FUNDAMENTALS OF MULTIPHASE FLOW

This book is targeted to graduate students and researchers at the cutting edge of investigations into the fundamental nature of multiphase flows. It is intended as a reference book for the basic methods used in the treatment of multiphase flows. The subject of multiphase flows encompasses a vast field, a host of different technological contexts, a wide spectrum of different scales, a broad range of engineering disciplines, and a multitude of different analytical approaches. The aim of *Fundamentals of Multiphase Flow* is to bring much of this fundamental understanding together into one book, presenting a unifying approach to the fundamental ideas of multiphase flows. The book summarizes those fundamental concepts with relevance to a broad spectrum of multiphase flows. It does not pretend to present a comprehensive review of the details of any one multiphase flow or technological context; references to such reviews are included where appropriate.

Christopher E. Brennen is professor of Mechanical Engineering in the Faculty of Engineering and Applied Science at the California Institute of Technology. He has published over 200 referred articles and is especially well known for his research on cavitation, turbomachinery flows, as well as multiphase flows. He is the author of *Cavitation and Bubble Dynamics* and *Hydrodynamics of Pumps* and has edited several other works.

Fundamentals of Multiphase Flow

CHRISTOPHER E. BRENNEN
California Institute of Technology

CAMBRIDGE UNIVERSITY PRESS
Cambridge, New York, Melbourne, Madrid, Cape Town, Singapore,
São Paulo, Delhi, Dubai, Tokyo

Cambridge University Press
32 Avenue of the Americas, New York, NY 10013-2473, USA

www.cambridge.org
Information on this title: www.cambridge.org/9780521139984

© Christopher E. Brennen 2005

This publication is in copyright. Subject to statutory exception
and to the provisions of relevant collective licensing agreements,
no reproduction of any part may take place without the written
permission of Cambridge University Press.

First published 2005
First paperback edition 2009
Reprinted 2006, 2009

Printed in the United States of America

A catalog record for this publication is available from the British Library.

Library of Congress Cataloging in Publication Data

Brennen, Christopher E. (Christopher Earls), 1941–
Fundamentals of multiphase flow / Christopher E. Brennen.
  p. cm.
Includes bibliographical references and index.
ISBN 0-521-84804-0 (hardback)
1. Multiphase flow. I. Title.
TA357.5.M84B76 2005
620.1'064 – dc22   2004020555

ISBN 978-0-521-84804-6 Hardback
ISBN 978-0-521-13998-4 Paperback

Cambridge University Press has no responsibility for the persistence or accuracy of URLs for external or
third-party Internet Web sites referred to in this publication and does not guarantee that any content on such
Web sites is, or will remain, accurate or appropriate.

Contents

Preface	*page* xiii
Nomenclature	xv

1	Introduction to Multiphase Flow	1
	1.1 Introduction	1
	1.1.1 Scope	1
	1.1.2 Multiphase Flow Models	2
	1.1.3 Multiphase Flow Notation	3
	1.1.4 Size Distribution Functions	6
	1.2 Equations of Motion	8
	1.2.1 Averaging	8
	1.2.2 Continuum Equations for Conservation of Mass	9
	1.2.3 Disperse Phase Number Continuity	10
	1.2.4 Fick's Law	11
	1.2.5 Continuum Equations for Conservation of Momentum	12
	1.2.6 Disperse Phase Momentum Equation	14
	1.2.7 Comments on Disperse Phase Interaction	15
	1.2.8 Equations for Conservation of Energy	16
	1.2.9 Heat Transfer between Separated Phases	19
	1.3 Interaction with Turbulence	21
	1.3.1 Particles and Turbulence	21
	1.3.2 Effect on Turbulence Stability	24
	1.4 Comments on the Equations of Motion	25
	1.4.1 Averaging	25
	1.4.2 Averaging Contributions to the Mean Motion	26
	1.4.3 Averaging in Pipe Flows	27
	1.4.4 Modeling with the Combined Phase Equations	28
	1.4.5 Mass, Force, and Energy Interaction Terms	28

2	Single-Particle Motion	30
	2.1 Introduction	30
	2.2 Flows Around a Sphere	31
	2.2.1 At High Reynolds Number	31
	2.2.2 At Low Reynolds Number	32
	2.2.3 Molecular Effects	37
	2.3 Unsteady Effects	38
	2.3.1 Unsteady Particle Motions	38
	2.3.2 Effect of Concentration on Added Mass	41
	2.3.3 Unsteady Potential Flow	41
	2.3.4 Unsteady Stokes Flow	44
	2.4 Particle Equation of Motion	48
	2.4.1 Equations of Motion	48
	2.4.2 Magnitude of Relative Motion	52
	2.4.3 Effect of Concentration on Particle Equation of Motion	53
	2.4.4 Effect of Concentration on Particle Drag	55
3	Bubble or Droplet Translation	60
	3.1 Introduction	60
	3.2 Deformation Due to Translation	60
	3.2.1 Dimensional analysis	60
	3.2.2 Bubble shapes and terminal velocities	61
	3.3 Marangoni Effects	66
	3.4 Bjerknes Forces	68
	3.5 Growing or Collapsing Bubbles	69
4	Bubble Growth and Collapse	73
	4.1 Introduction	73
	4.2 Bubble Growth and Collapse	73
	4.2.1 Rayleigh–Plesset Equation	73
	4.2.2 Bubble Contents	75
	4.2.3 In the Absence of Thermal Effects; Bubble Growth	78
	4.2.4 In the Absence of Thermal Effects; Bubble Collapse	81
	4.2.5 Stability of Vapor/Gas Bubbles	82
	4.3 Thermal Effects	84
	4.3.1 Thermal Effects on Growth	84
	4.3.2 Thermally Controlled Growth	85
	4.3.3 Cavitation and Boiling	89
	4.3.4 Bubble Growth by Mass Diffusion	89
	4.4 Oscillating Bubbles	91
	4.4.1 Bubble Natural Frequencies	91
	4.4.2 Nonlinear Effects	93
	4.4.3 Rectified Mass Diffusion	95

5	Cavitation	97
	5.1 Introduction	97
	5.2 Key Features of Bubble Cavitation	97
	5.2.1 Cavitation Inception	97
	5.2.2 Cavitation Bubble Collapse	99
	5.2.3 Shape Distortion during Bubble Collapse	101
	5.2.4 Cavitation Damage	104
	5.3 Cavitation Bubbles	106
	5.3.1 Observations of Cavitating Bubbles	106
	5.3.2 Cavitation Noise	109
	5.3.3 Cavitation Luminescence	115
6	Boiling and Condensation	116
	6.1 Introduction	116
	6.2 Horizontal Surfaces	117
	6.2.1 Pool Boiling	117
	6.2.2 Nucleate Boiling	119
	6.2.3 Film Boiling	120
	6.2.4 Leidenfrost Effect	121
	6.3 Vertical Surfaces	122
	6.3.1 Film Boiling	122
	6.4 Condensation	125
	6.4.1 Film Condensation	125
7	Flow Patterns	127
	7.1 Introduction	127
	7.2 Topologies of Multiphase Flow	127
	7.2.1 Multiphase Flow Patterns	127
	7.2.2 Examples of Flow Regime Maps	129
	7.2.3 Slurry Flow Regimes	131
	7.2.4 Vertical Pipe Flow	132
	7.2.5 Flow Pattern Classifications	134
	7.3 Limits of Disperse Flow Regimes	136
	7.3.1 Disperse Phase Separation and Dispersion	136
	7.3.2 Example: Horizontal Pipe Flow	138
	7.3.3 Particle Size and Particle Fission	140
	7.3.4 Examples of Flow-Determined Bubble Size	141
	7.3.5 Bubbly or Mist Flow Limits	142
	7.3.6 Other Bubbly Flow Limits	143
	7.3.7 Other Particle Size Effects	144
	7.4 Inhomogeneity Instability	144
	7.4.1 Stability of Disperse Mixtures	145
	7.4.2 Inhomogeneity Instability in Vertical Flows	148

Contents

    7.5 Limits on Separated Flow   151
        7.5.1 Kelvin–Helmoltz Instability   151
        7.5.2 Stratified Flow Instability   153
        7.5.3 Annular Flow Instability   154

8 Internal Flow Energy Conversion   155
    8.1 Introduction   155
    8.2 Frictional Loss in Disperse Flow   155
        8.2.1 Horizontal Flow   155
        8.2.2 Homogeneous Flow Friction   157
        8.2.3 Heterogeneous Flow Friction   159
        8.2.4 Vertical Flow   161
    8.3 Frictional Loss in Separated Flow   163
        8.3.1 Two-Component Flow   163
        8.3.2 Flow with Phase Change   168
    8.4 Energy Conversion in Pumps and Turbines   172
        8.4.1 Multiphase Flows in Pumps   172

9 Homogeneous Flows   176
    9.1 Introduction   176
    9.2 Equations of Homogeneous Flow   176
    9.3 Sonic Speed   177
        9.3.1 Basic Analysis   177
        9.3.2 Sonic Speeds at Higher Frequencies   181
        9.3.3 Sonic Speed with Change of Phase   182
    9.4 Barotropic Relations   186
    9.5 Nozzle Flows   187
        9.5.1 One-Dimensional Analysis   187
        9.5.2 Vapor/Liquid Nozzle Flow   192
        9.5.3 Condensation Shocks   195

10 Flows with Bubble Dynamics   199
    10.1 Introduction   199
    10.2 Basic Equations   200
    10.3 Acoustics of Bubbly Mixtures   200
        10.3.1 Analysis   200
        10.3.2 Comparison with Experiments   203
    10.4 Shock Waves in Bubbly Flows   205
        10.4.1 Shock-wave Analysis   205
        10.4.2 Shock-wave Structure   208
    10.5 Finite Bubble Clouds   210
        10.5.1 Natural Modes of a Spherical Cloud of Bubbles   210
        10.5.2 Response of a Spherical Bubble Cloud   214

Contents

11	**Flows with Gas Dynamics**	217
	11.1 Introduction	217
	11.2 Equations for a Dusty Gas	217
	11.2.1 Basic Equations	217
	11.2.2 Homogeneous Flow with Gas Dynamics	219
	11.2.3 Velocity and Temperature Relaxation	220
	11.3 Normal Shock Wave	221
	11.4 Acoustic Damping	224
	11.5 Other Linear Perturbation Analyses	227
	11.5.1 Stability of Laminar Flow	227
	11.5.2 Flow over a Wavy Wall	228
	11.6 Small Slip Perturbation	229
12	**Sprays**	232
	12.1 Introduction	232
	12.2 Types of Spray Formation	232
	12.3 Ocean Spray	233
	12.4 Spray Formation	234
	12.4.1 Spray Formation by Bubbling	234
	12.4.2 Spray Formation by Wind Shear	235
	12.4.3 Spray Formation by Initially Laminar Jets	237
	12.4.4 Spray Formation by Turbulent Jets	239
	12.5 Single-Droplet Mechanics	243
	12.5.1 Single-Droplet Evaporation	243
	12.5.2 Single-Droplet Combustion	245
	12.6 Spray Combustion	249
13	**Granular Flows**	252
	13.1 Introduction	252
	13.2 Particle Interaction Models	253
	13.2.1 Computer Simulations	255
	13.3 Flow Regimes	255
	13.3.1 Dimensional Analysis	255
	13.3.2 Flow Regime Rheologies	256
	13.3.3 Flow Regime Boundaries	259
	13.4 Slow Granular Flow	259
	13.4.1 Equations of Motion	259
	13.4.2 Mohr–Coulomb Models	260
	13.4.3 Hopper Flows	261
	13.5 Rapid Granular Flow	263
	13.5.1 Introduction	263
	13.5.2 Example of Rapid Flow Equations	264

	13.5.3 Boundary Conditions	267
	13.5.4 Computer Simulations	267
13.6	Effect of Interstitial Fluid	268
	13.6.1 Introduction	268
	13.6.2 Particle Collisions	268
	13.6.3 Classes of Interstitial Fluid Effects	270

14 Drift Flux Models — 272

- 14.1 Introduction — 272
- 14.2 Drift Flux Method — 273
- 14.3 Examples of Drift Flux Analyses — 274
  - 14.3.1 Vertical Pipe Flow — 274
  - 14.3.2 Fluidized Bed — 276
  - 14.3.3 Pool Boiling Crisis — 278
- 14.4 Corrections for Pipe Flows — 282

15 System Instabilities — 284

- 15.1 Introduction — 284
- 15.2 System Structure — 284
- 15.3 Quasistatic Stability — 286
- 15.4 Quasistatic Instability Examples — 288
  - 15.4.1 Turbomachine Surge — 288
  - 15.4.2 Ledinegg Instability — 288
  - 15.4.3 Geyser Instability — 289
- 15.5 Concentration Waves — 290
- 15.6 Dynamic Multiphase Flow Instabilities — 292
  - 15.6.1 Dynamic Instabilities — 292
  - 15.6.2 Cavitation Surge in Cavitating Pumps — 292
  - 15.6.3 Chugging and Condensation Oscillations — 293
- 15.7 Transfer Functions — 297
  - 15.7.1 Unsteady Internal Flow Methods — 297
  - 15.7.2 Transfer Functions — 298
  - 15.7.3 Uniform Homogeneous Flow — 300

16 Kinematic Waves — 302

- 16.1 Introduction — 302
- 16.2 Two-Component Kinematic Waves — 303
  - 16.2.1 Basic Analysis — 303
  - 16.2.2 Kinematic Wave Speed at Flooding — 304
  - 16.2.3 Kinematic Waves in Steady Flows — 305
- 16.3 Two-Component Kinematic Shocks — 306
  - 16.3.1 Kinematic Shock Relations — 306

16.3.2 Kinematic Shock Stability	308
16.3.3 Compressibility and Phase-Change Effects	309
16.4 Examples of Kinematic Wave Analyses	311
16.4.1 Batch Sedimentation	311
16.4.2 Dynamics of Cavitating Pumps	313
16.5 Two-Dimensional Kinematic Waves	318
Bibliography	321
Index	341

Preface

The subject of multiphase flows encompasses a vast field, a host of different technological contexts, a wide spectrum of different scales, a broad range of engineering disciplines, and a multitude of different analytical approaches. Not surprisingly, the number of books dealing with the subject is voluminous. For the student or researcher in the field of multiphase flow this broad spectrum presents a problem for the experimental or analytical methodologies that might be appropriate for his/her interests can be widely scattered and difficult to find. The aim of the present text is to try to bring much of this fundamental understanding together into one book and to present a unifying approach to the fundamental ideas of multiphase flows. Consequently the book summarizes those fundamental concepts with relevance to a broad spectrum of multiphase flows. It does not pretend to present a comprehensive review of the details of any one multiphase flow or technological context, although reference to books providing such reviews is included where appropriate. This book is targeted at graduate students and researchers at the cutting edge of investigations into the fundamental nature of multiphase flows; it is intended as a reference book for the basic methods used in the treatment of multiphase flows.

I am deeply grateful to all my many friends and fellow researchers in the field of multiphase flows whose ideas fill these pages. I am particularly indebted to my close colleagues Allan Acosta, Ted Wu, Rolf Sabersky, Melany Hunt, Tim Colonius, and the late Milton Plesset, all of whom made my professional life a real pleasure. This book grew out of many years of teaching and research at the California Institute of Technology. It was my privilege to have worked on multiphase flow problems with a group of marvelously talented students, including Hojin Ahn, Robert Bernier, Abhijit Bhattacharyya, David Braisted, Charles Campbell, Steven Ceccio, Luca d'Agostino, Fabrizio d'Auria, Mark Duttweiler, Ronald Franz, Douglas Hart, Steve Hostler, Gustavo Joseph, Joseph Katz, Yan Kuhn de Chizelle, Sanjay Kumar, Harri Kytomaa, Zhenhuan Liu, Beth McKenney, Sheung-Lip Ng, Tanh Nguyen, Kiam Oey, James Pearce, Garrett Reisman, Y.-C. Wang, Carl Wassgren, Roberto Zenit Camacho, and Steve Hostler. To them I owe a special debt. Also, Cecilia Lin devoted many selfless hours to the preparation of the illustrations.

A substantial fraction of the introductory material in this book is taken from my earlier book entitled "Cavitation and Bubble Dynamics" by Christopher Earls Brennen, © 1995 by Oxford University Press, Inc. It is reproduced here by permission of Oxford University Press, Inc.

The original hardback edition was dedicated to my mother, Muriel M. Brennen, whose love and encouragement inspired me throughout my life. This paperback edition is dedicated to another very special woman, my wife, Barbara, who gave me new life and love beyond measure.

> Christopher Earls Brennen
> California Institute of Technology
> December 2003.

Nomenclature

Roman Letters

a	Amplitude of wavelike disturbance
A	Cross-sectional area or cloud radius
$\mathcal{A}$	Attenuation
b	Power law index
Ba	Bagnold number, $\rho_S D^2 \dot{\gamma}/\mu_L$
c	Concentration
c	Speed of sound
c_κ	Phase velocity for wavenumber κ
c_p	Specific heat at constant pressure
c_s	Specific heat of solid or liquid
c_v	Specific heat at constant volume
C	Compliance
C	Damping coefficient
C_D	Drag coefficient
C_{ij}	Drag and lift coefficient matrix
C_L	Lift coefficient
C_p	Coefficient of pressure
C_{pmin}	Minimum coefficient of pressure
d	Diameter
d_j	Jet diameter
d_o	Hopper opening diameter
D	Particle, droplet or bubble diameter
D	Mass diffusivity
D_m	Volume (or mass) mean diameter
D_s	Sauter mean diameter
$D(T)$	Determinant of the transfer matrix $[T]$
$\mathcal{D}$	Thermal diffusivity
e	Specific internal energy

$\mathcal{E}$	Rate of exchange of energy per unit volume
f	Frequency in hertz
f	Friction factor
f_L, f_V	Liquid and vapor thermodynamic quantities
F_i	Force vector
Fr	Froude number
$\mathcal{F}$	Interactive force per unit volume
g	Acceleration due to gravity
g_L, g_V	Liquid and vapor thermodynamic quantities
G_{Ni}	Mass flux of component N in direction i
G_N	Mass flux of component N
h	Specific enthalpy
h	Height
H	Height
H	Total head, $p^T/\rho g$
He	Henry's law constant
Hm	Haberman–Morton number, normally $g\mu^4/\rho S^3$
i, j, k, m, n	Indices
i	Square root of -1
I	Acoustic impulse
$\mathcal{I}$	Rate of transfer of mass per unit volume
j_i	Total volumetric flux in direction i
j_{Ni}	Volumetric flux of component N in direction i
j_N	Volumetric flux of component N
k	Polytropic constant
k	Thermal conductivity
k	Boltzmann's constant
k_L, k_V	Liquid and vapor quantities
K	Constant
K^*	Cavitation compliance
Kc	Keulegan–Carpenter number
K_{ij}	Added mass coefficient matrix
K_n, K_s	Elastic spring constants in normal and tangential directions
Kn	Knudsen number, $\lambda/2R$
$\mathcal{K}$	Frictional constants
ℓ	Typical dimension
ℓ_t	Turbulent length scale
L	Inertance
$\mathcal{L}$	Latent heat of vaporization
m	Mass
$\dot{m}$	Mass flow rate
m_G	Mass of gas in bubble

m_p	Mass of particle
M	Mach number
M^*	Mass flow gain factor
M_{ij}	Added mass matrix
$\mathcal{M}$	Molecular weight
Ma	Martinelli parameter
n	Number of particles per unit volume
$\dot{n}$	Number of events per unit time
n_i	Unit vector in the i direction
$N(R)$, $N(D)$, $N(v)$	Particle size distribution functions
N^*	Number of sites per unit area
Nu	Nusselt number
p	Pressure
p^T	Total pressure
p_a	Radiated acoustic pressure
p_G	Partial pressure of gas
p_s	Sound pressure level
P	Perimeter
Pe	Peclet number, usually WR/α_C
Pr	Prandtl number, $\rho \nu c_p / k$
q	General variable
q_i	Heat flux vector
Q	General variable
$\mathcal{Q}$	Rate of heat transfer or release per unit mass
$\mathcal{Q}_\ell$	Rate of heat addition per unit length of pipe
r, r_i	Radial coordinate and position vector
r_d	Impeller discharge radius
R	Bubble, particle or droplet radius
R_k^*	Resistance of component, k
R_B	Equivalent volumetric radius, $(3\tau/4\pi)^{\frac{1}{3}}$
R_e	Equilibrium radius
Re	Reynolds number, usually $2WR/\nu_C$
$\mathcal{R}$	Gas constant
s	Coordinate measured along a streamline or pipe centerline
s	Laplace transform variable
s	Specific entropy
S	Surface tension
S_D	Surface of the disperse phase
St	Stokes number
Str	Strouhal number
t	Time

t_c	Binary collision time
t_u	Relaxation time for particle velocity
t_T	Relaxation time for particle temperature
T	Temperature
T	Granular temperature
T_{ij}	Transfer matrix
u_i	Velocity vector
u_{Ni}	Velocity of component N in direction i
u_r, u_θ	Velocity components in polar coordinates
u_s	Shock velocity
u^*	Friction velocity
U, U_i	Fluid velocity and velocity vector in absence of particle
U_∞	Velocity of upstream uniform flow
v	Volume of particle, droplet or bubble
V, V_i	Absolute velocity and velocity vector of particle
V	Volume
V	Control volume
$\dot{V}$	Volume flow rate
w	Dimensionless relative velocity, W/W_∞
W, W_i	Relative velocity of particle and relative velocity vector
W_∞	Terminal velocity of particle
W_p	Typical phase separation velocity
W_t	Typical phase mixing velocity
We	Weber number, $2\rho W^2 R/S$
$\mathcal{W}$	Rate of work done per unit mass
x, y, z	Cartesian coordinates
x_i	Position vector
x	Mass fraction
$\mathcal{X}$	Mass quality
z	Coordinate measured vertically upward

Greek Letters

α	Volume fraction
β	Volume quality
γ	Ratio of specific heats of gas
$\dot{\gamma}$	Shear rate
Γ	Rate of dissipation of energy per unit volume
δ	Boundary layer thickness
δ_d	Damping coefficient
δm	Fractional mass
δ_T	Thermal boundary layer thickness
δ_2	Momentum thickness of the boundary layer

Nomenclature

δ_{ij}	Kronecker delta: $\delta_{ij} = 1$ for $i = j$; $\delta_{ij} = 0$ for $i \neq j$
ϵ	Fractional volume
ϵ	Coefficient of restitution
ϵ	Rate of dissipation of energy per unit mass
ζ	Attenuation or amplification rate
η	Bubble population per unit liquid volume
θ	Angular coordinate or direction of velocity vector
θ	Reduced frequency
θ_w	Hopper opening half-angle
κ	Wavenumber
κ	Bulk modulus of compressibility
κ_L, κ_G	Shape constants
λ	Wavelength
λ	Mean free path
λ	Kolmogorov length scale
Λ	Integral length scale of the turbulence
μ	Dynamic viscosity
μ^*	Coulomb friction coefficient
ν	Kinematic viscosity
ν	Mass-based stoichiometric coefficient
ξ	Particle loading
ρ	Density
σ	Cavitation number
σ_i	Inception cavitation number
σ_{ij}	Stress tensor
σ_{ij}^D	Deviatoric stress tensor
$\Sigma(T)$	Thermodynamic parameter
τ	Kolmogorov time scale
τ_i	Interfacial shear stress
τ_n	Normal stress
τ_s	Shear stress
τ_w	Wall shear stress
ψ	Stokes stream function
ψ	Head coefficient, $\Delta p^T / \rho \Omega^2 r_d^2$
ϕ	Velocity potential
ϕ	Internal friction angle
ϕ	Flow coefficient, $j / \Omega r_d$
$\phi_L^2, \phi_G^2, \phi_{L0}^2$	Martinelli pressure gradient ratios
φ	Fractional perturbation in bubble radius
ω	Radian frequency
ω_a	Acoustic mode frequency
ω_i	Instability frequency
ω_n	Natural frequency

ω_m	Cloud natural frequencies
ω_m	Manometer frequency
ω_p	Peak frequency
Ω	Rotating frequency (radians per second)

Subscripts

On any variable, Q:

Q_o	Initial value, upstream value or reservoir value
Q_1, Q_2, Q_3	Components of Q in three Cartesian directions
Q_1, Q_2	Values upstream and downstream of a component or flow structure
Q_∞	Value far from the particle or bubble
Q_*	Throat values
Q_A	Pertaining to a general phase or component, A
Q_b	Pertaining to the bulk
Q_B	Pertaining to a general phase or component, B
Q_B	Value in the bubble
Q_C	Pertaining to the continuous phase or component, C
Q_c	Critical values and values at the critical point
Q_D	Pertaining to the disperse phase or component, D
Q_e	Equilibrium value or value on the saturated liquid/vapor line
Q_e	Effective value or exit value
Q_G	Pertaining to the gas phase or component
Q_i	Components of vector Q
Q_{ij}	Components of tensor Q
Q_L	Pertaining to the liquid phase or component
Q_m	Maximum value of Q
Q_N	Pertaining to a general phase or component, N
Q_O	Pertaining to the oxidant
Q_r	Component in the r direction
Q_s	A surface, system or shock value
Q_S	Pertaining to the solid particles
Q_V	Pertaining to the vapor phase or component
Q_w	Value at the wall
Q_θ	Component in the θ direction

Superscripts and Other Qualifiers

On any variable, Q:

Q', Q'', Q^*	Used to differentiate quantities similar to Q
$\bar{Q}$	Mean value of Q or complex conjugate of Q

Nomenclature

$\dot{Q}$	Small perturbation in Q
$\tilde{Q}$	Complex amplitude of oscillating Q
$\dot{Q}$	Time derivative of Q
$\ddot{Q}$	Second time derivative of Q
$\hat{Q}(s)$	Laplace transform of $Q(t)$
$\check{Q}$	Coordinate with origin at image point
δQ	Small change in Q
$\text{Re}\{Q\}$	Real part of Q
$\text{Im}\{Q\}$	Imaginary part of Q

Notes

Notation

The reader is referred to Section 1.1.3 for a more complete description of the multiphase flow notation employed in this book. Note also that a few symbols that are only used locally in the text have been omitted from the above lists.

Units

In most of this book, the emphasis is placed on the nondimensional parameters that govern the phenomenon being discussed. However, there are also circumstances in which we shall utilize dimensional thermodynamic and transport properties. In such cases the International System of Units will be employed using the basic units of mass (kg), length (m), time (s), and absolute temperature (K).

1

Introduction to Multiphase Flow

1.1 Introduction

1.1.1 Scope

In the context of this book, the term *multiphase flow* is used to refer to any fluid flow consisting of more than one phase or component. For brevity and because they are covered in other texts, we exclude those circumstances in which the components are well mixed above the molecular level. Consequently, the flows considered here have some level of phase or component separation at a scale well above the molecular level. This still leaves an enormous spectrum of different multiphase flows. One could classify them according to the state of the different phases or components and therefore refer to gas/solids flows or liquid/solids flows or gas/particle flows or bubbly flows and so on; many texts exist that limit their attention in this way. Some treatises are defined in terms of a specific type of fluid flow and deal with low-Reynolds-number suspension flows, dusty gas dynamics, and so on. Others focus attention on a specific application such as slurry flows, cavitating flows, aerosols, debris flows, fluidized beds, and so on; again, there are many such texts. In this book we attempt to identify the basic fluid mechanical phenomena and to illustrate those phenomena with examples from a broad range of applications and types of flow.

Parenthetically, it is valuable to reflect on the diverse and ubiquitous challenges of multiphase flow. Virtually every processing technology must deal with multiphase flow, from cavitating pumps and turbines to electrophotographic processes to papermaking to the pellet form of almost all raw plastics. The amount of granular material, coal, grain, ore, and so on that is transported every year is enormous and, at many stages, that material is required to flow. Clearly the ability to predict the fluid flow behavior of these processes is central to the efficiency and effectiveness of those processes. For example, the effective flow of toner is a major factor in the quality and speed of electrophotographic printers. Multiphase flows are also a ubiquitous feature of our environment whether one considers rain, snow, fog, avalanches, mud slides, sediment transport, debris flows, and countless other natural phenomena, to say nothing of what happens beyond our planet. Very critical biological and medical flows are also multiphase, from blood flow to semen to the bends to lithotripsy to laser surgery cavitation and so on. No single list can adequately illustrate the diversity and ubiquity; consequently, any attempt at a comprehensive treatment of multiphase flows is flawed unless it focuses on common phenomenological themes and avoids the temptation to digress into lists of observations.

Two general topologies of multiphase flow can be usefully identified at the outset, namely *disperse flows* and *separated flows*. By *disperse flows* we mean those consisting of finite particles, drops, or bubbles (the disperse phase) distributed in a connected volume of the continuous phase. *Separated flows* consist of two or more continuous streams of different fluids separated by interfaces.

1.1.2 Multiphase Flow Models

A persistent theme throughout the study of multiphase flows is the need to model and predict the detailed behavior of those flows and the phenomena that they manifest. There are three ways in which such models are explored: (1) experimentally, through laboratory-sized models equipped with appropriate instrumentation; (2) theoretically, using mathematical equations and models for the flow; and (3) computationally, using the power and size of modern computers to address the complexity of the flow. Clearly there are some applications in which full-scale laboratory models are possible. But, in many instances, the laboratory model must have a very different scale from the prototype and then a reliable theoretical or computational model is essential for confident extrapolation to the scale of the prototype. There are also cases in which a laboratory model is impossible for a wide variety of reasons.

Consequently, the predictive capability and physical understanding must rely heavily on theoretical and/or computational models and here the complexity of most multiphase flows presents a major hurdle. It may be possible at some distant time in the future to code the Navier–Stokes equations for each of the phases or components and to compute every detail of a multiphase flow, the motion of all the fluid around and inside every particle or drop, the position of every interface. But the computer power and speed required to do this are far beyond present capability for most of the flows that are commonly experienced. When one or both of the phases become turbulent (as often happens), the magnitude of the challenge becomes truly astronomical. Therefore, simplifications are essential in realistic models of most multiphase flows.

In disperse flows two types of models are prevalent, *trajectory models* and *two-fluid models*. In trajectory models, the motion of the disperse phase is assessed by following the motion of either the actual particles or larger, representative *particles*. The details of the flow around each of the particles are subsumed into assumed drag, lift, and moment forces acting on and altering the trajectory of those particles. The thermal history of the particles can also be tracked if it is appropriate to do so. Trajectory models have been very useful in studies of the rheology of granular flows (see Chapter 13) primarily

1.1 Introduction

because the effects of the interstitial fluid are small. In the alternative approach, two-fluid models, the disperse phase is treated as a second continuous phase intermingled and interacting with the continuous phase. Effective conservation equations (of mass, momentum, and energy) are developed for the two fluid flows; these include interaction terms modeling the exchange of mass, momentum, and energy between the two flows. These equations are then solved either theoretically or computationally. Thus, two-fluid models neglect the discrete nature of the disperse phase and approximate its effects on the continuous phase. Inherent in this approach are averaging processes necessary to characterize the properties of the disperse phase; these involve significant difficulties. The boundary conditions appropriate in two-fluid models also pose difficult modeling issues.

In contrast, separated flows present many fewer issues. In theory one must solve the single-phase fluid-flow equations in the two streams, coupling them through appropriate kinematic and dynamic conditions at the interface. Free streamline theory (see, for example, Birkhoff and Zarantonello 1957, Tulin 1964, Woods 1961, Wu 1972) is an example of a successful implementation of such a strategy, though the interface conditions used in that context are particularly simple.

In the first part of this book, the basic tools for both trajectory and two-fluid models are developed and discussed. In the remainder of this first chapter, a basic notation for multiphase flow is developed and this leads naturally into a description of the mass, momentum, and energy equations applicable to multiphase flows and, particularly in two-fluid models. In Chapters 2, 3, and 4, we examine the dynamics of individual particles, drops, and bubbles. In Chapter 7 we address the different topologies of multiphase flows and, in the subsequent chapters, we examine phenomena in which *particle* interactions and the particle/fluid interactions modify the flow.

1.1.3 Multiphase Flow Notation

The notation that is used is close to the standard described by Wallis (1969). It has, however, been slightly modified to permit more ready adoption to the Cartesian tensor form. In particular the subscripts that can be attached to a property consist of a group of uppercase subscripts followed by lowercase subscripts. The lowercase subscripts (i, ij, etc.) are used in the conventional manner to denote vector or tensor components. A single uppercase subscript (N) refers to the property of a specific phase or component. In some contexts generic subscripts N = A, B are used for generality. However, other letters such as N = C (continuous phase), N = D (disperse phase), N = L (liquid), N = G (gas), N = V (vapor), or N = S (solid) are used for clarity in other contexts. Finally two uppercase subscripts imply the difference between the two properties for the two single uppercase subscripts.

Specific properties frequently used are as follows. *Volumetric fluxes* (volume flow per unit area) of individual components are denoted by j_{Ai}, j_{Bi} ($i = 1$, 2, or 3 in three-dimensional flow). These are sometimes referred to as superficial component velocities. The *total volumetric flux*, j_i, is then given by the following:

$$j_i = j_{Ai} + j_{Bi} + \cdots = \sum_N j_{Ni}. \tag{1.1}$$

Mass fluxes are similarly denoted by G_{Ai}, G_{Bi}, or G_i. Thus if the densities of individual components are denoted by ρ_A, ρ_B it follows that

$$G_{Ai} = \rho_A j_{Ai}; \quad G_{Bi} = \rho_B j_{Bi}; \quad G_i = \sum_N \rho_N j_{Ni}. \tag{1.2}$$

Velocities of the specific phases are denoted by u_{Ai}, u_{Bi} or, in general, by u_{Ni}. The relative velocity between the two phases A and B is denoted by u_{ABi} such that

$$u_{Ai} - u_{Bi} = u_{ABi}. \tag{1.3}$$

The volume fraction of a component or phase is denoted by α_N and, in the case of two components or phases, A and B, it follows that $\alpha_B = 1 - \alpha_A$. Though this is clearly a well-defined property for any finite volume in the flow, there are some substantial problems associated with assigning a value to an infinitesimal volume or point in the flow. Provided these can be resolved, it follows that the volumetric flux of a component, N, and its velocity are related by

$$j_{Ni} = \alpha_N u_{Ni} \tag{1.4}$$

and that

$$j_i = \alpha_A u_{Ai} + \alpha_B u_{Bi} + \cdots = \sum_N \alpha_N u_{Ni}. \tag{1.5}$$

Two other fractional properties are relevant only in the context of one-dimensional flows. The *volumetric quality*, β_N, is the ratio of the volumetric flux of the component, N, to the total volumetric flux, that is,

$$\beta_N = j_N/j, \tag{1.6}$$

where the index i has been dropped from j_N and j because β is only used in the context of one-dimensional flows and the j_N, j refer to cross-sectionally averaged quantities.

The *mass fraction*, x_A, of a phase or component, A, is simply given by $\rho_A \alpha_A / \rho$ [see Eq. (1.8) for ρ]. Conversely, the *mass quality*, $\mathcal{X}_A$, is often referred to simply as *the quality* and is the ratio of the mass flux of component A to the total mass flux or

$$\mathcal{X}_A = \frac{G_A}{G} = \frac{\rho_A j_A}{\sum_N \rho_N j_N}. \tag{1.7}$$

Furthermore, when only two components or phases are present it is often redundant to use subscripts on the volume fraction and the qualities because $\alpha_A = 1 - \alpha_B$, $\beta_A = 1 - \beta_B$, and $\mathcal{X}_A = 1 - \mathcal{X}_B$. Thus unsubscripted quantities α, β, and $\mathcal{X}$ are often used in these circumstances.

1.1 Introduction

It is clear that a multiphase mixture has certain *mixture* properties, of which the most readily evaluated is the *mixture* density denoted by ρ and given by the following:

$$\rho = \sum_N \alpha_N \rho_N. \tag{1.8}$$

Conversely, the specific enthalpy, h, and specific entropy, s, being defined as per unit mass rather than per unit volume, are weighted according to the following:

$$\rho h = \sum_N \rho_N \alpha_N h_N; \quad \rho s = \sum_N \rho_N \alpha_N s_N. \tag{1.9}$$

Other properties such as the *mixture* viscosity or thermal conductivity cannot be reliably obtained from such simple weighted means.

Aside from the relative velocities between phases that were described earlier, there are two other measures of relative motion that are frequently used. The *drift velocity* of a component is defined as the velocity of that component in a frame of reference moving at a velocity equal to the total volumetric flux, j_i, and is therefore given by u_{NJi}, where

$$u_{NJi} = u_{Ni} - j_i. \tag{1.10}$$

Even more frequent use will be made of the *drift flux* of a component, which is defined as the volumetric flux of a component in the frame of reference moving at j_i. Denoted by j_{NJi}, this is given by the following:

$$j_{NJi} = j_{Ni} - \alpha_N j_i = \alpha_N (u_{Ni} - j_i) = \alpha_N u_{NJi}. \tag{1.11}$$

It is particularly important to notice that the sum of all the drift fluxes must be zero because from Eq. (1.11) we have the following:

$$\sum_N j_{NJi} = \sum_N j_{Ni} - j_i \sum_N \alpha_N = j_i - j_i = 0. \tag{1.12}$$

When only two phases or components, A and B, are present it follows that $j_{AJi} = -j_{BJi}$ and hence it is convenient to denote both of these drift fluxes by the vector j_{ABi}, where we have the following:

$$j_{ABi} = j_{AJi} = -j_{BJi}. \tag{1.13}$$

Moreover it follows from Eq. (1.11) that we have the following:

$$j_{ABi} = \alpha_A \alpha_B u_{ABi} = \alpha_A (1 - \alpha_A) u_{ABi}, \tag{1.14}$$

and hence the drift flux, j_{ABi}, and the relative velocity, u_{ABi}, are simply related.

Finally, it is clear that certain basic relations follow from the preceding definitions and it is convenient to identify these here for later use. First the relations between the volume and mass qualities that follow from Eqs. (1.6) and (1.7) only involve ratios of

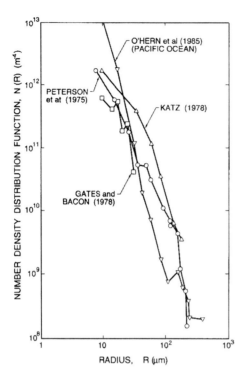

Figure 1.1. Measured size distribution functions for small bubbles in three different water tunnels (Peterson et al. 1975, Gates and Bacon 1978, Katz 1978) and in the ocean off Los Angeles, California (O'Hern et al. 1985).

the densities of the following components:

$$X_A = \beta_A / \sum_N \left(\frac{\rho_N}{\rho_A}\right) \beta_N; \quad \beta_A = X_A / \sum_N \left(\frac{\rho_A}{\rho_N}\right) X_N. \tag{1.15}$$

Conversely, the relation between the volume fraction and the volume quality necessarily involves some measure of the relative motion between the phases (or components). The following useful results for two-phase (or two-component) one-dimensional flows can readily be obtained from Eqs. (1.11) and (1.6) as follows:

$$\beta_N = \alpha_N + \frac{j_{NJ}}{j}; \quad \beta_A = \alpha_A + \frac{j_{AB}}{j}; \quad \beta_B = \alpha_B - \frac{j_{AB}}{j}, \tag{1.16}$$

which demonstrate the importance of the drift flux as a measure of the relative motion.

1.1.4 Size Distribution Functions

In many multiphase flow contexts we shall make the simplifying assumption that all the disperse phase particles (bubbles, droplets, or solid particles) have the same size. However, in many natural and technological processes it is necessary to consider the distribution of particle size. One fundamental measure of this is the size distribution function, $N(v)$, defined such that the number of particles in a unit volume of the multiphase mixture with volume between v and $v + dv$ is $N(v)dv$. For convenience,

1.1 Introduction

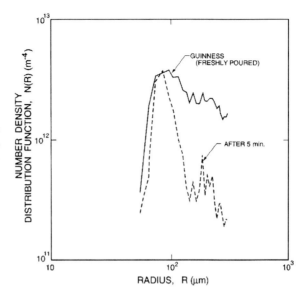

Figure 1.2. Size distribution functions for bubbles in freshly poured Guinness and after five minutes. Adapted from Kawaguchi and Maeda (2003).

it is often assumed that the particle size can be represented by a single linear dimension (for example, the diameter, D, or radius, R, in the case of spherical particles) so that alternative size distribution functions, $N'(D)$ or $N''(R)$, may be used. Examples of size distribution functions based on radius are shown in Figures 1.1 and 1.2.

Often such information is presented in the form of cumulative number distributions. For example, the cumulative distribution, $N^*(v^*)$, defined as

$$N^*(v^*) = \int_0^{v^*} N(v)dv, \qquad (1.17)$$

is the total number of particles of volume less than v^*. Examples of cumulative distributions (in this case for coal slurries) are shown in Figure 1.3.

In these disperse flows, the evaluation of global quantities or characteristics of the disperse phase will clearly require integration over the full range of particle sizes using the size distribution function. For example, the volume fraction of the disperse phase,

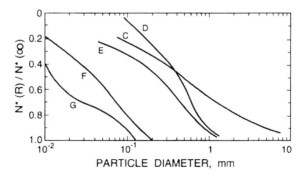

Figure 1.3. Cumulative size distributions for various coal slurries. Adapted from Shook and Roco (1991).

α_D, is given by the following:

$$\alpha_D = \int_0^\infty vN(v)dv = \frac{\pi}{6}\int_0^\infty D^3 N'(D)dD, \qquad (1.18)$$

where the last expression clearly applies to spherical particles. Other properties of the disperse phase or of the interactions between the disperse and continuous phases can involve other moments of the size distribution function (see, for example, Friedlander 1977). This leads to a series of mean diameters (or sizes in the case of nonspherical particles) of the form, D_{jk}, where

$$D_{jk} = \left[\frac{\int_0^\infty D^j N'(D)dD}{\int_0^\infty D^k N'(D)dD}\right]^{\frac{1}{j-k}}. \qquad (1.19)$$

A commonly used example is the *mass mean* diameter, D_{30}. Conversely, processes that are controlled by particle surface area would be characterized by the *surface area mean* diameter, D_{20}. The surface area mean diameter would be important, for example, in determining the exchange of heat between the phases or the rates of chemical interaction at the disperse phase surface. Another measure of the average size that proves useful in characterizing many disperse particulates is the Sauter mean diameter, D_{32}. This is a measure of the ratio of the particle volume to the particle surface area and, as such, is often used in characterizing particulates (see, for example, Chapter 14).

1.2 Equations of Motion

1.2.1 Averaging

In Section 1.1.3 it was implicitly assumed that there existed an *infinitesimal* volume of dimension, ϵ, such that ϵ was not only very much smaller than the typical distance over which the flow properties varied significantly but also very much larger than the size of the individual phase elements (the disperse phase particles, drops, or bubbles). The first condition is necessary to define derivatives of the flow properties within the flow field. The second is necessary so that each *averaging* volume (of volume ϵ^3) contains representative samples of each of the components or phases. In the sections that follow (Sections 1.2.2 to 1.2.9), we proceed to develop the effective differential equations of motion for multiphase flow assuming that these conditions hold.

However, one of the more difficult hurdles in treating multiphase flows is that the preceding two conditions are rarely both satisfied. As a consequence the averaging volumes contain a finite number of finite-sized particles and therefore flow properties such as the continuous phase velocity vary significantly from point to point within these averaging volumes. These variations pose the challenge of how to define appropriate average quantities in the averaging volume. Moreover, the gradients of those averaged

1.2 Equations of Motion

flow properties appear in the equations of motion that follow and the mean of the gradient is not necessarily equal to the gradient of the mean. These difficulties are addressed in Section 1.4 after we have explored the basic structure of the Equations in the absence of such complications.

1.2.2 Continuum Equations for Conservation of Mass

Consider now the construction of the effective differential equations of motion for a disperse multiphase flow (such as might be used in a two-fluid model) assuming that an appropriate elemental volume can be identified. For convenience this elemental volume is chosen to be a *unit* cube with edges parallel to the x_1, x_2, x_3 directions. The mass flow of component N through one of the faces perpendicular to the i direction is given by $\rho_N j_{Ni}$ and therefore the net outflow of mass of component N from the cube is given by the divergence of $\rho_N j_{Ni}$ or

$$\frac{\partial(\rho_N j_{Ni})}{\partial x_i}. \tag{1.20}$$

The rate of increase of the mass of component N stored in the elemental volume is $\partial(\rho_N \alpha_N)/\partial t$ and hence conservation of mass of component N requires that

$$\frac{\partial}{\partial t}(\rho_N \alpha_N) + \frac{\partial(\rho_N j_{Ni})}{\partial x_i} = \mathcal{I}_N, \tag{1.21}$$

where $\mathcal{I}_N$ is the rate of transfer of mass to the phase N from the other phases per unit total volume. Such mass exchange would result from a phase change or chemical reaction. This is the first of several phase interaction terms that are identified and, for ease of reference, the quantities $\mathcal{I}_N$ are termed the *mass interaction terms*.

Clearly there is a continuity equation such as Eq. (1.21) for each phase or component present in the flow. They are referred to as the individual phase continuity equations (IPCE). However, because mass as a whole must be conserved whatever phase changes or chemical reactions are happening, it follows that

$$\sum_N \mathcal{I}_N = 0 \tag{1.22}$$

and hence the sum of all the IPCEs results in a combined phase continuity equation (CPCE) that does not involve $\mathcal{I}_N$ as follows:

$$\frac{\partial}{\partial t}\left(\sum_N \rho_N \alpha_N\right) + \frac{\partial}{\partial x_i}\left(\sum_N \rho_N j_{Ni}\right) = 0 \tag{1.23}$$

or, using Eqs. (1.4) and (1.8),

$$\frac{\partial \rho}{\partial t} + \frac{\partial}{\partial x_i}\left(\sum_N \rho_N \alpha_N u_{Ni}\right) = 0. \tag{1.24}$$

Notice that only under the conditions of *zero* relative velocity in which $u_{Ni} = u_i$ does this reduce to the mixture continuity equation (MCE), which is identical to that for an equivalent single-phase flow of density ρ as follows:

$$\frac{\partial \rho}{\partial t} + \frac{\partial}{\partial x_i}(\rho u_i) = 0. \tag{1.25}$$

We also record that for one-dimensional duct flow the individual phase continuity equation [Eq. (1.21)] becomes the following:

$$\frac{\partial}{\partial t}(\rho_N \alpha_N) + \frac{1}{A}\frac{\partial}{\partial x}(A \rho_N \alpha_N u_N) = \mathcal{I}_N, \tag{1.26}$$

where x is measured along the duct; $A(x)$ is the cross-sectional area; u_N, α_N are cross-sectionally averaged quantities, and $A\mathcal{I}_N$ is the rate of transfer of mass to the phase N per unit length of the duct. The sum over the constituents yields the following combined phase continuity equation

$$\frac{\partial \rho}{\partial t} + \frac{1}{A}\frac{\partial}{\partial x}\left(A \sum_N \rho_N \alpha_N u_n\right) = 0. \tag{1.27}$$

When all the phases travel at the same speed, $u_N = u$, this reduces to the following:

$$\frac{\partial \rho}{\partial t} + \frac{1}{A}\frac{\partial}{\partial x}(\rho A u) = 0. \tag{1.28}$$

Finally, we should make note of the form of the equations when the two components or species are intermingled rather than separated because we analyze several situations with gases diffusing through one another. Then both components occupy the entire volume and the void fractions are effectively unity so that the continuity equation [Eq. (1.21)] becomes the following:

$$\frac{\partial \rho_N}{\partial t} + \frac{\partial (\rho_N u_{Ni})}{\partial x_i} = \mathcal{I}_N. \tag{1.29}$$

1.2.3 Disperse Phase Number Continuity

Complementary to the equations of conservation of mass are the equations governing the conservation of the number of bubbles, drops, particles, and so on that constitute a disperse phase. If no such particles are created or destroyed within the elemental volume and if the number of particles of the disperse component, D, per unit total volume is denoted by n_D, it follows that

$$\frac{\partial n_D}{\partial t} + \frac{\partial}{\partial x_i}(n_D u_{Di}) = 0. \tag{1.30}$$

This is referred to as the disperse phase number equation (DPNE).

If the volume of the particles of component D is denoted by v_D it follows that

$$\alpha_D = n_D v_D \tag{1.31}$$

1.2 Equations of Motion

and substituting this into Eq. (1.21) one obtains the following:

$$\frac{\partial}{\partial t}(n_D \rho_D v_D) + \frac{\partial}{\partial x_i}(n_D u_{Di} \rho_D v_D) = \mathcal{I}_D. \tag{1.32}$$

Expanding this equation using Eq. (1.30) leads to the following relation for $\mathcal{I}_D$:

$$\mathcal{I}_D = n_D \left(\frac{\partial(\rho_D v_D)}{\partial t} + u_{Di} \frac{\partial(\rho_D v_D)}{\partial x_i} \right) = n_D \frac{D_D}{D_D t}(\rho_D v_D), \tag{1.33}$$

where $D_D/D_D t$ denotes the Lagrangian derivative following the disperse phase. This demonstrates a result that could, admittedly, be assumed a priori. Namely that the rate of transfer of mass to the component D in each particle, $\mathcal{I}_D/n_D$, is equal to the Lagrangian rate of increase of mass, $\rho_D v_D$, of each particle.

It is sometimes convenient in the study of bubbly flows to write the bubble number conservation equation in terms of a population, η, of bubbles per unit *liquid* volume rather than the number per unit total volume, n_D. Note that if the bubble volume is v and the volume fraction is α then

$$\eta = \frac{n_D}{(1-\alpha)}; \quad n_D = \frac{\eta}{(1+\eta v)}; \quad \alpha = \eta \frac{v}{(1+\eta v)} \tag{1.34}$$

and the bubble number conservation equation can be written as follows:

$$\frac{\partial u_{Di}}{\partial x_i} = -\frac{(1+\eta v)}{\eta} \frac{D_D}{D_D t}\left(\frac{\eta}{1+\eta v}\right). \tag{1.35}$$

If the number population, η, is assumed uniform and constant (which requires neglect of slip and the assumption of liquid incompressibility) then Eq. (1.35) can be written as follows:

$$\frac{\partial u_{Di}}{\partial x_i} = \frac{\eta}{1+\eta v} \frac{D_D v}{D_D t}. \tag{1.36}$$

In other words, the divergence of the velocity field is directly related to the Lagrangian rate of change in the volume of the bubbles.

1.2.4 Fick's Law

We digress briefly to complete the kinematics of two interdiffusing gases. Equation (1.29) represents the conservation of mass for the two gases in these circumstances. The kinematics are then completed by a statement of Fick's law that governs the interdiffusion. For the gas, A, this law is as follows:

$$u_{Ai} = u_i - \frac{\rho D}{\rho_A} \frac{\partial}{\partial x_i}\left(\frac{\rho_A}{\rho}\right), \tag{1.37}$$

where D is the diffusivity.

1.2.5 Continuum Equations for Conservation of Momentum

Continuing with the development of the differential equations, the next step is to apply the momentum principle to the elemental volume. Prior to doing so we make some minor modifications to that control volume to avoid some potential difficulties. Specifically, we deform the bounding surfaces so that they never cut through disperse phase particles but everywhere are within the continuous phase. Because it is already assumed that the dimensions of the particles are very small compared with the dimensions of the control volume, the required modification is correspondingly small. It is possible to proceed without this modification but several complications arise. For example, if the boundaries cut through particles, it would then be necessary to determine what fraction of the control volume surface is acted on by tractions within each of the phases and to face the difficulty of determining the tractions within the particles. Moreover, we need to evaluate the interacting force between the phases within the control volume and this is complicated by the issue of dealing with the parts of particles intersected by the boundary.

Now proceeding to the application of the momentum theorem for either the disperse (N = D) or continuous phase (N = C), the flux of momentum of the N component in the k direction through a side perpendicular to the i direction is $\rho_N j_{Ni} u_{Nk}$ and hence the net flux of momentum (in the k direction) out of the elemental volume is $\partial(\rho_N \alpha_N u_{Ni} u_{Nk})/\partial x_i$. The rate of increase of momentum of component N in the k direction within the elemental volume is $\partial(\rho_N \alpha_N u_{Nk})/\partial t$. Thus using the momentum conservation principle, the net force in the k direction acting on the component N in the control volume (of unit volume), $\mathcal{F}_{Nk}^T$, must be given by the following:

$$\mathcal{F}_{Nk}^T = \frac{\partial}{\partial t}(\rho_N \alpha_N u_{Nk}) + \frac{\partial}{\partial x_i}(\rho_N \alpha_N u_{Ni} u_{Nk}). \tag{1.38}$$

It is more difficult to construct the forces, $\mathcal{F}_{Nk}^T$, to complete the equations of motion. We must include body forces acting within the control volume, the force due to the pressure and viscous stresses on the exterior of the control volume, and, most particularly, the force that each component imposes on the other components within the control volume.

The first contribution is that due to an external force field on the component N within the control volume. In the case of gravitational forces, this is clearly given by the following:

$$\alpha_N \rho_N g_k, \tag{1.39}$$

where g_k is the component of the gravitational acceleration in the k direction (the direction of g is considered vertically downward).

The second contribution, namely that due to the tractions on the control volume, differs for the two phases because of the small deformation previously discussed. It is zero for the disperse phase. For the continuous phase we define the stress tensor,

1.2 Equations of Motion

σ_{Cki}, so that the contribution from the surface tractions to the force on that phase is as follows:

$$\frac{\partial \sigma_{Cki}}{\partial x_i}. \tag{1.40}$$

For future purposes it is also convenient to decompose σ_{Cki} into a pressure, $p_C = p$, and a deviatoric stress, σ_{Cki}^D, as follows:

$$\sigma_{Cki} = -p\delta_{ki} + \sigma_{Cki}^D, \tag{1.41}$$

where δ_{ki} is the Kronecker delta such that $\delta_{ki} = 1$ for $k = i$ and $\delta_{ij} = 0$ for $k \neq i$.

The third contribution to $\mathcal{F}_{Nk}^T$ is the force (per unit total volume) imposed *on* the component N *by* the other components within the control volume. We write this as $\mathcal{F}_{Nk}$ so that the individual phase momentum equation (IPME) becomes the following:

$$\frac{\partial}{\partial t}(\rho_N \alpha_N u_{Nk}) + \frac{\partial}{\partial x_i}(\rho_N \alpha_N u_{Ni} u_{Nk}) = \alpha_N \rho_N g_k + \mathcal{F}_{Nk} - \delta_N \left\{ \frac{\partial p}{\partial x_k} - \frac{\partial \sigma_{Cki}^D}{\partial x_i} \right\}, \tag{1.42}$$

where $\delta_D = 0$ for the disperse phase and $\delta_C = 1$ for the continuous phase.

Thus we identify the second of the interaction terms, namely the *force interaction*, $\mathcal{F}_{Nk}$. Note that, as in the case of the mass interaction $\mathcal{I}_N$, it must follow that

$$\sum_N \mathcal{F}_{Nk} = 0. \tag{1.43}$$

In disperse flows it is often useful to separate $\mathcal{F}_{Nk}$ into two components, one due to the pressure gradient in the continuous phase, $-\alpha_D \partial p / \partial x_k$, and the remainder, $\mathcal{F}'_{Dk}$, because of other effects such as the relative motion between the phases. Then we obtain the following:

$$\mathcal{F}_{Dk} = -\mathcal{F}_{Ck} = -\alpha_D \frac{\partial p}{\partial x_k} + \mathcal{F}'_{Dk}. \tag{1.44}$$

The IPME [Eq. (1.42)] is frequently used in a form in which the terms on the left-hand side are expanded and use is made of the continuity equation [Eq. (1.21)]. In single-phase flow this yields a Lagrangian time derivative of the velocity on the left-hand side. In the present case the use of the continuity equation results in the appearance of the mass interaction, $\mathcal{I}_N$. Specifically, one obtains the following:

$$\rho_N \alpha_N \left\{ \frac{\partial u_{Nk}}{\partial t} + u_{Ni} \frac{\partial u_{Nk}}{\partial x_i} \right\} = \alpha_N \rho_N g_k + \mathcal{F}_{Nk} - \mathcal{I}_N u_{Nk} - \delta_N \left\{ \frac{\partial p}{\partial x_k} - \frac{\partial \sigma_{Cki}^D}{\partial x_i} \right\}. \tag{1.45}$$

Viewed from a Lagrangian perspective, the left-hand side is the normal rate of increase of the momentum of the component N; the term $\mathcal{I}_N u_{Nk}$ is the rate of increase of the momentum in the component N due to the gain of mass by that phase.

If the momentum equations [Eq. (1.42)] for each of the components are added together, the resulting combined phase momentum equation (CPME) becomes the following:

$$\frac{\partial}{\partial t}\left(\sum_N \rho_N \alpha_N u_{Nk}\right) + \frac{\partial}{\partial x_i}\left(\sum_N \rho_N \alpha_N u_{Ni} u_{Nk}\right) = \rho g_k - \frac{\partial p}{\partial x_k} + \frac{\partial \sigma^D_{Cki}}{\partial x_i}. \quad (1.46)$$

Note that this equation reduces to the equation of motion for a single-phase flow only in the absence of relative motion, $u_{Ck} = u_{Dk}$. Note also that, in the absence of any motion (when the deviatoric stress is zero), Eq. (1.46) yields the appropriate hydrostatic pressure gradient $\partial p/\partial x_k = \rho g_k$ based on the mixture density, ρ.

Another useful limit is the case of uniform and constant sedimentation of the disperse component (volume fraction, $\alpha_D = \alpha = 1 - \alpha_C$) through the continuous phase under the influence of gravity. Then Eq. (1.42) yields the following:

$$0 = \alpha \rho_D g_k + \mathcal{F}_{Dk}$$
$$0 = \frac{\partial \sigma_{Cki}}{\partial x_i} + (1-\alpha)\rho_C g_k + \mathcal{F}_{Ck}. \quad (1.47)$$

But $\mathcal{F}_{Dk} = -\mathcal{F}_{Ck}$ and, in this case, the deviatoric part of the continuous phase stress should be zero (because the flow is a simple uniform stream) so that $\sigma_{Ckj} = -p$. It follows from Eq. (1.47) that

$$\mathcal{F}_{Dk} = -\mathcal{F}_{Ck} = -\alpha \rho_D g_k \quad \text{and} \quad \partial p/\partial x_k = \rho g_k \quad (1.48)$$

or, in words, the pressure gradient is hydrostatic.

Finally, note that the equivalent one-dimensional or duct flow form of the IPME is as follows:

$$\frac{\partial}{\partial t}(\rho_N \alpha_N u_N) + \frac{1}{A}\frac{\partial}{\partial x}(A \rho_N \alpha_N u_N^2) = -\delta_N \left\{\frac{\partial p}{\partial x} + \frac{P\tau_w}{A}\right\} + \alpha_N \rho_N g_x + \mathcal{F}_{Nx}, \quad (1.49)$$

where, in the usual pipe flow notation, $P(x)$ is the perimeter of the cross section and τ_w is the wall shear stress. In this equation, $A\mathcal{F}_{Nx}$ is the force imposed on the component N in the x direction by the other components per unit length of the duct. A sum over the constituents yields the combined phase momentum equation for duct flow, namely

$$\frac{\partial}{\partial t}\left(\sum_N \rho_N \alpha_N u_N\right) + \frac{1}{A}\frac{\partial}{\partial x}\left(A \sum_N \rho_N \alpha_N u_N^2\right) = -\frac{\partial p}{\partial x} - \frac{P\tau_w}{A} + \rho g_x \quad (1.50)$$

and, when all phases travel at the same velocity, $u = u_N$, this reduces to the following:

$$\frac{\partial}{\partial t}(\rho u) + \frac{1}{A}\frac{\partial}{\partial x}(A \rho u^2) = -\frac{\partial p}{\partial x} - \frac{P\tau_w}{A} + \rho g_x. \quad (1.51)$$

1.2.6 Disperse Phase Momentum Equation

At this point we should consider the relation between the equation of motion for an individual particle of the disperse phase and the disperse phase momentum equation (DPME) delineated in the last section. This relation is analogous to that between

1.2 Equations of Motion

the number continuity equation and the disperse phase continuity equation (DPCE). The construction of the equation of motion for an individual particle in an infinite fluid medium is discussed at some length in Chapter 2. It is sufficient at this point to recognize that we may write Newton's equation of motion for an individual particle of volume v_D in the following form:

$$\frac{D_D}{D_D t}(\rho_D v_D u_{Dk}) = F_k + \rho_D v_D g_k, \quad (1.52)$$

where $D_D/D_D t$ is the Lagrangian time derivative following the particle so that

$$\frac{D_D}{D_D t} \equiv \frac{\partial}{\partial t} + u_{Di}\frac{\partial}{\partial x_i} \quad (1.53)$$

and $\mathcal{F}_k$ is the force that the surrounding continuous phase imparts to the particle in the direction k. Note that F_k includes not only the force due to the velocity and acceleration of the particle relative to the fluid but also the *buoyancy* forces due to pressure gradients within the continuous phase. Expanding Eq. (1.52) and using Eq. (1.33) for the mass interaction, $\mathcal{I}_D$, one obtains the following form of the DPME:

$$\rho_D v_D \left\{ \frac{\partial u_{Dk}}{\partial t} + u_{Di}\frac{\partial u_{Dk}}{\partial x_i} \right\} + u_{Dk}\frac{\mathcal{I}_D}{n_D} = F_k + \rho_D v_D g_k. \quad (1.54)$$

Now examine the implication of this relation when considered alongside the IPME [Eq. (1.45)] for the disperse phase. Setting $\alpha_D = n_D v_D$ in Eq. (1.45), expanding and comparing the result with Eq. (1.54) (using the continuity equation [Eq. (1.21)] one observes the following:

$$\mathcal{F}_{Dk} = n_D F_k. \quad (1.55)$$

Hence the appropriate force interaction term in the disperse phase momentum equation is simply the sum of the fluid forces acting on the individual particles in a unit volume, namely $n_D F_k$. As an example, note that the steady, uniform sedimentation interaction force $\mathcal{F}_{Dk}$ given by Eq. (1.48), when substituted into Eq. (1.55), leads to the result $F_k = -\rho_D v_D g_k$ or, in words, a fluid force on an individual particle that precisely balances the weight of the particle.

1.2.7 Comments on Disperse Phase Interaction

In Section 1.2.6 the relation between the force interaction term, $\mathcal{F}_{Dk}$, and the force, F_k, acting on an individual particle of the disperse phase was established. In Chapter 2 we include extensive discussions of the forces acting on a single particle moving in an infinite fluid. Various forms of the fluid force, F_k, acting *on* the particle are presented [for example, Eqs. (2.47), (2.49), (2.50), (2.67), (2.71), and (3.20)] in terms of (a) the particle velocity, $V_k = u_{Dk}$, (b) the fluid velocity $U_k = u_{Ck}$ that would have existed at the center of the particle in the latter's absence, and (c) the relative velocity $W_k = V_k - U_k$.

Downstream of some disturbance that creates a relative velocity, W_k, the drag will tend to reduce that difference. It is useful to characterize the rate of equalization of the particle (mass, m_p, and radius, R) and fluid velocities by defining a velocity *relaxation* time, t_u. For example, it is common in dealing with gas flows laden with small droplets or particles to assume that the equation of motion can be approximated by just two terms, namely the particle inertia and a Stokes drag, which for a spherical particle is $6\pi \mu_C R W_k$ (see Section 2.2.2). It follows that the relative velocity decays exponentially with a time constant, t_u, given by the following:

$$t_u = m_p/6\pi R \mu_C. \tag{1.56}$$

This is known as the velocity relaxation time. A more complete treatment that includes other parametric cases and other fluid mechanical effects is contained in Sections 2.4.1 and 2.4.2.

There are many issues with the equation of motion for the disperse phase that have yet to be addressed. Many of these are delayed until Section 1.4 and others are addressed later in the book, for example in Sections 2.3.2, 2.4.3, and 2.4.4.

1.2.8 Equations for Conservation of Energy

The third fundamental conservation principle that is utilized in developing the basic equations of fluid mechanics is the principle of conservation of energy. Even in single-phase flow the general statement of this principle is complicated when energy transfer processes such as heat conduction and viscous dissipation are included in the analysis. Fortunately it is frequently possible to show that some of these complexities have a negligible effect on the results. For example, one almost always neglects viscous and heat conduction effects in preliminary analyses of gas dynamic flows. In the context of multiphase flows the complexities involved in a general statement of energy conservation are so numerous that it is of little value to attempt such generality. Thus we present only a simplified version that neglects, for example, viscous heating and the global conduction of heat (though not the heat transfer from one phase to another).

However, these limitations are often minor compared with other difficulties that arise in constructing an energy equation for multiphase flows. In single-phase flows it is usually adequate to assume that the fluid is in an equilibrium thermodynamic state at all points in the flow and that an appropriate thermodynamic constraint (for example, *constant* and *locally uniform* entropy or temperature) may be used to relate the pressure, density, temperature, entropy, and so on. In many multiphase flows the different phases and/or components are often *not* in equilibrium and consequently thermodynamic equilibrium arguments that might be appropriate for single-phase flows are no longer valid. Under those circumstances it is important to evaluate the heat and mass transfer occurring between the phases and/or components; this is discussed in Section 1.2.9.

1.2 Equations of Motion

In single-phase flow application of the principle of energy conservation to the control volume (CV) uses the following statement of the first law of thermodynamics:

> Rate of heat addition to the CV, Q
> + Rate of work done on the CV, W
> =
> Net flux of total internal energy out of CV
> + Rate of increase of total internal energy in CV.

In chemically nonreacting flows the total internal energy per unit mass, e^*, is the sum of the internal energy, e, the kinetic energy $u_i u_i /2$ (u_i are the velocity components), and the potential energy gz (where z is a coordinate measured in the vertically upward direction) as follows:

$$e^* = e + \frac{1}{2} u_i u_i + gz. \qquad (1.57)$$

Consequently, the energy equation in single-phase flow becomes the following:

$$\frac{\partial}{\partial t}(\rho e^*) + \frac{\partial}{\partial x_i}(\rho e^* u_i) = Q + W - \frac{\partial}{\partial x_j}(u_i \sigma_{ij}), \qquad (1.58)$$

where σ_{ij} is the stress tensor. Then, if there is no heat addition to ($Q = 0$) or external work done on ($W = 0$) the CV and if the flow is steady with no viscous effects (no deviatoric stresses), the energy equation for single-phase flow becomes the following:

$$\frac{\partial}{\partial x_i} \left\{ \rho u_i \left(e^* + \frac{p}{\rho} \right) \right\} = \frac{\partial}{\partial x_i} \{\rho u_i h^*\} = 0, \qquad (1.59)$$

where $h^* = e^* + p/\rho$ is the total enthalpy per unit mass. Thus, when the total enthalpy of the incoming flow is uniform, h^* is constant everywhere.

Now examine the task of constructing an energy equation for each of the components or phases in a multiphase flow. First, it is necessary to define a total internal energy density, e_N^*, for each component N such that

$$e_N^* = e_N + \frac{1}{2} u_{Ni} u_{Ni} + gz. \qquad (1.60)$$

Then an appropriate statement of the first law of thermodynamics for each phase (the individual phase energy equation, IPEE) is as follows:

> Rate of heat addition to N from outside CV, Q_N
> + Rate of work done to N by the exterior surroundings, $\mathcal{WA}_N$
> + Rate of heat transfer to N within the CV, $\mathcal{QI}_N$
> + Rate of work done to N by other components in CV, $\mathcal{WI}_N$
> =
> Rate of increase of total kinetic energy of N in CV
> + Net flux of total internal energy of N out of the CV,

where each of the terms is conveniently evaluated for a unit total volume.

First note that the last two terms can be written as follows:

$$\frac{\partial}{\partial t}(\rho_N \alpha_N e_N^*) + \frac{\partial}{\partial x_i}(\rho_N \alpha_N e_N^* u_{Ni}). \tag{1.61}$$

Turning then to the upper part of the equation, the first term due to external heating and to conduction of heat from the surroundings into the control volume is left as $\mathcal{Q}_N$. The second term contains two contributions: (1) minus the rate of work done by the stresses acting on the component N on the surface of the control volume and (2) the rate of external *shaft work*, $\mathcal{W}_N$, done on the component N. In evaluating the first of these, we make the same modification to the control volume as was discussed in the context of the momentum equation; specifically, we make small deformations to the control volume so that its boundaries lie wholly within the continuous phase. Then using the continuous phase stress tensor, σ_{Cij}, as defined in Eq. (1.41) the expressions for $\mathcal{WA}_N$ become the following:

$$\mathcal{WA}_C = \mathcal{W}_C + \frac{\partial}{\partial x_j}(u_{Ci}\sigma_{Cij}) \quad \text{and} \quad \mathcal{WA}_D = \mathcal{W}_D. \tag{1.62}$$

The individual phase energy equation may then be written as follows:

$$\frac{\partial}{\partial t}(\rho_N \alpha_N e_N^*) + \frac{\partial}{\partial x_i}(\rho_N \alpha_N e_N^* u_{Ni}) = \mathcal{Q}_N + \mathcal{W}_N + \mathcal{QI}_N + \mathcal{WI}_N + \delta_N \frac{\partial}{\partial x_j}(u_{Ci}\sigma_{Cij}). \tag{1.63}$$

Note that the two terms involving internal exchange of energy between the phases may be combined into an *energy interaction* term given by $\mathcal{E}_N = \mathcal{QI}_N + \mathcal{WI}_N$. It follows that

$$\sum_N \mathcal{QI}_N = 0 \quad \text{and} \quad \sum_N \mathcal{WI}_N = 0 \quad \text{and} \quad \sum_N \mathcal{E}_N = 0. \tag{1.64}$$

Moreover, the work done terms, $\mathcal{WI}_N$, may clearly be related to the interaction forces, $\mathcal{F}_{Nk}$. In a two-phase system with one disperse phase we have the following:

$$\mathcal{QI}_C = -\mathcal{QI}_D \quad \text{and} \quad \mathcal{WI}_C = -\mathcal{WI}_D = -u_{Di}\mathcal{F}_{Di} \quad \text{and} \quad \mathcal{E}_C = -\mathcal{E}_D. \tag{1.65}$$

As with the continuity and momentum equations, the individual phase energy equations can be summed to obtain the combined phase energy equation (CPEE). Then, denoting the total rate of external heat added (per unit total volume) by $\mathcal{Q}$ and the total rate of external shaft work done (per unite total volume) by $\mathcal{W}$ where

$$\mathcal{Q} = \sum_N \mathcal{Q}_N \quad \text{and} \quad \mathcal{W} = \sum_N \mathcal{W}_N \tag{1.66}$$

the CPEE becomes the following:

$$\frac{\partial}{\partial t}\left(\sum_N \rho_N \alpha_N e_N^*\right) + \frac{\partial}{\partial x_i}\left(-u_{Cj}\sigma_{Cij} + \sum_N \rho_N \alpha_N u_{Ni} e_N^*\right) = \mathcal{Q} + \mathcal{W}. \tag{1.67}$$

1.2 Equations of Motion

When the left-hand sides of the individual or combined phase equations [Eqs. (1.63) and (1.67)] are expanded and use is made of the continuity equation [Eq. (1.21)] and the momentum equation [Eq. (1.42)] (in the absence of deviatoric stresses), the results are known as the *thermodynamic* forms of the energy equations. Using equation Eq. (1.65) and the relation

$$e_N = c_{vN} T_N + \text{constant} \qquad (1.68)$$

between the internal energy, e_N, the specific heat at constant volume, c_{vN}, and the temperature, T_N, of each phase, the thermodynamic form of the IPEE can be written as follows:

$$\rho_N \alpha_N c_{vN} \left\{ \frac{\partial T_N}{\partial t} + u_{Ni} \frac{\partial T_N}{\partial x_i} \right\} = \delta_N \sigma_{Cij} \frac{\partial u_{Ci}}{\partial x_j} + \mathcal{Q}_N + \mathcal{W}_N + \mathcal{QI}_N \\ + \mathcal{F}_{Ni}(u_{Di} - u_{Ni}) - (e_N^* - u_{Ni}u_{Ni})\mathcal{I}_N \qquad (1.69)$$

and, summing these, the thermodynamic form of the CPEE is

$$\sum_N \left\{ \rho_N \alpha_N c_{vN} \left(\frac{\partial T_N}{\partial t} + u_{Ni} \frac{\partial T_N}{\partial x_i} \right) \right\} = \sigma_{Cij} \frac{\partial u_{Ci}}{\partial x_j} - \mathcal{F}_{Di}(u_{Di} - u_{Ci}) \\ - \mathcal{I}_D(e_D^* - e_C^*) + \sum_N u_{Ni}u_{Ni}\mathcal{I}_N. \qquad (1.70)$$

In Eqs. (1.69) and (1.70), it has been assumed that the specific heats, c_{vN}, are constant and uniform.

Finally, we note that the one-dimensional duct flow version of the IPEE, Eq. (1.63), is as follows:

$$\frac{\partial}{\partial t}(\rho_N \alpha_N e_N^*) + \frac{1}{A}\frac{\partial}{\partial x}(A\rho_N \alpha_N e_N^* u_N) = \mathcal{Q}_N + \mathcal{W}_N + \mathcal{E}_N - \delta_N \frac{\partial}{\partial x}(pu_C), \qquad (1.71)$$

where $A\mathcal{Q}_N$ is the rate of external heat addition to the component N per unit length of the duct, $A\mathcal{W}_N$ is the rate of external work done on component N per unit length of the duct, $A\mathcal{E}_N$ is the rate of energy transferred to the component N from the other phases per unit length of the duct, and p is the pressure in the continuous phase neglecting deviatoric stresses. The CPEE, Eq. (1.67), becomes the following:

$$\frac{\partial}{\partial t}\left(\sum_N \rho_N \alpha_N e_N^*\right) + \frac{1}{A}\frac{\partial}{\partial x}\left(\sum_N A\rho_N \alpha_N e_N^* u_N\right) = \mathcal{Q} + \mathcal{W} - \frac{\partial}{\partial x}(pu_C), \qquad (1.72)$$

where $A\mathcal{Q}$ is the total rate of external heat addition to the flow per unit length of the duct and $A\mathcal{W}$ is the total rate of external work done on the flow per unit length of the duct.

1.2.9 Heat Transfer between Separated Phases

In the preceding section, the rate of heat transfer, $\mathcal{QI}_N$, to each phase, N, from the other phases was left undefined. Now we address the functional form of this rate of heat transfer in the illustrative case of a two-phase flow consisting of a disperse solid particle or liquid droplet phase and a gaseous continuous phase.

In Section 1.2.7, we defined a relaxation time that typifies the natural attenuation of velocity differences between the phases. In an analogous manner, the temperatures of the phases might be different downstream of a flow disturbance and consequently there would be a second *relaxation* time associated with the equilibration of temperatures through the process of heat transfer between the phases. This temperature relaxation time is denoted by t_T and can be obtained by equating the rate of heat transfer from the continuous phase to the particle with the rate of increase of heat stored in the particle. The heat transfer to the particle can occur as a result of conduction, convection, or radiation and there are practical flows in which each of these mechanisms is important. For simplicity, we neglect the radiation component. Then, if the relative motion between the particle and the gas is sufficiently small, the only contributing mechanism is conduction and it will be limited by the thermal conductivity, k_C, of the gas (because the thermal conductivity of the particle is usually much greater). Then the rate of heat transfer to a particle (radius R) is given approximately by $2\pi R k_C (T_C - T_D)$, where T_C and T_D are representative temperatures of the gas and particle respectively.

Now we add in the component of heat transfer by the convection caused by relative motion. To do so we define the Nusselt number, Nu, as twice the ratio of the rate of heat transfer with convection to that without convection. Then the rate of heat transfer becomes Nu times the above result for conduction. Typically, the Nusselt number is a function of both the Reynolds number of the relative motion, $\text{Re} = 2WR/\nu_C$ [where W is the typical magnitude of $(u_{Di} - u_{Ci})$], and the Prandtl number, $\text{Pr} = \rho_C \nu_C c_{pC}/k_C$. One frequently used expression for Nu (see Ranz and Marshall 1952) is as follows:

$$\text{Nu} = 2 + 0.6 \text{Re}^{\frac{1}{2}} \text{Pr}^{\frac{1}{3}} \tag{1.73}$$

and, of course, this reduces to the pure conduction result, Nu = 2, when the second term on the right-hand side is small.

Assuming that the particle temperature has a roughly uniform value of T_D, it follows that

$$\mathcal{QI}_D = 2\pi R k_C \text{Nu}(T_C - T_D) n_D = \rho_D \alpha_D c_{sD} \frac{DT_D}{Dt}, \tag{1.74}$$

where the material derivative, D/Dt, follows the particle. This provides the equation that must be solved for T_D namely

$$\frac{DT_D}{Dt} = \frac{\text{Nu}}{2} \frac{(T_C - T_D)}{t_T}, \tag{1.75}$$

where

$$t_T = c_{sD} \rho_D R^2 / 3k_C. \tag{1.76}$$

Clearly t_T represents a typical time for equilibration of the temperatures in the two phases and is referred to as the *temperature relaxation time*.

The preceding construction of the temperature relaxation time and the equation for the particle temperature represents perhaps the simplest formulation that retains the essential ingredients. Many other effects may become important and require modification of the equations. Examples are the rarefied gas effects and turbulence effects. Moreover, the preceding was based on a uniform particle temperature and steady-state heat transfer correlations; in many flows heat transfer to the particles is highly transient and a more accurate heat transfer model is required. For a discussion of these effects the reader is referred to Rudinger (1969) and Crowe *et al.* (1998).

1.3 Interaction with Turbulence

1.3.1 Particles and Turbulence

Turbulent flows of a single Newtonian fluid, even those of quite simple external geometry such as a fully developed pipe flow, are very complex and their solution at high Reynolds numbers requires the use of empirical models to represent the unsteady motions. It is self-evident that the addition of particles to such a flow will result in the following:

1. Complex unsteady motions of the particles that may result in nonuniform spatial distribution of the particles and, perhaps, particle segregation. It can also result in particle agglomeration or in particle fission, especially if the particles are bubbles or droplets.
2. Modifications of the turbulence itself caused by the presence and motions of the particles. One can visualize that the turbulence could be damped by the presence of particles, or it could be enhanced by the wakes and other flow disturbances that the motion of the particles may introduce.

Since the mid-1970s, a start has been made in the understanding of these complicated issues, though many aspects remain to be understood. The advent of laser Doppler velocimetry resulted in the first measurements of these effects, and the development of direct numerical simulation allowed the first calculations of these complex flows, albeit at rather low Reynolds numbers. Here we are confined to a brief summary of these complex issues. The reader is referred to the early review of Hetsroni (1989) and the text by Crowe *et al.* (1998) for a summary of the current understanding.

To set the stage, recall that turbulence is conveniently characterized at any point in the flow by the Kolmogorov length and time scales, λ and τ, given by

$$\lambda = \left(\frac{\nu^3}{\epsilon}\right)^{\frac{1}{4}} \quad \text{and} \quad \tau = \left(\frac{\nu}{\epsilon}\right)^{\frac{1}{2}} \tag{1.77}$$

where ν is the kinematic viscosity and ϵ is the mean rate of dissipation per unit mass of fluid. Because ϵ is proportional to U^3/ℓ, where U and ℓ are the typical velocity and

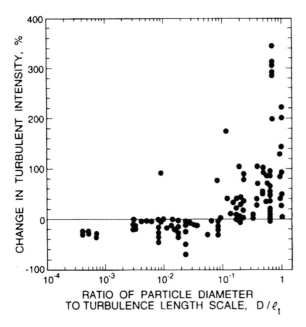

Figure 1.4. The percentage change in the turbulence intensity as a function of the ratio of particle size to turbulence length scale, D/ℓ_t, from a wide range of experiments. Adapted from Gore and Crowe (1989).

dimension of the flow, it follows that

$$\lambda/\ell \propto \mathrm{Re}^{-\frac{3}{4}} \quad \text{and} \quad U\tau/\ell \propto \mathrm{Re}^{-\frac{1}{2}} \qquad (1.78)$$

and the difficulties in resolving the flow either by measurement or by computation increase as Re increases.

Gore and Crowe (1989) collected data from a wide range of turbulent pipe and jet flows (all combinations of gas, liquid, and solid flows; volume fractions from 2.5×10^{-6} to 0.2; density ratios from 0.001 to 7500; and Reynolds numbers from 8000 to 100,000) and constructed Figure 1.4, which plots the fractional change in the turbulence intensity (defined as the rms fluctuating velocity) as a result of the introduction of the disperse phase against the ratio of the particle size to the turbulent length scale, D/ℓ_t. They judge that the most appropriate turbulent length scale, ℓ_t, is the size of the most energetic eddy. Single-phase experiments indicate that ℓ_t is about 0.2 times the radius in a pipe flow and 0.039 times the distance from the exit in a jet flow. To explain Figure 1.4 Gore and Crowe argue that when the particles are small compared with the turbulent length scale, they tend to follow the turbulent fluid motions and in doing so absorb energy from them, thus reducing the turbulent energy. It appears that the turbulence reduction is a strong function of Stokes number, $\mathrm{St} = m_p/6\pi R\mu\tau$, the ratio of the particle relaxation time, $m_p/6\pi R\mu$, to the Kolmogorov time scale, τ. A few experiments (Eaton 1994, Kulick *et al.* 1994) suggest that the maximum reduction occurs at St values of the order of unity though other features of the flow may also influence the effect. Of course, the change in the turbulence intensity also depends on the particle concentration. Figure 1.5, from Paris and Eaton (2001), shows one example of how the turbulent kinetic energy and the rate of viscous dissipation depend on the mass fraction of particles for a case in which D/ℓ_t is small.

1.3 Interaction with Turbulence

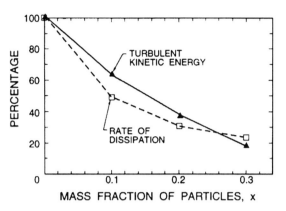

Figure 1.5. The percentage change in the turbulent kinetic energy and the rate of viscous dissipation with mass fraction for a channel flow of 150-μm glass spheres suspended in air (from Paris and Eaton 2001).

Conversely, large particles do not follow the turbulent motions and the relative motion produces wakes that tend to add to the turbulence (see, for example, Parthasarathy and Faeth 1990). Under these circumstances, when the response times of the particles are comparable with or greater than the typical times associated with the fluid motion, the turbulent flow with particles is more complex due to the effects of relative motion. Particles in a gas tend to be centrifuged out of the more intense vortices and accumulate in the shear zones in between. Figure 1.6 is a photograph of a turbulent flow of a gas loaded with particles showing the accumulation of particles in shear zones between strong vortices. Conversely, bubbles in a liquid flow tend to accumulate in the center of the vortices.

Analyses of turbulent flows with particles or bubbles are currently the subject of active research, and many issues remain. The literature includes a number of heuristic and approximate quantitative analyses of the enhancement of turbulence due to particle relative motion. Examples are the work of Yuan and Michaelides (1992) and Kenning

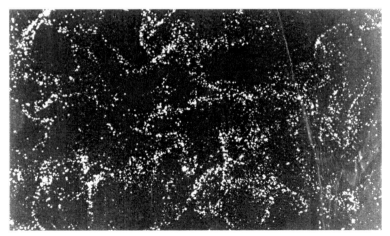

Figure 1.6. Image of the centerplane of a fully developed, turbulent channel flow of air loaded with 28-μm particles. The area is 50 × 30 mm. Reproduced from Fessler *et al.* (1994) with the authors' permission.

and Crowe (1997). The latter relate the percentage change in the turbulence intensity due to the particle wakes; this yields a percentage change that is a function not only of D/ℓ_t but also of the mean relative motion and the density ratio. They show qualitative agreement with some of the data included in Figure 1.4.

An alternative to these heuristic methodologies is the use of direct numerical simulations (DNS) to examine the details of the interaction between the turbulence and the particles or bubbles. Such simulations have been carried out both for solid particles (for example, Squires and Eaton 1990, Elghobashi and Truesdell 1993) and for bubbles (for example, Pan and Banarejee 1997). Because each individual simulation is so time consuming and leads to complex consequences, it is not yet possible to draw general conclusions over a wide parameter range. However, the kinds of particle segregation mentioned earlier are readily apparent in the simulations.

1.3.2 Effect on Turbulence Stability

The issue of whether particles promote or delay transition to turbulence is somewhat distinct from their effect on developed turbulent flows. Saffman (1962) investigated the effect of dust particles on the stability of parallel flows and showed theoretically that if the relaxation time of the particles, t_u, is small compared with ℓ/U, the characteristic time of the flow, then the dust destablizes the flow. Conversely if $t_u \gg \ell/U$ the dust destablizes the flow.

In a somewhat similar investigation of the effect of bubbles on the stability of parallel liquid flows, d'Agostino *et al.* (1997) found that the effect depends on the relative magnitude of the most unstable frequency, ω_m, and the natural frequency of the bubbles, ω_n (see Section 4.4.1). When the ratio, $\omega_m/\omega_n \ll 1$, the primary effect of the bubbles is to increase the effective compressibility of the fluid, and because increased compressibility causes increased stability, the bubbles are stabilizing. Conversely, at or near resonance, when ω_m/ω_n is of order unity, there are usually bands of frequencies in which the flow is less stable and the bubbles are therefore destabilizing.

In summary, when the response times of the particles or bubbles (both the relaxation time and the natural period of volume oscillation) are short compared with the typical times associated with the fluid motion, the particles simply alter the effective properties of the fluid, its effective density, viscosity, and compressibility. It follows that under these circumstances the stability is governed by the effective Reynolds number and effective Mach number. Saffman considered dusty gases at low volume concentrations, α, and low Mach numbers; under those conditions the net effect of the dust is to change the density by $(1 + \alpha \rho_S/\rho_G)$ and the viscosity by $(1 + 2.5\alpha)$. The effective Reynolds number therefore varies like $(1 + \alpha \rho_S/\rho_G)/(1 + 2.5\alpha)$. Because $\rho_S \gg \rho_G$ the effective Reynolds number is increased and the dust is therefore destabilizing. In the case of d'Agostino *et al.* the primary effect of the bubbles (when $\omega_m \gg \omega_n$) is to change the compressibility of the mixture. Because such a change is stabilizing in a single-phase flow, the result is that the bubbles tend to stabilize the flow.

1.4 Comments on the Equations of Motion

Conversely, when the response times are comparable with or greater than the typical times associated with the fluid motion, the particles do not follow the motions of the continuous phase. The disturbances caused by this relative motion tend to generate unsteady motions and promote instability in the continuous phase.

1.4 Comments on the Equations of Motion

In Sections 1.2.2 through 1.2.8 we assembled the basic form for the equations of motion for a multiphase flow that would be used in a two-fluid model. However, these only provide the initial framework; there are many additional complications that must be addressed. The relative importance of these complications varies greatly from one type of multiphase flow to another. Consequently the level of detail with which they must be addressed varies enormously. In this general introduction we can indicate only the various types of complications that can arise.

1.4.1 Averaging

As discussed in Section 1.2.1, when the ratio of the particle size, D, to the typical dimension of the averaging volume (estimated as the typical length, ϵ, over which there is significant change in the averaged flow properties) becomes significant, several issues arise (see Hinze 1959, Vernier and Delhaye 1968, Nigmatulin 1979, Reeks 1992). The reader is referred to Slattery (1972) or Crowe et al. (1997) for a systematic treatment of these issues; only a summary is presented here. Clearly an appropriate volume average of a property, Q_C, of the continuous phase is given by $\langle Q_C \rangle$ where

$$\langle Q_C \rangle = \frac{1}{V_C} \int_{V_C} Q_C dV, \tag{1.79}$$

where V_C denotes the volume of the continuous phase within the control volume, V. For present purposes, it is also convenient to define an average

$$\overline{Q_C} = \frac{1}{V} \int_{V_C} Q_C dV = \alpha_C \langle Q_C \rangle \tag{1.80}$$

over the whole of the control volume.

Because the conservation equations discussed in the preceding sections contain derivatives in space and time and because the leading order set of equations we seek are versions in which all the terms are averaged over some local volume, the equations contain averages of spatial gradients and time derivatives. For these terms to be evaluated they must be converted to derivatives of the volume-averaged properties. Those relations take the form (Crowe et al. 1997):

$$\overline{\frac{\partial Q_C}{\partial x_i}} = \frac{\partial \overline{Q_C}}{\partial x_i} - \frac{1}{V} \int_{S_D} Q_C n_i dS, \tag{1.81}$$

where S_D is the total surface area of the particles within the averaging volume. With regard to the time derivatives, if the volume of the particles is not changing with time then

$$\frac{\overline{\partial Q_C}}{\partial t} = \frac{\partial \overline{Q_C}}{\partial t}, \tag{1.82}$$

but if the location of a point on the surface of a particle relative to its center is given by r_i and if r_i is changing with time (for example, growing bubbles) then

$$\frac{\overline{\partial Q_C}}{\partial t} = \frac{\partial \overline{Q_C}}{\partial t} + \frac{1}{V} \int_{S_D} Q_C \frac{Dr_i}{Dt} dS. \tag{1.83}$$

When the Eqs. (1.81) and (1.83) are employed in the development of appropriate averaged conservation equations, the integrals over the surface of the disperse phase introduce additional terms that might not have been anticipated (see Crowe et al. 1997 for specific forms of those equations). Here it is of value to observe that the magnitude of the additional surface integral term in Eq. (1.81) is of the order of $(D/\epsilon)^2$. Consequently these additional terms are small as long as D/ϵ is sufficiently small.

1.4.2 Averaging Contributions to the Mean Motion

Thus far we have discussed only those additional terms introduced as a result of the fact that the gradient of the average may differ from the average of the gradient. Inspection of the form of the basic equations (for example, the continuity equation [Eq. (1.21)] or the momentum equation [Eq. (1.42)]) readily demonstrates that additional averaging terms are introduced because the average of a product is different from the product of averages. In single-phase flows, the *Reynolds stress terms* in the averaged equations of motion for turbulent flows are a prime example of this phenomenon. We use the name *quadratic rectification terms* to refer to the appearance in the averaged equations of motion of the mean of two fluctuating components of velocity and/or volume fraction. Of course multiphase flows also exhibit conventional Reynolds stress terms when they become turbulent (see Section 1.3 for more on the complicated subject of turbulence in multiphase flows). But even multiphase flows that are not turbulent in the strictest sense exhibit variations in the velocities due to the flows around particles, and these variations yield quadratic rectification terms. These must be recognized and modeled when considering the effects of locally nonuniform and unsteady velocities on the equations of motion. Much more has to be learned of both the laminar and turbulent quadratic rectification terms before these can be confidently incorporated in model equations for multiphase flow. Both experiments and computer simulation are valuable in this regard.

One simpler example in which the fluctuations in velocity have been measured and considered is the case of concentrated granular flows in which direct particle/particle

1.4 Comments on the Equations of Motion

interactions create particle velocity fluctuations. These particle velocity fluctuations and the energy associated with them (the so-called granular temperature) have been studied both experimentally and computationally (see Chapter 13) and their role in the effective continuum equations of motion is better understood than in more complex multiphase flows.

With two interacting phases or components, the additional terms that emerge from an averaging process can become extremely complex. In recent decades a number of valiant efforts have been made to codify these issues and establish at least the forms of the important terms that result from these interactions. For example, Wallis (1991) has devoted considerable effort to identify the inertial coupling of spheres in inviscid, locally irrotational flow. Arnold, Drew, and Lahey (1989) and Drew (1991) have focused on the application of cell methods (see Section 2.4.3) to interacting multiphase flows. Both these authors as well as Sangani and Didwania (1993) and Zhang and Prosperetti (1994) have attempted to include the fluctuating motions of the particles (as in granular flows) in the construction of equations of motion for the multiphase flow; Zhang and Prosperetti also provide a useful comparative summary of these various averaging efforts. However, it is also clear that these studies have some distance to go before they can be incorporated into any real multiphase flow prediction methodology.

1.4.3 Averaging in Pipe Flows

One specific example of a quadratic rectification term (in this case a discrepancy between the product of an average and the average of a product) is that recognized by Zuber and Findlay (1965). To account for the variations in velocity and volume fraction over the cross section of a pipe in constructing the one-dimensional equations of pipe flow, they found it necessary to introduce a distribution parameter, C_0, defined by the following:

$$C_0 = \frac{\overline{\alpha j}}{\overline{\alpha}\, \overline{j}}, \tag{1.84}$$

where the overbar now represents an average over the cross section of the pipe. The importance of C_0 is best demonstrated by observing that it follows from Eq. (1.16) that the cross-sectionally averaged volume fraction, $\overline{\alpha_A}$, is now related to the volume fluxes, $\overline{j_A}$ and $\overline{j_B}$ by the following:

$$\overline{\alpha_A} = \frac{1}{C_0} \frac{\overline{j_A}}{(\overline{j_A} + \overline{j_B})}. \tag{1.85}$$

Values of C_0 of the order of 1.13 (Zuber and Findlay 1965) or 1.25 (Wallis 1969) appear necessary to match the experimental observations.

1.4.4 Modeling with the Combined Phase Equations

One of the simpler approaches is to begin by modeling the combined phase equations [Eqs. (1.24), (1.46), and (1.67)] and hence avoid having to codify the mass, force, and energy interaction terms. By defining mixture properties such as the density, ρ, and the total volumetric flux, j_i, one can begin to construct equations of motion in terms of those properties. But none of the summation terms (equivalent to various weighted averages) in the combined phase equations can be written accurately in terms of these mixture properties. For example, the summations

$$\sum_N \rho_N \alpha_N u_{Ni} \quad \text{and} \quad \sum_N \rho_N \alpha_N u_{Ni} u_{Nk} \tag{1.86}$$

are not necessarily given with any accuracy by ρj_i and $\rho j_i j_k$. Indeed, the discrepancies are additional rectification terms that would require modeling in such an approach. Thus any effort to avoid addressing the mass, force, and energy interaction terms by focusing exclusively on the mixture equations of motion immediately faces difficult modeling questions.

1.4.5 Mass, Force, and Energy Interaction Terms

Most multiphase flow modeling efforts concentrate on the individual phase equations of motion and must therefore face the issues associated with construction of $\mathcal{I}_N$, the mass interaction term, $\mathcal{F}_{Nk}$, the force interaction term, and $\mathcal{E}_N$, the energy interaction term. These represent the core of the problem in modeling multiphase flows and there exist no universally applicable methodologies that are independent of the topology of the flow, the flow pattern. Indeed, efforts to find systems of model equations that would be applicable to a range of flow patterns would seem fruitless. Therein lies the main problem for the user who may not be able to predict the flow pattern and therefore has little hope of finding an accurate and reliable method to predict flow rates, pressure drops, temperatures, and other flow properties.

The best that can be achieved with the present state of knowledge is to attempt to construct heuristic models for $\mathcal{I}_N$, $\mathcal{F}_{Nk}$, and $\mathcal{E}_N$ given a particular flow pattern. Substantial efforts have been made in this direction particularly for dispersed flows; the reader is directed to the excellent reviews by Hinze (1961), Drew (1983), Gidaspow (1994), and Crowe et al. (1998) among others. Both direct experimentation and computer simulation have been used to create data from which heuristic expressions for the interaction terms could be generated. Computer simulations are particularly useful not only because high-fidelity instrumentation for the desired experiments is often very difficult to develop but also because one can selectively incorporate a range of different effects and thereby evaluate the importance of each.

It is important to recognize that there are several constraints to which any mathematical model must adhere. Any violation of those constraints is likely to produce strange and physically inappropriate results (see Garabedian 1964). Thus, the system

1.4 Comments on the Equations of Motion

of equations must have appropriate frame-indifference properties (see, for example, Ryskin and Rallison 1980). It must also have real characteristics; Prosperetti and Jones (1987) show that some models appearing in the literature do have real characteristics, whereas others do not.

In this book Chapters 2, 3, and 4 review what is known of the behavior of individual particles, bubbles, and drops, with a view to using this information to construct $\mathcal{I}_N$, $\mathcal{F}_{Nk}$, and $\mathcal{E}_N$ and therefore the equations of motion for particular forms of multiphase flow.

2

Single-Particle Motion

2.1 Introduction

This chapter briefly reviews the issues and problems involved in constructing the equations of motion for individual particles, drops, or bubbles moving through a fluid. For convenience we use the generic name *particle* to refer to the finite pieces of the disperse phase or component. The analyses are implicitly confined to those circumstances in which the interactions between neighboring particles are negligible. In very dilute multiphase flows in which the particles are very small compared with the global dimensions of the flow and are very far apart compared with the particle size, it is often sufficient to solve for the velocity and pressure, $u_i(x_i, t)$ and $p(x_i, t)$, of the continuous suspending fluid while ignoring the particles or disperse phase. Given this solution one could then solve an equation of motion for the particle to determine its trajectory. This chapter focuses on the construction of such a particle or bubble equation of motion.

The body of fluid mechanical literature on the subject of flows around particles or bodies is very large indeed. Here we present a summary that focuses on a spherical particle of radius R and employs the following common notation. The components of the translational velocity of the center of the particle is denoted by $V_i(t)$. The velocity that the fluid would have had at the location of the particle center in the absence of the particle is denoted by $U_i(t)$. Note that such a concept is difficult to extend to the case of interactive multiphase flows. Finally, the velocity of the particle relative to the fluid is denoted by $W_i(t) = V_i - U_i$.

Frequently the approach used to construct equations for $V_i(t)$ (or $W_i(t)$) given $U_i(x_i, t)$ is to individually estimate all the fluid forces acting on the particle and to equate the total fluid force, F_i, to $m_p dV_i/dt$ (where m_p is the particle mass, assumed constant). These fluid forces may include forces due to buoyancy, added mass, drag, and so on. In the absence of fluid acceleration ($dU_i/dt = 0$) such an approach can be made unambiguously; however, in the presence of fluid acceleration, this kind of heuristic approach can be misleading. Hence we concentrate in the next few sections on a fundamental fluid mechanical approach that minimizes possible ambiguities. The

2.2 Flows Around a Sphere

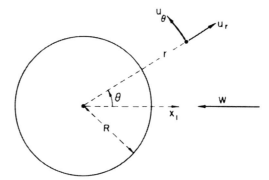

Figure 2.1. Notation for a spherical particle.

classical results for a spherical particle or bubble are reviewed first. The analysis is confined to a suspending fluid that is incompressible and Newtonian so that the basic equations to be solved are the continuity equation.

$$\frac{\partial u_j}{\partial x_j} = 0 \tag{2.1}$$

and the Navier–Stokes equations

$$\rho_C \left\{ \frac{\partial u_i}{\partial t} + u_j \frac{\partial u_i}{\partial x_j} \right\} = -\frac{\partial p}{\partial x_i} + \rho_C \nu_C \frac{\partial^2 u_i}{\partial x_j \partial x_j} \tag{2.2}$$

where ρ_C and ν_C are the density and kinematic viscosity of the suspending fluid. It is assumed that the only external force is that due to gravity, g. Then the actual pressure is $p' = p - \rho_C g z$, where z is a coordinate measured vertically upward. Furthermore, to maintain clarity we confine our attention to rectilinear relative motion in a direction conveniently chosen to be the x_1 direction.

2.2 Flows Around a Sphere

2.2.1 At High Reynolds Number

For steady flows about a sphere in which $dU_i/dt = dV_i/dt = dW_i/dt = 0$, it is convenient to use a coordinate system, x_i, fixed in the particle as well as polar coordinates (r, θ) and velocities u_r, u_θ as defined in Figure 2.1.

Then Eqs. (2.1) and (2.2) become the following:

$$\frac{1}{r^2}\frac{\partial}{\partial r}(r^2 u_r) + \frac{1}{r \sin\theta}\frac{\partial}{\partial \theta}(u_\theta \sin\theta) = 0 \tag{2.3}$$

and

$$\rho_C \left\{ \frac{\partial u_r}{\partial t} + u_r \frac{\partial u_r}{\partial r} + \frac{u_\theta}{r}\frac{\partial u_r}{\partial \theta} - \frac{u_\theta^2}{r} \right\} \tag{2.4}$$
$$= -\frac{\partial p}{\partial r} + \rho_C \nu_C \left\{ \frac{1}{r^2}\frac{\partial}{\partial r}\left(r^2 \frac{\partial u_r}{\partial r}\right) + \frac{1}{r^2 \sin\theta}\frac{\partial}{\partial \theta}\left(\sin\theta \frac{\partial u_r}{\partial \theta}\right) - \frac{2 u_r}{r^2} - \frac{2}{r^2}\frac{\partial u_\theta}{\partial \theta} \right\}$$

$$\rho_C \left\{ \frac{\partial u_\theta}{\partial t} + u_r \frac{\partial u_\theta}{\partial r} + \frac{u_\theta}{r} \frac{\partial u_\theta}{\partial \theta} + \frac{u_r u_\theta}{r} \right\} = -\frac{1}{r} \frac{\partial p}{\partial \theta} \qquad (2.5)$$
$$+ \rho_C \nu_C \left\{ \frac{1}{r^2} \frac{\partial}{\partial r} \left(r^2 \frac{\partial u_\theta}{\partial r} \right) + \frac{1}{r^2 \sin \theta} \frac{\partial}{\partial \theta} \left(\sin \theta \frac{\partial u_\theta}{\partial \theta} \right) + \frac{2}{r^2} \frac{\partial u_r}{\partial \theta} - \frac{u_\theta}{r^2 \sin^2 \theta} \right\}.$$

The Stokes streamfunction, ψ, is defined to satisfy continuity automatically:

$$u_r = \frac{1}{r^2 \sin \theta} \frac{\partial \psi}{\partial \theta}; \quad u_\theta = -\frac{1}{r \sin \theta} \frac{\partial \psi}{\partial r} \qquad (2.6)$$

and the inviscid potential flow solution is as follows:

$$\psi = -\frac{W r^2}{2} \sin^2 \theta - \frac{D}{r} \sin^2 \theta \qquad (2.7)$$
$$u_r = -W \cos \theta - \frac{2D}{r^3} \cos \theta \qquad (2.8)$$
$$u_\theta = +W \sin \theta - \frac{D}{r^3} \sin \theta \qquad (2.9)$$
$$\phi = -W r \cos \theta + \frac{D}{r^2} \cos \theta, \qquad (2.10)$$

where, because of the boundary condition $(u_r)_{r=R} = 0$, it follows that $D = -WR^3/2$. In potential flow one may also define a velocity potential, ϕ, such that $u_i = \partial \phi / \partial x_i$. The classic problem with such solutions is the fact that the drag is zero, a circumstance termed D'Alembert's paradox. The flow is symmetric about the $x_2 x_3$ plane through the origin and there is no wake.

The real viscous flows around a sphere at large Reynolds numbers, $Re = 2WR/\nu_C > 1$, are well documented. In the range from $\sim 10^3$ to 3×10^5, laminar boundary layer separation occurs at $\theta \cong 84°$ and a large wake is formed behind the sphere (see Figure 2.2). Close to the sphere the *near-wake* is laminar; further downstream transition and turbulence occurring in the shear layers spreads to generate a turbulent *far-wake*. As the Reynolds number increases the shear layer transition moves forward until, quite abruptly, the turbulent shear layer reattaches to the body, resulting in a major change in the final position of separation ($\theta \cong 120°$) and in the form of the turbulent wake (Figure 2.2). Associated with this change in flow pattern is a dramatic decrease in the drag coefficient, C_D (defined as the drag force on the body in the negative x_1 direction divided by $\frac{1}{2} \rho_C W^2 \pi R^2$), from a value of ~ 0.5 in the laminar separation regime to a value of ~ 0.2 in the turbulent separation regime (Figure 2.3). At values of Re less than $\sim 10^3$ the flow becomes quite unsteady with periodic shedding of vortices from the sphere.

2.2.2 At Low Reynolds Number

At the other end of the Reynolds number spectrum is the classic Stokes solution for flow around a sphere. In this limit the terms on the left-hand side of Eq. (2.2) are

2.2 Flows Around a Sphere

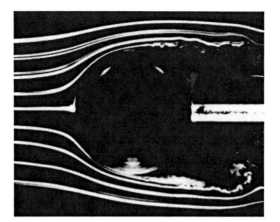

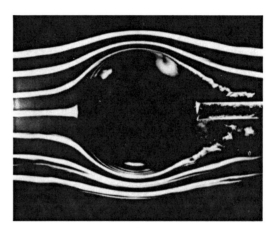

Figure 2.2. Smoke visualization of the nominally steady flows (from left to right) past a sphere showing, at the top, laminar separation at Re $= 2.8 \times 10^5$ and, on the bottom, turbulent separation at Re $= 3.9 \times 10^5$. Photographs by F.N.M. Brown, reproduced with the permission of the University of Notre Dame.

neglected and the viscous term is retained. This solution has the following form:

$$\psi = \sin^2\theta \left\{ -\frac{Wr^2}{2} + \frac{A}{r} + Br \right\} \quad (2.11)$$

$$u_r = \cos\theta \left\{ -W + \frac{2A}{r^3} + \frac{2B}{r} \right\} \quad (2.12)$$

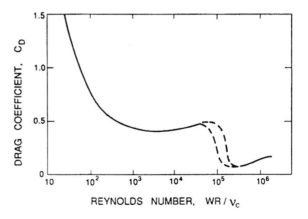

Figure 2.3. Drag coefficient on a sphere as a function of Reynolds number. Dashed curves indicate the drag crisis regime in which the drag is very sensitive to other factors such as the free stream turbulence.

$$u_\theta = -\sin\theta \left\{ -W - \frac{A}{r^3} + \frac{B}{r} \right\}, \qquad (2.13)$$

where A and B are constants to be determined from the boundary conditions on the surface of the sphere. The force, F, on the *particle* in the x_1 direction is as follows:

$$F_1 = \frac{4}{3}\pi R^2 \rho_C \nu_C \left\{ -\frac{4W}{R} + \frac{8A}{R^4} + \frac{2B}{R^2} \right\}. \qquad (2.14)$$

Several subcases of this solution are of interest in the present context. The first is the classic Stokes (1851) solution for a solid sphere in which the no-slip boundary condition, $(u_\theta)_{r=R} = 0$, is applied [in addition to the kinematic condition $(u_r)_{r=R} = 0$]. This set of boundary conditions, referred to as the Stokes boundary conditions, leads to the following:

$$A = -\frac{WR^3}{4}, \quad B = +\frac{3WR}{4}, \quad \text{and} \quad F_1 = -6\pi \rho_C \nu_C WR. \qquad (2.15)$$

The second case originates with Hadamard (1911) and Rybczynski (1911) who suggested that, in the case of a bubble, a condition of zero shear stress on the sphere surface would be more appropriate than a condition of zero tangential velocity, u_θ. Then it transpires that

$$A = 0, \quad B = +\frac{WR}{2}, \quad \text{and} \quad F_1 = -4\pi \rho_C \nu_C WR. \qquad (2.16)$$

Real bubbles may conform to either the Stokes or Hadamard–Rybczynski solutions depending on the degree of contamination of the bubble surface, as discussed in more detail in section 3.3. Finally, it is of interest to observe that the potential flow solution given in Eqs. (2.7) to (2.10) is also a subcase with

$$A = +\frac{WR^3}{2}, \quad B = 0, \quad \text{and} \quad F_1 = 0. \qquad (2.17)$$

However, another paradox, known as the Whitehead paradox, arises when the validity of these Stokes flow solutions at small (rather than zero) Reynolds numbers is considered. The nature of this paradox can be demonstrated by examining the magnitude of the neglected term, $u_j \partial u_i/\partial x_j$, in the Navier–Stokes equations relative to the magnitude of the retained term $\nu_C \partial^2 u_i/\partial x_j \partial x_j$. As is evident from Eq. (2.11), far from the sphere the former is proportional to $W^2 R/r^2$, whereas the latter behaves like $\nu_C WR/r^3$. It follows that although the retained term dominates close to the body (provided $\text{Re} = 2WR/\nu_C \ll 1$), there will always be a radial position, r_c, given by $R/r_c = \text{Re}$ beyond which the neglected term will exceed the retained viscous term. Hence, even if $\text{Re} \ll 1$, the Stokes solution is not uniformly valid. Recognizing this limitation, Oseen (1910) attempted to correct the Stokes solution by retaining in the basic equation an approximation to $u_j \partial u_i/\partial x_j$ that would be valid in

2.2 Flows Around a Sphere

the far field, $-W\partial u_i/\partial x_1$. Thus the Navier–Stokes equations are approximated by the following:

$$-W\frac{\partial u_i}{\partial x_1} = -\frac{1}{\rho_C}\frac{\partial p}{\partial x_i} + \nu_C\frac{\partial^2 u_i}{\partial x_j \partial x_j}. \qquad (2.18)$$

Oseen was able to find a closed form solution to this equation that satisfies the Stokes boundary conditions approximately as follows:

$$\psi = -WR^2 \left\{ \frac{r^2\sin^2\theta}{2R^2} + \frac{R\sin^2\theta}{4r} + \frac{3\nu_C(1+\cos\theta)}{2WR}\left(1 - e^{\frac{Wr}{2\nu_C(1-\cos\theta)}}\right) \right\}, \qquad (2.19)$$

which yields a drag force as follows:

$$F_1 = -6\pi\rho_C\nu_C WR\left\{1 + \frac{3}{16}\mathrm{Re}\right\}. \qquad (2.20)$$

It is readily shown that Eq. (2.19) reduces to Eq. (2.11) as Re → 0. The corresponding solution for the Hadamard–Rybczynski boundary conditions is not known to the author; its validity would be more questionable because, unlike the case of Stokes' boundary conditions, the inertial terms $u_j\partial u_i/\partial x_j$ are not identically zero on the surface of the bubble.

Proudman and Pearson (1957) and Kaplun and Lagerstrom (1957) showed that Oseen's solution is, in fact, the first term obtained when the method of matched asymptotic expansions is used in an attempt to patch together consistent asymptotic solutions of the full Navier–Stokes equations for both the near field close to the sphere and the far field. They also obtained the next term in the expression for the drag force:

$$F_1 = -6\pi\rho_C\nu_C WR\left\{1 + \frac{3}{16}\mathrm{Re} + \frac{9}{160}\mathrm{Re}^2\ln\left(\frac{Re}{2}\right) + 0(\mathrm{Re}^2)\right\}. \qquad (2.21)$$

The additional term leads to an error of 1% at Re $= 0.3$ and does not, therefore, have much practical consequence.

The most notable feature of the Oseen solution is that the geometry of the streamlines depends on the Reynolds number. The downstream flow is *not* a mirror image of the upstream flow as in the Stokes or potential flow solutions. Indeed, closer examination of the Oseen solution reveals that, downstream of the sphere, the streamlines are further apart and the flow is slower than in the equivalent upstream location. Furthermore, this effect increases with Reynolds number. These features of the Oseen solution are entirely consistent with experimental observations and represent the initial development of a wake behind the body.

The flow past a sphere at Reynolds numbers between ∼ 0.5 and several thousand has proven intractable to analytical methods though numerical solutions are numerous. Experimentally, it is found that a recirculating zone (or vortex ring) develops close to the rear stagnation point at about Re $= 30$ (see Taneda 1956 and Figure 2.4). With

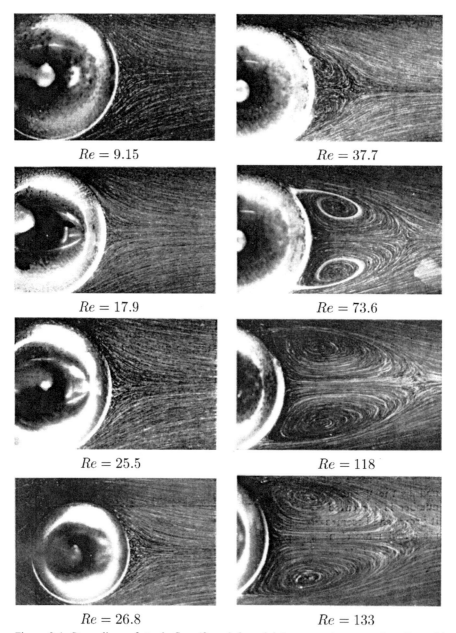

Figure 2.4. Streamlines of steady flow (from left to right) past a sphere at various Reynolds numbers (from Taneda 1956, reproduced by permission of the author).

further increase in the Reynolds number this recirculating zone or wake expands. Defining locations on the surface by the angle from the front stagnation point, the separation point moves forward from $\sim$130° at Re = 100 to $\sim$115° at Re = 300. In the process the wake reaches a diameter comparable to that of the sphere when Re $\approx$ 130. At this point the flow becomes unstable and the ring vortex that makes up

2.2 Flows Around a Sphere

the wake begins to oscillate (Taneda 1956). However, it continues to be attached to the sphere until ~Re = 500 (Torobin and Gauvin 1959).

At Reynolds numbers above about 500, vortices begin to be shed and then convected downstream. The frequency of vortex shedding has not been studied as extensively as in the case of a circular cylinder and seems to vary more with Reynolds number. In terms of the conventional Strouhal number, Str, defined as follows:

$$\text{Str} = 2fR/W \qquad (2.22)$$

the vortex shedding frequencies, f, that Moller (1938) observed correspond to a range of Str varying from 0.3 at $Re = 1000$ to ~ 1.8 at Re = 5000. Furthermore, as Re increases above 500 the flow develops a fairly steady near-wake behind which vortex shedding forms an unsteady and increasingly turbulent far-wake. This process continues until, at a value of Re of the order of 1000, the flow around the sphere and in the near-wake again becomes quite steady. A recognizable boundary layer has developed on the front of the sphere and separation settles down to a position about 84° from the front stagnation point. Transition to turbulence occurs on the free shear layer (which defines the boundary of the near-wake) and moves progressively forward as the Reynolds number increases. The flow is similar to that of the top picture in Figure 2.2. Then the events described in the previous section occur with further increase in the Reynolds number.

Because the Reynolds number range between 0.5 and several hundred can often pertain in multiphase flows, one must resort to an empirical formula for the drag force in this regime. A number of empirical results are available; for example, Klyachko (1934) recommends the following:

$$F_1 = -6\pi \rho_C \nu_C WR \left\{ 1 + \frac{\text{Re}^{\frac{2}{3}}}{6} \right\}, \qquad (2.23)$$

which fits the data fairly well up to Re $\approx$ 1000. At Re = 1 the factor in the square brackets is 1.167, whereas the same factor in Eq. (2.20) is 1.187. Conversely, at Re = 1000, the two factors are respectively 17.7 and 188.5.

2.2.3 Molecular Effects

When the mean free path of the molecules in the surrounding fluid, λ, becomes comparable with the size of the particles, the flow will clearly deviate from the continuum models, which are relevant only when $\lambda \ll R$. The Knudsen number, Kn = $\lambda/2R$, is used to characterize these circumstances, and Cunningham (1910) showed that the first-order correction for small but finite Knudsen number leads to an additional factor, $(1 + 2A\text{Kn})$, in the Stokes drag for a spherical particle. The numerical factor, A, is roughly a constant of order unity (see, for example, Green and Lane 1964).

When the impulse generated by the collision of a single fluid molecule with the particle is large enough to cause significant change in the particle velocity, the resulting random motions of the particle are called *Brownian motion* (Einstein 1956). This leads to diffusion of solid particles suspended in a fluid. Einstein showed that the diffusivity, D, of this process is given by the following:

$$D = kT/6\pi\mu_C R, \tag{2.24}$$

where k is Boltzmann's constant. It follows that the typical rms displacement of the particle in a time, t, is given by $(kTt/3\pi\mu_C R)^{\frac{1}{2}}$. Brownian motion is usually significant only for micron- and sub-micron-sized particles. The example quoted by Einstein is that of a 1-μm-diameter particle in water at 17°C for which the typical displacement during 1s is 0.8 μm.

A third, related, phenomenon is the response of a particle to the collisions of molecules when there is a significant temperature gradient in the fluid. Then the impulses imparted to the particle by molecular collisions on the hot side of the particle will be larger than the impulses on the cold side. The particle will therefore experience a net force diving it in the direction of the colder fluid. This phenomenon is known as *thermophoresis* (see, for example, Davies 1966). A similar phenomenon known as *photophoresis* occurs when a particle is subjected to nonuniform radiation. One could include in this list the Bjerknes forces described in section 3.4 because they constitute *sonophoresis*, namely forces acting on a particle in a sound field.

2.3 Unsteady Effects

2.3.1 Unsteady Particle Motions

Having reviewed the steady motion of a particle relative to a fluid, we must now consider the consequences of unsteady relative motion in which the particle, the fluid, or both are accelerating. The complexities of fluid acceleration are delayed until the next section. First we consider the simpler circumstance in which the fluid either is at rest or has a steady uniform streaming motion ($U =$ constant) far from the particle. Clearly the second case is readily reduced to the first by a simple Galilean transformation and it is assumed that this has been accomplished.

In the ideal case of unsteady inviscid potential flow, it can then be shown by using the concept of the total kinetic energy of the fluid that the force on a rigid particle in an incompressible flow is given by F_i as follows:

$$F_i = -M_{ij}\frac{dV_j}{dt}, \tag{2.25}$$

where M_{ij} is called the *added mass matrix* (or tensor), although the name *induced inertia tensor* used by Batchelor (1967) is, perhaps, more descriptive. The reader is referred to Sarpkaya and Isaacson (1981), Yih (1969), or Batchelor (1967) for detailed

2.3 Unsteady Effects

descriptions of such analyses. The above-mentioned methods also show that M_{ij} for any finite particle can be obtained from knowledge of several *steady* potential flows. In fact,

$$M_{ij} = \frac{\rho_C}{2} \int_{\text{volume of fluid}} u_{ik} u_{jk} \, d(\text{volume}), \qquad (2.26)$$

where the integration is performed over the entire volume of the fluid. The velocity field, u_{ij}, is the fluid velocity in the i direction caused by the *steady* translation of the particle with unit velocity in the j direction. Note that this means that M_{ij} is necessarily a symmetric matrix. Furthermore, it is clear that particles with planes of symmetry will not experience a force perpendicular to that plane when the direction of acceleration is parallel to that plane. Hence if there is a plane of symmetry perpendicular to the k direction, then for $i \neq k$, $M_{ki} = M_{ik} = 0$, and the only off-diagonal matrix elements that can be nonzero are M_{ij}, $j \neq k$, $i \neq k$. In the special case of the sphere *all* the off-diagonal terms will be zero.

Tables of some available values of the diagonal components of M_{ij} are given by Sarpkaya and Isaacson (1981), who also summarize the experimental results, particularly for planar flows past cylinders. Other compilations of added mass results can be found in Kennard (1967), Patton (1965), and Brennen (1982). Some typical values for three-dimensional particles are listed in Table 2.1. The uniform diagonal value for a sphere (often referred to simply as the added mass of a sphere) is $2\rho_C \pi R^3 / 3$ or one-half the displaced mass of fluid. This value can readily be obtained from Eq. (2.26) using the steady flow results given in Eqs. (2.7) to (2.10). In general, of course, there is *no* special relation between the added mass and the displaced mass. Consider, for example, the case of the infinitely thin plate or disk with zero displaced mass that has a finite added mass in the direction normal to the surface. Finally, it should be noted that the literature contains little, if any, information on off-diagonal components of added mass matrices.

Now consider the application of these potential flow results to real viscous flows at high Reynolds numbers (the case of low Reynolds number flows is discussed in section 2.3.4). Significant doubts about the applicability of the added masses calculated from potential flow analysis would be justified because of the experience of D'Alembert's paradox for steady potential flows and the substantial difference between the streamlines of the potential and actual flows. Furthermore, analyses of experimental results requires the separation of the *added mass* forces from the viscous drag forces. Usually this is accomplished by heuristic summation of the two forces so that

$$F_i = -M_{ij} \frac{dV_j}{dt} - \frac{1}{2} \rho_C A C_{ij} |V_j| V_j, \qquad (2.27)$$

where C_{ij} is a lift and drag coefficient matrix and A is a typical cross-sectional area for the body. This is known as Morison's equation (see Morison *et al.* 1950).

Actual unsteady high Reynolds number flows are more complicated and not necessarily compatible with such simple superposition. This is reflected in the fact that

Table 2.1. *Added masses (diagonal terms in M_{ij}) for some three-dimensional bodies (particles)*

Particle	Matrix element	Value			
Sphere (T)	M_{ii}	$\frac{2}{3}\rho_C \pi R^3$			
Disk (T)	M_{11}	$\frac{8}{3}\rho_C R^3$			
Ellipsoids (T)	$M_{ii} = K_{ii}\frac{4}{3}\rho_C \pi ab^2$	a/b	K_{11}	$K_{22}(K_{33})$	
		2	0.209	0.702	
		5	0.059	0.895	
		10	0.021	0.960	
Sphere near plane wall (T) ($R/H \ll 1$)	$M_{ii} = K_{ii}\frac{4}{3}\rho_C \pi R^3$	$K_{11} = \frac{1}{2}\left(1 + \frac{3}{8}\frac{R^3}{H^3} + \cdots\right)$			
		$K_{22} = \frac{1}{2}\left(1 + \frac{3}{16}\frac{R^3}{H^3} + \cdots\right)$			
		$K_{33} = K_{22}$			
Sphere near free surface (E) ($R/H \ll 1$)	$M_{ii} = K_{ii}\frac{4}{3}\rho_C \pi R^3$	H/R	K_{11}		
		8.0	0.52		
		4.0	0.59		
		2.0	0.54		
		1.0	0.44		
		0.0	0.25		

Note: (T), Potential flow calculations; (E), experimental data from Patton (1965).

the coefficients M_{ij} and C_{ij} appear from the experimental results to be functions not only of Re but also of the reduced time or frequency of the unsteady motion. Typically, experiments involve either oscillation of a body in a fluid or acceleration from rest. The most extensively studied case involves planar flow past a cylinder (for example, Keulegan and Carpenter 1958), and a detailed review of this data is included in Sarpkaya and Isaacson (1981). For oscillatory motion of the cylinder with velocity amplitude U_M and period t^*, the coefficients are functions of both the Reynolds number, $\text{Re} = 2U_M R/\nu_C$, and the reduced period or Keulegan–Carpenter number, $\text{Kc} = U_M t^*/2R$. When the amplitude, $U_M t^*$, is less than about $10R(\text{Kc} < 5)$, the inertial effects dominate and M_{ii} is only a little less than its potential flow value over a wide range of Reynolds numbers ($10^4 < \text{Re} < 10^6$). However, for larger values of

2.3 Unsteady Effects

Kc, M_{ii} can be substantially smaller than this and, in some range of Re and Kc, may actually be negative. The values of C_{ii} (the drag coefficient) that are deduced from experiments are also a complicated function of Re and Kc. The behavior of the coefficients is particularly pathological when the reduced period, Kc, is close to that of vortex shedding (Kc of the order of 10). Large transverse or *lift* forces can be generated under these circumstances. To the author's knowledge, detailed investigations of this kind have not been made for a spherical body, but one might expect the same qualitative phenomena to occur.

2.3.2 Effect of Concentration on Added Mass

Although most multiphase flow effects are delayed until later chapters it is convenient at this point to address the issue of the effect on the added mass of the particles in the surrounding mixture. It is to be expected that the added mass coefficient for an individual particle would depend on the void fraction of the surrounding medium. Zuber (1964) first addressed this issue using a *cell method* and found that the added mass, M_{ii}, for spherical bubbles increased with volume fraction, α, as follows:

$$\frac{M_{ii}(\alpha)}{M_{ii}(0)} = \frac{(1+2\alpha)}{(1-\alpha)} = 1 + 3\alpha + O(\alpha^2). \tag{2.28}$$

The simplistic geometry assumed in the cell method (a concentric spherical shell of fluid surrounding each spherical particle) caused later researchers to attempt improvements to Zuber's analysis; for example, van Wijngaarden (1976) used an improved geometry (and the assumption of potential flow) to study the $O(\alpha)$ term and found that

$$\frac{M_{ii}(\alpha)}{M_{ii}(0)} = 1 + 2.76\alpha + O(\alpha^2), \tag{2.29}$$

which is close to Zuber's result. However, even more accurate and more recent analyses by Sangani *et al.* (1991) have shown that Zuber's original result is, in fact, remarkably accurate even up to volume fractions as large as 50% (see also Zhang and Prosperetti 1994).

2.3.3 Unsteady Potential Flow

In general, a particle moving in any flow other than a steady uniform stream will experience fluid accelerations, and it is therefore necessary to consider the structure of the equation governing the particle motion under these circumstances. Of course, this will include the special case of acceleration of a particle in a fluid at rest (or with a steady streaming motion). As in the earlier sections we confine the detailed solutions to those for a spherical particle or bubble. Furthermore, we consider only those circumstances in which both the particle and fluid acceleration are in one direction, chosen for convenience to be the x_1 direction. The effect of an external force field such as gravity is omitted; it can readily be inserted into any of the solutions that follow by the addition of the conventional buoyancy force.

All the solutions discussed are obtained in an accelerating frame of reference *fixed* in the center of the fluid particle. Therefore, if the velocity of the particle in some original, noninertial coordinate system, x_i^*, was $V(t)$ in the x_1^* direction, the Navier–Stokes equations in the new frame, x_i, fixed in the particle center are as follows:

$$\frac{\partial u_i}{\partial t} + u_j \frac{\partial u_i}{\partial x_j} = -\frac{1}{\rho_C} \frac{\partial P}{\partial x_i} + \nu_C \frac{\partial^2 u_i}{\partial x_j \partial x_j}, \qquad (2.30)$$

where the pseudo-pressure, P, is related to the actual pressure, p, by the following:

$$P = p + \rho_C x_1 \frac{dV}{dt}. \qquad (2.31)$$

Here the conventional time derivative of $V(t)$ is denoted by d/dt, but it should be noted that in the original x_i^* frame it implies a Lagrangian derivative following the particle. As before, the fluid is assumed incompressible (so that continuity requires $\partial u_i/\partial x_i = 0$) and Newtonian. The velocity that the fluid would have at the x_i origin in the absence of the particle is then $W(t)$ in the x_1 direction. It is also convenient to define the quantities r, θ, u_r, u_θ as shown in Figure 2.1 and the Stokes streamfunction as in Eq. (2.6). In some cases we shall also be able to consider the unsteady effects due to growth of the bubble so the radius is denoted by $R(t)$.

First consider inviscid potential flow for which Eq. (2.30) may be integrated to obtain the Bernoulli equation as follows:

$$\frac{\partial \phi}{\partial t} + \frac{P}{\rho_C} + \frac{1}{2}\left(u_\theta^2 + u_r^2\right) = \text{constant} \qquad (2.32)$$

where ϕ is a velocity potential ($u_i = \partial \phi/\partial x_i$) and ψ must satisfy the following equation:

$$L\psi = 0 \quad \text{where} \quad L \equiv \frac{\partial^2}{\partial r^2} + \frac{\sin\theta}{r^2}\frac{\partial}{\partial \theta}\left(\frac{1}{\sin\theta}\frac{\partial}{\partial \theta}\right). \qquad (2.33)$$

This is of course the same equation as in steady flow and has harmonic solutions, only five of which are necessary for present purposes:

$$\psi = \sin^2\theta \left\{-\frac{Wr^2}{2} + \frac{D}{r}\right\} + \cos\theta \sin^2\theta \left\{\frac{2Ar^3}{3} - \frac{B}{r^2}\right\} + E\cos\theta \qquad (2.34)$$

$$\phi = \cos\theta \left\{-Wr + \frac{D}{r^2}\right\} + \left(\cos^2\theta - \frac{1}{3}\right)\left\{Ar^2 + \frac{B}{r^3}\right\} + \frac{E}{r} \qquad (2.35)$$

$$u_r = \cos\theta\left\{-W - \frac{2D}{r^3}\right\} + \left(\cos^2\theta - \frac{1}{3}\right)\left\{2Ar - \frac{3B}{r^4}\right\} - \frac{E}{r^2} \qquad (2.36)$$

$$u_\theta = -\sin\theta\left\{-W + \frac{D}{r^3}\right\} - 2\cos\theta\sin\theta\left\{Ar + \frac{B}{r^4}\right\}. \qquad (2.37)$$

2.3 Unsteady Effects

The first part, which involves W and D, is identical to that for steady translation. The second, involving A and B, will provide the fluid velocity gradient in the x_1 direction, and the third, involving E, permits a time-dependent particle (bubble) radius. The W and A terms represent the fluid flow in the absence of the particle, and the D, B, and E terms allow the following boundary condition:

$$(u_r)_{r=R} = \frac{dR}{dt} \tag{2.38}$$

to be satisfied provided the following hold:

$$D = -\frac{WR^3}{2}, \quad B = \frac{2AR^5}{3}, \quad E = -R^2 \frac{dR}{dt}. \tag{2.39}$$

In the absence of the particle the velocity of the fluid at the origin, $r = 0$, is simply $-W$ in the x_1 direction and the gradient of the velocity $\partial u_1/\partial x_1 = 4A/3$. Hence A is determined from the fluid velocity gradient in the original frame as the following:

$$A = \frac{3}{4} \frac{\partial U}{\partial x_1^*}. \tag{2.40}$$

Now the force, F_1, on the bubble in the x_1 direction is given by the following:

$$F_1 = -2\pi R^2 \int_0^\pi p \sin\theta \cos\theta \, d\theta, \tag{2.41}$$

which upon using Eqs. (2.31), (2.32), and (2.35) to (2.37) can be integrated to yield the following:

$$\frac{F_1}{2\pi R^2 \rho_C} = -\frac{D}{Dt}(WR) - \frac{4}{3}RWA + \frac{2}{3}R\frac{dV}{dt}. \tag{2.42}$$

Reverting to the original coordinate system and using v as the sphere volume for convenience ($v = 4\pi R^3/3$), one obtains the following:

$$F_1 = -\frac{1}{2}\rho_C v \frac{dV}{dt^*} + \frac{3}{2}\rho_C v \frac{DU}{Dt^*} + \frac{1}{2}\rho_C (U-V) \frac{dv}{dt^*}, \tag{2.43}$$

where the two Lagrangian time derivatives are defined by the following:

$$\frac{D}{Dt^*} \equiv \frac{\partial}{\partial t^*} + U\frac{\partial}{\partial x_1^*} \tag{2.44}$$

$$\frac{d}{dt^*} \equiv \frac{\partial}{\partial t^*} + V\frac{\partial}{\partial x_1^*}. \tag{2.45}$$

Equation (2.43) is an important result, and care must be taken not to confuse the different time derivatives contained in it. Note that in the absence of bubble growth, viscous drag, and body forces, the equation of motion that results from setting $F_1 = m_p dV/dt^*$ is as follows:

$$\left(1 + \frac{2m_p}{\rho_C v}\right)\frac{dV}{dt^*} = 3\frac{DU}{Dt^*}, \tag{2.46}$$

where m_p is the mass of the *particle*. Thus for a massless bubble the acceleration of the bubble is three times the fluid acceleration.

In a more comprehensive study of unsteady potential flows Symington (1978) has shown that the result for more general (i.e., noncolinear) accelerations of the fluid and particle is merely the vector equivalent of Eq. (2.43):

$$F_i = -\frac{1}{2}\rho_C v \frac{dV_i}{dt^*} + \frac{3}{2}\rho_C v \frac{DU_i}{Dt^*} + \frac{1}{2}\rho_C (U_i - V_i) \frac{dv}{dt^*}, \quad (2.47)$$

where

$$\frac{d}{dt^*} = \frac{\partial}{\partial t^*} + V_j \frac{\partial}{\partial x_j^*}; \quad \frac{D}{Dt^*} = \frac{\partial}{\partial t^*} + U_j \frac{\partial}{\partial x_j^*}. \quad (2.48)$$

The first term in Eq. (2.47) represents the conventional added mass effect due to the particle acceleration. The factor 3/2 in the second term due to the fluid acceleration may initially seem surprising. However, it is made up of two components: (1) $\frac{1}{2}\rho_C dV_i/dt^*$, which is the added mass effect of the fluid acceleration, and (2) $\rho_C v DU_i/Dt^*$, which is a buoyancy-like force due to the pressure gradient associated with the fluid acceleration. The last term in Eq. (2.47) is caused by particle (bubble) volumetric growth, dv/dt^*, and is similar in form to the force on a source in a uniform stream.

Now it is necessary to ask how this force given by Eq. (2.47) should be used in the practical construction of an equation of motion for a particle. Frequently, a viscous drag force F_i^D, is quite arbitrarily added to F_i to obtain some total *effective* force on the particle. Drag forces, F_i^D, with the following conventional forms:

$$F_i^D = \frac{C_D}{2}\rho_C |U_i - V_i|(U_i - V_i)\pi R^2 \quad (\text{Re} \gg 1) \quad (2.49)$$

$$F_i^D = 6\pi\mu_C (U_i - V_i) R \quad (\text{Re} \ll 1) \quad (2.50)$$

have both been employed in the literature. It is, however, important to recognize that there is no fundamental analytical justification for such superposition of these forces. At high Reynolds numbers, we noted in the last section that experimentally observed added masses are indeed quite close to those predicted by potential flow within certain parametric regimes, and hence the superposition has some experimental justification. At low Reynolds numbers, it is improper to use the results of the potential flow analysis. The appropriate analysis under these circumstances is examined in the next section.

2.3.4 Unsteady Stokes Flow

To elucidate some of the issues raised in the last section, it is instructive to examine solutions for the unsteady flow past a sphere in low Reynolds number Stokes flow. In the asymptotic case of zero Reynolds number, the solution of Section 2.2.2 is unchanged by unsteadiness, and hence the solution at any instant in time is identical to the steady-flow solution for the same particle velocity. In other words, because the

2.3 Unsteady Effects

fluid has no inertia, it is always in static equilibrium. Thus the instantaneous force is identical to that for the steady flow with the same $V_i(t)$.

The next step is therefore to investigate the effects of small but nonzero inertial contributions. The Oseen solution provides some indication of the effect of the *convective* inertial terms, $u_j \partial u_i / \partial x_j$, in steady flow. Here we investigate the effects of the *unsteady* inertial term, $\partial u_i / \partial t$. Ideally it would be best to include *both* the $\partial u_i / \partial t$ term *and* the Oseen approximation to the convective term, $U \partial u_i / \partial x$. However, the resulting *unsteady* Oseen flow is sufficiently difficult that only small-time expansions for the impulsively started motions of droplets and bubbles exist in the literature (Pearcey and Hill 1956).

Consider, therefore, the unsteady Stokes equations in the absence of the convective inertial terms:

$$\rho_C \frac{\partial u_i}{\partial t} = -\frac{\partial P}{\partial x_i} + \mu_C \frac{\partial^2 u_i}{\partial x_j \partial x_j}. \tag{2.51}$$

Because both the equations and the boundary conditions used below are linear in u_i, we need only consider colinear particle and fluid velocities in one direction, say x_1. The solution to the general case of noncolinear particle and fluid velocities and accelerations may then be obtained by superposition. As in Section 2.3.3 the colinear problem is solved by first transforming to an accelerating coordinate frame, x_i, fixed in the center of the particle so that $P = p + \rho_C x_1 dV/dt$. Elimination of P by taking the curl of Eq. (2.51) leads to the following:

$$\left(L - \frac{1}{\nu_C} \frac{\partial}{\partial t} \right) L \psi = 0, \tag{2.52}$$

where L is the same operator as defined in Eq. (2.33). Guided by both the steady Stokes flow and the unsteady potential flow solution, one can anticipate a solution of the following form:

$$\psi = \sin^2 \theta \, f(r, t) + \cos \theta \sin^2 \theta \, g(r, t) + \cos \theta \, h(t), \tag{2.53}$$

plus other spherical harmonic functions. The first term has the form of the steady Stokes flow solution; the last term would be required if the particle were a growing spherical bubble. After substituting Eq. (2.53) into Eq. (2.52), the equations for f, g, h are as follows:

$$\left(L_1 - \frac{1}{\nu_C} \frac{\partial}{\partial t} \right) L_1 f = 0 \quad \text{where} \quad L_1 \equiv \frac{\partial^2}{\partial r^2} - \frac{2}{r^2} \tag{2.54}$$

$$\left(L_2 - \frac{1}{\nu_C} \frac{\partial}{\partial t} \right) L_2 g = 0 \quad \text{where} \quad L_2 \equiv \frac{\partial^2}{\partial r^2} - \frac{6}{r^2} \tag{2.55}$$

$$\left(L_0 - \frac{1}{\nu_C} \frac{\partial}{\partial t} \right) L_0 h = 0 \quad \text{where} \quad L_0 \equiv \frac{\partial^2}{\partial r^2}. \tag{2.56}$$

Moreover, the form of the expression for the force, F_1, on the spherical particle (or bubble) obtained by evaluating the stresses on the surface and integrating is as follows:

$$\frac{F_1}{\frac{4}{3}\rho_C \pi R^3} = \frac{dV}{dt} + \left\{\frac{1}{r}\frac{\partial^2 f}{\partial r \partial t} + \frac{\nu_C}{r}\left(\frac{2}{r^2}\frac{\partial f}{\partial r} + \frac{2}{r}\frac{\partial^2 f}{\partial r^2} - \frac{\partial^3 f}{\partial r^3}\right)\right\}_{r=R}. \quad (2.57)$$

It transpires that this is *independent* of g or h. Hence only the solution to Eq. (2.54) for $f(r, t)$ need be sought to find the force on a spherical particle, and the other spherical harmonics that might have been included in Eq. (2.53) are now seen to be unnecessary.

Fourier or Laplace transform methods may be used to solve Eq. (2.54) for $f(r, t)$, and we choose Laplace transforms. The Laplace transforms for the relative velocity, $W(t)$, and the function $f(r, t)$ are denoted by $\hat{W}(s)$ and $\hat{f}(r, s)$ as follows:

$$\hat{W}(s) = \int_0^\infty e^{-st} W(t) dt; \quad \hat{f}(r, s) = \int_0^\infty e^{-st} f(r, t) dt. \quad (2.58)$$

Then Eq. (2.54) becomes the following:

$$\left(L_1 - \xi^2\right) L_1 \hat{f} = 0, \quad (2.59)$$

where $\xi = (s/\nu_C)^{\frac{1}{2}}$, and the solution after application of the condition that $\hat{u}_1(s, t)$ far from the particle be equal to $\hat{W}(s)$ is as follows:

$$\hat{f} = -\frac{\hat{W}r^2}{2} + \frac{A(s)}{r} + B(s)\left(\frac{1}{r} + \xi\right)e^{-\xi r}, \quad (2.60)$$

where A and B are functions of s whose determination requires application of the boundary conditions on $r = R$. In terms of A and B the Laplace transform of the force $\hat{F}_1(s)$ is as follows:

$$\frac{\hat{F}_1}{\frac{4}{3}\rho_C \pi R^3} = \frac{d\hat{V}}{dt} + \left\{\frac{s}{r}\frac{\partial \hat{f}}{\partial r} + \frac{\nu_C}{R}\left(-\frac{4\hat{W}}{r} + \frac{8A}{r^4} + CBe^{-\xi r}\right)\right\}_{r=R}, \quad (2.61)$$

where

$$C = \xi^4 + \frac{3\xi^3}{r} + \frac{3\xi^2}{r^2} + \frac{8\xi}{r^3} + \frac{8}{r^4}. \quad (2.62)$$

The classical solution (see Landau and Lifshitz 1959) is for a solid sphere (i.e., constant R) using the no-slip (Stokes) boundary condition for which

$$f(R, t) = \left.\frac{\partial f}{\partial r}\right|_{r=R} = 0 \quad (2.63)$$

and hence

$$A = +\frac{\hat{W}R^3}{2} + \frac{3\hat{W}R\nu_C}{2s}\{1 + \xi R\}; \quad B = -\frac{3\hat{W}R\nu_C}{2s}e^{\xi R} \quad (2.64)$$

2.3 Unsteady Effects

so that

$$\frac{\hat{F}_1}{\frac{4}{3}\rho_C \pi R^3} = \frac{d\hat{V}}{dt} - \frac{3}{2}s\hat{W} - \frac{9\nu_C \hat{W}}{2R^2} - \frac{9\nu_C^{\frac{1}{2}}}{2R}s^{\frac{1}{2}}\hat{W}. \qquad (2.65)$$

For a motion starting at rest at $t = 0$ the inverse Laplace transform of this yields the following:

$$\frac{F_1}{\frac{4}{3}\rho_C \pi R^3} = \frac{dV}{dt} - \frac{3}{2}\frac{dW}{dt} - \frac{9\nu_C}{2R^2}W - \frac{9}{2R}\left(\frac{\nu_C}{\pi}\right)^{\frac{1}{2}} \int_0^t \frac{dW(\tilde{t})}{d\tilde{t}} \frac{d\tilde{t}}{(t-\tilde{t})^{\frac{1}{2}}}, \qquad (2.66)$$

where $\tilde{t}$ is a dummy time variable. This result must then be written in the original coordinate framework with $W = V - U$ and can be generalized to the noncolinear case by superposition so that

$$F_i = -\frac{1}{2}\upsilon\rho_C \frac{dV_i}{dt^*} + \frac{3}{2}\upsilon\rho_C \frac{dU_i}{dt^*} + \frac{9\upsilon\mu_C}{2R^2}(U_i - V_i)$$
$$+ \frac{9\upsilon\rho_C}{2R}\left(\frac{\nu_C}{\pi}\right)^{\frac{1}{2}} \int_0^{t^*} \frac{d(U_i - V_i)}{d\tilde{t}} \frac{d\tilde{t}}{(t^* - \tilde{t})^{\frac{1}{2}}}, \qquad (2.67)$$

where d/dt^* is the Lagrangian time derivative following the particle. This is then the general force on the particle or bubble in unsteady Stokes flow when the Stokes boundary conditions are applied.

Compare this result with that obtained from the potential flow analysis, Eq. (2.47) with υ taken as constant. It is striking to observe that the coefficients of the added mass terms involving dV_i/dt^* and dU_i/dt^* are identical to those of the potential flow solution. On superficial examination it might be noted that dU_i/dt^* appears in Eq. (2.67), whereas DU_i/Dt^* appears in Eq. (2.47); the difference is, however, of order $W_j \partial U_i/\partial x_j$ and terms of this order have already been dropped from the equation of motion on the basis that they were negligible compared with the temporal derivatives like $\partial W_i/\partial t$. Hence it is inconsistent with the initial assumption to distinguish between d/dt^* and D/Dt^* in the present unsteady Stokes flow solution.

The term $9\nu_C W/2R^2$ in Eq. (2.67) is, of course, the steady Stokes drag. The new phenomenon introduced by this analysis is contained in the last term of Eq. (2.67). This is a fading memory term that is often named the Basset term after one of its identifiers (Basset 1888). It results from the fact that additional vorticity created at the solid particle surface due to relative acceleration diffuses into the flow and creates a temporary perturbation in the flow field. Like all diffusive effects it produces an $\omega^{\frac{1}{2}}$ term in the equation for oscillatory motion.

Before we conclude this section, comment should be included on three other analytical results. Morrison and Stewart (1976) considered the case of a spherical bubble for which the Hadamard–Rybczynski boundary conditions rather than the Stokes conditions are applied. Then, instead of the conditions of Eq. (2.63), the conditions for zero normal velocity and zero shear stress on the surface require that

$$f(R, t) = \left\{\frac{\partial^2 f}{\partial r^2} - \frac{2}{r}\frac{\partial f}{\partial r}\right\}_{r=R} = 0 \quad (2.68)$$

and hence in this case (see Morrison and Stewart 1976)

$$A(s) = +\frac{\hat{W}R^3}{2} + \frac{3\hat{W}R(1+\xi R)}{\xi^2(3+\xi R)}; \quad B(s) = -\frac{3\hat{W}Re^{+\xi R}}{\xi^2(3+\xi R)} \quad (2.69)$$

so that

$$\frac{\hat{F}_1}{\frac{4}{3}\pi\rho_C R^3} = \frac{d\hat{V}}{dt} - \frac{9\hat{W}\nu_C}{R^2} - \frac{3}{2}\hat{W}s + \frac{6\nu_C \hat{W}}{R^2\left\{1 + s^{\frac{1}{2}}R/3\nu_C^{\frac{1}{2}}\right\}}. \quad (2.70)$$

The inverse Laplace transform of this for motion starting at rest at $t = 0$ is as follows:

$$\frac{F_1}{\frac{4}{3}\rho_C \pi R^3} = \frac{dV}{dt} - \frac{3}{2}\frac{dW}{dt} - \frac{3\nu_C W}{R^2} - \frac{6\nu_C}{R^2}\int_0^t \frac{dW(\tilde{t})}{d\tilde{t}}$$

$$\times \exp\left\{\frac{9\nu_C(t-\tilde{t})}{R^2}\right\} \text{erfc}\left\{\left(\frac{9\nu_C(t-\tilde{t})}{R^2}\right)^{\frac{1}{2}}\right\} d\tilde{t}. \quad (2.71)$$

Comparing this with the solution for the Stokes conditions, we note that the first two terms are unchanged and the third term is the expected Hadamard–Rybczynski steady drag term [see Eq. (2.16)]. The last term is significantly different from the Basset term in Eq. (2.67) but still represents a fading memory.

More recently, Magnaudet and Legendre (1998) have extended these results further by obtaining an expression for the force on a particle (bubble) whose radius is changing with time.

Another interesting case is that for unsteady Oseen flow, which essentially consists of attempting to solve the Navier–Stokes equations with the convective inertial terms approximated by $U_j \partial u_i/\partial x_j$. Pearcey and Hill (1956) have examined the small-time behavior of droplets and bubbles started from rest when this term is included in the equations.

2.4 Particle Equation of Motion

2.4.1 Equations of Motion

In a multiphase flow with a very dilute discrete phase the fluid forces discussed in Sections 2.1 to 2.3.4 determine the motion of the particles that constitute that discrete phase. In this section we discuss the implications of some of the fluid force terms. The equation that determines the particle velocity, V_i, is generated by equating the total force, F_i^T, on the particle to $m_p dV_i/dt^*$. Consider the motion of a spherical particle

2.4 Particle Equation of Motion

(or bubble) of mass m_p and volume v (radius R) in a *uniformly* accelerating fluid. The simplest example of this is the vertical motion of a particle under gravity, g, in a pool of otherwise quiescent fluid. Thus the results are written in terms of the buoyancy force. However, the same results apply to motion generated by any uniform acceleration of the fluid, and hence g can be interpreted as a general uniform fluid acceleration (dU/dt). This will also allow some tentative conclusions to be drawn concerning the relative motion of a particle in the nonuniformly accelerating fluid situations that can occur in general multiphase flow. For the motion of a sphere at small relative Reynolds number, Re $\ll 1$ (where Re $= 2WR/\nu_C$ and W is the typical magnitude of the relative velocity), only the forces due to buoyancy and the weight of the particle need be added to F_i as given by Eqs. (2.67) or (2.71) to obtain F_i^T. This addition is simply given by $(\rho_C v - m_p)g_i$ where g is a vector in the vertically upward direction with magnitude equal to the acceleration due to gravity. Conversely, at high relative Reynolds numbers, Re $\gg 1$, one must resort to a more heuristic approach in which the fluid forces given by Eq. (2.47) are supplemented by drag (and lift) forces given by $\frac{1}{2}\rho_C A C_{ij}|W_j|W_j$ as in Eq. (2.27). In either case it is useful to nondimensionalize the resulting equation of motion so that the pertinent nondimensional parameters can be identified.

Examine first the case in which the relative velocity, W (defined as positive in the direction of the acceleration, g, and therefore positive in the vertically upward direction of the rising bubble or sedimenting particle), is sufficiently small so that the relative Reynolds number is much less than unity. Then, using the Stokes boundary conditions, the equation governing W may be obtained from Eq. (2.66) as follows:

$$w + \frac{dw}{dt_*} + \left\{\frac{9}{\pi(1 + 2m_p/\rho_C v)}\right\}^{\frac{1}{2}} \int_0^{t_*} \frac{dw}{d\tilde{t}} \frac{d\tilde{t}}{(t_* - \tilde{t})^{\frac{1}{2}}} = 1 \qquad (2.72)$$

where the dimensionless time, $t_* = t/t_u$ and the relaxation time, t_u, is given by the following:

$$t_u = R^2 (1 + 2m_p/\rho_C v)/9\nu_C \qquad (2.73)$$

and

$$w = W/W_\infty,$$

where W_∞ is the steady terminal velocity given by the following:

$$W_\infty = 2R^2 g (1 - m_p/\rho_C v)/9\nu_C. \qquad (2.74)$$

In the absence of the Basset term the solution of Eq. (2.72) is simply the following:

$$w = 1 - e^{-t/t_u} \qquad (2.75)$$

and therefore the typical response time is given by the relaxation time, t_u (see, for example, Rudinger 1969 and Section 1.2.7). In the general case that includes the Basset

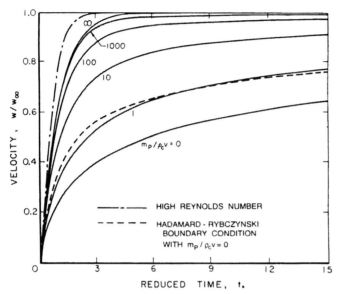

Figure 2.5. The velocity, W, of a particle released from rest at $t_* = 0$ in a quiescent fluid and its approach to terminal velocity, W_∞. Horizontal axis is a dimensionless time defined in text. Solid lines represent the low-Reynolds-number solutions for various particle mass/displaced mass ratios, $m_p/\rho_C v$, and the Stokes boundary condition. The dashed line is for the Hadamard–Rybczynski boundary condition and $m_p/\rho_C v = 0$. The dash-dot line is the high-Reynolds-number result; note that t_* is nondimensionalized differently in that case.

term the dimensionless solution, $w(t_*)$, of Eq. (2.72) depends only on the parameter $m_p/\rho_C v$ (particle mass/displaced fluid mass) appearing in the Basset term. Indeed, the dimensionless Eq. (2.72) clearly illustrates the fact that the Basset term is much less important for solid particles in a gas where $m_p/\rho_C v \gg 1$ than it is for bubbles in a liquid where $m_p/\rho_C v \ll 1$. Note also that for initial conditions of zero relative velocity ($w(0) = 0$) the small-time solution of equation 2.72 takes the following form:

$$\omega = t_* - \frac{2}{\pi^{\frac{1}{2}} \{1 + 2m_p/\rho_C v\}^{\frac{1}{2}}} t_*^{\frac{3}{2}} + \cdots. \tag{2.76}$$

Hence the initial acceleration at $t = 0$ is given dimensionally by the following:

$$2g \left(1 - m_p/\rho_C v\right) / \left(1 + 2m_p/\rho_C v\right)$$

or $2g$ in the case of a massless bubble and $-g$ in the case of a heavy solid particle in a gas where $m_p \gg \rho_C v$. Note also that the effect of the Basset term is to *reduce* the acceleration of the relative motion, thus increasing the time required to achieve terminal velocity.

Numerical solutions of the form of $w(t_*)$ for various $m_p/\rho_C v$ are shown in Figure 2.5 where the delay caused by the Basset term can be clearly seen. In fact in the later stages of approach to the terminal velocity the Basset term dominates over the added mass term, (dw/dt_*). The integral in the Basset term becomes

2.4 Particle Equation of Motion

approximately $2t_*^{1/2} dw/dt_*$ so that the final approach to $w = 1$ can be approximated by the following:

$$w = 1 - C \exp\left\{-t_*^{\frac{1}{2}} \Big/ \left(\frac{9}{\pi\{1 + 2m_p/\rho_C v\}}\right)^{\frac{1}{2}}\right\}, \tag{2.77}$$

where C is a constant. As can be seen in Figure 2.5, the result is a much slower approach to W_∞ for small $m_p/\rho_C v$ than for larger values of this quantity.

The case of a bubble with Hadamard–Rybczynski boundary conditions is very similar except that

$$W_\infty = R^2 g \left(1 - m_p/\rho_C v\right)/3v_C \tag{2.78}$$

and the equation $w(t_*)$ is as follows:

$$w + \frac{3}{2}\frac{dw}{dt_*} + 2\int_0^{t_*} \frac{dw}{d\tilde{t}}\Gamma\left(t_* - \tilde{t}\right) d\tilde{t} = 1, \tag{2.79}$$

where the function, $\Gamma(\xi)$, is given by the following:

$$\Gamma(\xi) = \exp\left\{\left(1 + \frac{2m_p}{\rho_C v}\right)\xi\right\} \operatorname{erfc}\left\{\left(\left(1 + \frac{2m_p}{\rho_C v}\right)\xi\right)^{\frac{1}{2}}\right\}. \tag{2.80}$$

For the purposes of comparison the form of $w(t_*)$ for the Hadamard–Rybczynski boundary condition with $m_p/\rho_C v = 0$ is also shown in Figure 2.5. Though the altered Basset term leads to a more rapid approach to terminal velocity than occurs for the Stokes boundary condition, the difference is not qualitatively significant.

If the terminal Reynolds number is much greater than unity then, in the absence of particle growth, Eq. (2.47) heuristically supplemented with a drag force of the form of Eq. (2.49) leads to the following equation of motion for unidirectional motion:

$$w^2 + \frac{dw}{dt_*} = 1, \tag{2.81}$$

where $w = W/W_\infty$, $t_* = t/t_u$, and the relaxation time, t_u, is now given by the following:

$$t_u = \left(1 + 2m_p/\rho_C v\right)\left(2R/3C_D g(1 - m_p/v\rho_C)\right)^{\frac{1}{2}} \tag{2.82}$$

and

$$W_\infty = \{8Rg\left(1 - m_p/\rho_C v\right)/3C_D\}^{\frac{1}{2}}. \tag{2.83}$$

The solution to Eq. (2.81) for $w(0) = 0$,

$$w = \tanh t_*, \tag{2.84}$$

is also shown in Figure 2.5 though, of course, t_* has a different definition in this case.

The relaxation times given by the Eqs. (2.73) and (2.82) are particularly valuable in assessing relative motion in disperse multiphase flows. When this time is short compared with the typical time associated with the fluid motion, the particle will essentially follow the fluid motion and the techniques of homogeneous flow (see Chapter 9) are applicable. Otherwise the flow is more complex and special effort is needed to evaluate the relative motion and its consequences.

For the purposes of reference in Section 3.2 note that, if we define a Reynolds number, Re, and a Froude number, Fr, by the following:

$$\text{Re} = \frac{2W_\infty R}{\nu_C}; \quad \text{Fr} = \frac{W_\infty}{\{2Rg(1 - m_p/\rho_C v)\}^{\frac{1}{2}}}, \quad (2.85)$$

then the expressions for the terminal velocities, W_∞, given by Eqs. (2.74), (2.78), and (2.83), can be written as follows:

$$\text{Fr} = (\text{Re}/18)^{\frac{1}{2}}, \quad \text{Fr} = (\text{Re}/12)^{\frac{1}{2}}, \quad \text{and} \quad \text{Fr} = (4/3C_D)^{\frac{1}{2}} \quad (2.86)$$

respectively. Indeed, dimensional analysis of the governing Navier–Stokes equations requires that the general expression for the terminal velocity can be written as follows:

$$F(\text{Re}, \text{Fr}) = 0 \quad (2.87)$$

or, alternatively, if C_D is defined as $4/3\text{Fr}^2$, then it could be written as follows:

$$F^*(\text{Re}, C_D) = 0. \quad (2.88)$$

2.4.2 Magnitude of Relative Motion

Qualitative estimates of the magnitude of the relative motion in multiphase flows can be made from the analyses of the last section. Consider a general steady fluid flow characterized by a velocity, U, and a typical dimension, ℓ; it may, for example, be useful to visualize the flow in a converging nozzle of length, ℓ, and mean axial velocity, U. A particle in this flow will experience a typical fluid acceleration (or effective g) of U^2/ℓ for a typical time given by ℓ/U and hence will develop a velocity, W, relative to the fluid. In many practical flows it is necessary to determine the maximum value of W (denoted by W_m) that could develop under these circumstances. To do so, one must first consider whether the available time, ℓ/U, is large or small compared with the typical time, t_u, required for the particle to reach its terminal velocity as given by Eq. (2.73) or (2.82). If $t_u \ll \ell/U$ then W_m is given by Eqs. (2.74), (2.78), or (2.83) for W_∞ and qualitative estimates for W_m/U would be as follows:

$$\left(1 - \frac{m_p}{\rho_C v}\right)\left(\frac{UR}{\nu_C}\right)\left(\frac{R}{\ell}\right) \quad \text{and} \quad \left(1 - \frac{m_p}{\rho_C v}\right)^{\frac{1}{2}} \frac{1}{C_D^{\frac{1}{2}}}\left(\frac{R}{\ell}\right)^{\frac{1}{2}}, \quad (2.89)$$

2.4 Particle Equation of Motion

When $WR/\nu_C \ll 1$ and $WR/\nu_C \gg 1$ respectively. We refer to this as the *quasistatic regime*. Conversely, if $t_u \gg \ell/U$, W_m can be estimated as $W_\infty \ell/Ut_u$ so that W_m/U is of the order of

$$\frac{2\left(1 - m_p/\rho_C v\right)}{\left(1 + 2m_p/\rho_C v\right)} \tag{2.90}$$

for all WR/ν_C. This is termed the *transient regime*.

In practice, WR/ν_C will not be known in advance. The most meaningful quantities that can be evaluated prior to any analysis are a Reynolds number, UR/ν_C, based on flow velocity and particle size, a size parameter

$$X = \frac{R}{\ell}\left|1 - \frac{m_p}{\rho_C v}\right| \tag{2.91}$$

and the parameter

$$Y = \left|1 - \frac{m_p}{\rho_C v}\right| \bigg/ \left(1 + \frac{2m_p}{\rho_C v}\right). \tag{2.92}$$

The resulting regimes of relative motion are displayed graphically in Figure 2.6. The transient regime in the upper right-hand sector of the graph is characterized by large relative motion, as suggested by Eq. (2.90). The quasistatic regimes for $WR/\nu_C \gg 1$ and $WR/\nu_C \ll 1$ are in the lower right- and left-hand sectors respectively. The shaded boundaries between these regimes are, of course, approximate and are functions of the parameter Y, that must have a value in the range $0 < Y < 1$. As one proceeds deeper into either of the quasistatic regimes, the magnitude of the relative velocity, W_m/U, becomes smaller and smaller. Thus, homogeneous flows (see Chapter 9) in which the relative motion is neglected require that *either $X \ll Y^2$ or $X \ll Y/(UR/\nu_C)$*. Conversely, if either of these conditions is violated, relative motion must be included in the analysis.

2.4.3 Effect of Concentration on Particle Equation of Motion

When the concentration of the disperse phase in a multiphase flow is small (less than, say, 0.01% by volume) the particles have little effect on the motion of the continuous phase and analytical or computational methods are much simpler. Quite accurate solutions are then obtained by solving a single-phase flow for the continuous phase (perhaps with some slightly modified density) and inputting those fluid velocities into equations of motion for the particles. This is known as *one-way coupling*.

As the concentration of the disperse phase is increased a whole spectrum of complications can arise. These may effect both the continuous phase flow and the disperse phase motions and flows with this *two-way coupling* pose many modeling challenges. A few examples are appropriate. The particle motions may initiate or alter the turbulence in the continuous phase flow; this particularly challenging issue is briefly

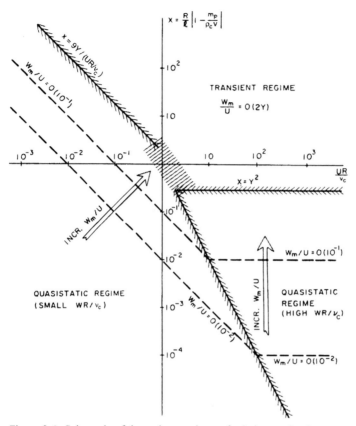

Figure 2.6. Schematic of the various regimes of relative motion between a particle and the surrounding flow.

addressed in Section 1.3. Moreover, particles may begin to collide with one another, altering their effective equation of motion and introducing random particle motions that may need to be accounted for; Chapter 13 is devoted to flows dominated by such collisions. These collisions and random motions may generate additional turbulent motions in the continuous phase. Often the interactions of particles become important even if they do not actually collide. Fortes *et al.* (1987) have shown that in flows with high relative Reynolds numbers there are several important mechanisms of particle/particle interactions that occur when a particle encounters the wake of another particle. The following particle drafts the leading particle and impacts it when it catches up with it and the pair then begin tumbling. In packed beds these interactions result in the development of lateral bands of higher concentration separated by regions of low, almost zero, volume fraction. How these complicated interactions could be incorporated into a two-fluid model (short of complete and direct numerical simulation) is unclear.

At concentrations that are sufficiently small so that the complications of the preceding paragraph do not arise, there are still effects on the coefficients in the particle

2.4 Particle Equation of Motion

equation of motion that may need to be accounted for. For example, the drag on a particle or the added mass of a particle may be altered by the presence of neighboring particles. These issues are somewhat simpler to deal with than those of the preceding paragraph and we cover them in this chapter. The effect on the added mass was addressed earlier in Section 2.3.2. In the next section we address the issue of the effect of concentration on the particle drag.

2.4.4 Effect of Concentration on Particle Drag

Section 2.2 reviewed the dependence of the drag coefficient on the Reynolds number for a single particle in a fluid and the effect on the sedimentation of that single particle in an otherwise quiescent fluid was examined as a particular example in Section 2.4. Such results would be directly applicable to the evaluation of the relative velocity between the disperse phase (the particles) and the continuous phase in a very dilute multiphase flow. However, at higher concentrations, the interactions between the flow fields around individual particles alter the force experienced by those particles and therefore change the velocity of sedimentation. Furthermore, the volumetric flux of the disperse phase is no longer negligible because of the finite concentration and, depending on the boundary conditions in the particular problem, this may cause a nonnegligible volumetric flux of the continuous phase. For example, particles sedimenting in a containing vessel with a downward particle volume flux, $-j_S$ (upward is deemed the positive direction), at a concentration, α, will have a mean velocity as follows:

$$-u_S = -j_S/\alpha \tag{2.93}$$

and will cause an equal and opposite upward flux of the suspending liquid, $j_L = -j_S$, so that the mean velocity of the liquid is as follows:

$$u_L = j_L/(1-\alpha) = -j_S/(1-\alpha). \tag{2.94}$$

Hence the relative velocity is as follows:

$$u_{SL} = u_S - u_L = j_S/\alpha(1-\alpha) = u_S/(1-\alpha). \tag{2.95}$$

Thus care must be taken to define the terminal velocity and here we focus on the more fundamental quantity, namely the relative velocity, u_{SL}, rather than on quantities such as the sedimentation velocity, u_S, that are dependent on the boundary conditions.

Barnea and Mizrahi (1973) have reviewed the experimental, theoretical and empirical data on the sedimentation of particles in otherwise quiescent fluids at various concentrations, α. The experimental data of Mertes and Rhodes (1955) on the ratio of the relative velocity, u_{SL}, to the sedimentation velocity for a single particle, $(u_{SL})_0$ (equal to the value of u_{SL} as $\alpha \to 0$), are presented in Figure 2.7. As one might anticipate, the relative motion is hindered by the increasing concentration. It can also be seen that $u_{SL}/(u_{SL})_0$ is not only a function of α but varies systematically with the Reynolds

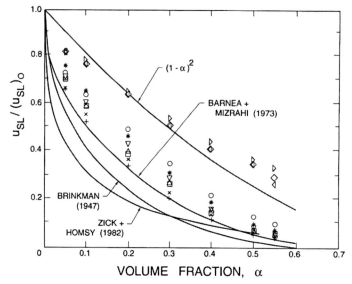

Figure 2.7. Relative velocity of sedimenting particles, u_{SL} [normalized by the velocity as $\alpha \rightarrow 0$, $(u_{SL})_0$] as a function of the volume fraction, α. Experimental data from Mertes and Rhodes (1955) are shown for various Reynolds numbers, Re, as follows: Re = 0.003 (+), 0.019 (×), 0.155 (□), 0.98 (△), 1.45 (▽), 4.8 (∗), 16 (○), 641, (◇), 020 (▷), and 2180 (◁). Also shown are the analytical results of Brinkman [Eq. (2.97)] and Zick and Homsy and the empirical results of Wallis [Eq. (2.100)] and Barnea and Mizrahi [Eq. (2.98)].

number, $2R(u_{SL})_0/\nu_L$, where ν_L is the kinematic viscosity of the suspending medium. Specifically, $u_{SL}/(u_{SL})_0$ increases significantly with Re so that the rate of decrease of $u_{SL}/(u_{SL})_0$ with increasing α is lessened as the Reynolds number increases. One might intuitively expect this decrease in the interactions between the particles because the far field effects of the flow around a single particle decline as the Reynolds number increases.

We also note that complementary to the data of Figure 2.7 is extensive data on the flow through packed beds of particles. The classical analyses of that data by Kozeny (1927) and, independently, by Carman (1937) led to the widely used expression for the pressure drop in the low Reynolds number flow of a fluid of viscosity, μ_C, and superficial velocity, j_{CD}, through a packed bed of spheres of diameter, D, and solids volume fraction, α, namely

$$\frac{dp}{ds} = \frac{180\alpha^3 \mu_C j_{CD}}{(1-\alpha)^3 D^2}, \qquad (2.96)$$

where the 180 and the powers on the functions of α were empirically determined. This expression, known as the Carman–Kozeny equation, is used shortly.

Several curves that are representative of the analytical and empirical results are also shown in Figure 2.7 (and in Figure 2.8). One of the first approximate, analytical

2.4 Particle Equation of Motion

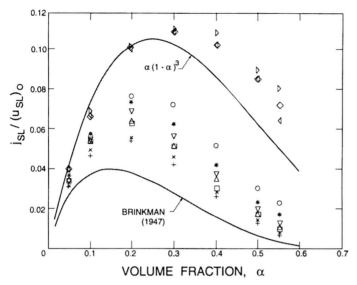

Figure 2.8. The drift flux, j_{SL} (normalized by the velocity $(u_{SL})_0$) corresponding to the relative velocities of Figure 2.7 (see that caption for codes).

models to include the interactions between particles was that of Brinkman (1947) for spherical particles at asymptotically small Reynolds numbers who obtained the following:

$$\frac{u_{SL}}{(u_{SL})_0} = \frac{(2-3\alpha)^2}{4 + 3\alpha + 3(8\alpha - 3\alpha^2)^{\frac{1}{2}}} \quad (2.97)$$

and this result is included in Figures 2.7 and 2.8. Other researchers (see, for example, Tam 1969 and Brady and Bossis 1988) have studied this low Reynolds number limit quite closely. Exact solutions for the sedimentation velocity of a various regular arrays of spheres at asymptotically low Reynolds number were obtained by Zick and Homsy (1982) and the particular result for a simple cubic array is included in Figure 2.7. Clearly, these results deviate significantly from the experimental data and it is currently thought that the sedimentation process cannot be modeled by a regular array because the fluid mechanical effects are dominated by the events that occur when particles happen to come close to one another.

Switching attention to particle Reynolds numbers greater than unity, it was mentioned earlier that the work of Fortes *et al.* (1987) and others has illustrated that the interactions between particles become very complex because they result, primarily, from the interactions of particles with the wakes of the particles ahead of them. Fortes *et al.* (1987) have shown this results in a variety of behaviors they term *drafting, kissing*, and *tumbling* that can be recognized in fluidized beds. As yet, these behaviors have not been amenable to theoretical analyses.

58   Single-Particle Motion

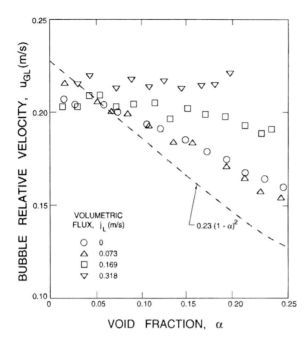

Figure 2.9. Data indicating the variation in the bubble relative velocity, u_{GL}, with the void fraction, α, and the overall flow rate (as represented by j_L) in a vertical, 10.2-cm diameter tube. The dashed line is the correlation of Wallis [Eq. (2.100)]. Adapted from Bernier (1982).

The literature contains numerous empirical correlations but three will suffice for present purposes. At small Reynolds numbers, Barnea and Mizrahi (1973) show that the experimental data closely follow an expression of the following form:

$$\frac{u_{SL}}{(u_{SL})_0} \approx \frac{(1-\alpha)}{(1+\alpha^{\frac{1}{3}})e^{5\alpha/3(1-\alpha)}} \qquad (2.98)$$

By way of comparison the Carman–Kozeny Eq. (2.96) implies that a sedimenting packed bed would have a terminal velocity given by the following:

$$\frac{u_{SL}}{(u_{SL})_0} = \frac{1}{80}\frac{(1-\alpha)^2}{\alpha^2}, \qquad (2.99)$$

which has magnitudes comparable to the Eq. (2.98) at the volume fractions of packed beds.

At large rather than small Reynolds number, the ratio $u_{SL}/(u_{SL})_0$ seems to be better approximated by the following empirical relation:

$$\frac{u_{SL}}{(u_{SL})_0} \approx (1-\alpha)^{b-1}, \qquad (2.100)$$

where Wallis (1969) suggests a value of $b = 3$. Both of these empirical formulae are included in Figure 2.7.

In later chapters discussing sedimentation phenomena, we use the drift flux, j_{SL}, more frequently than the relative velocity, u_{SL}. Recalling that $j_{SL} = \alpha(1-\alpha)u_{SL}$, the data from Figure 2.7 are replotted in Figure 2.8 to display $j_{SL}/(u_{SL})_0$.

2.4 Particle Equation of Motion

It is appropriate to end by expressing some reservations regarding the generality of the experimental data presented in Figures 2.7 and 2.8. At the higher concentrations, vertical flows of this type often develop instabilities that produce large-scale mixing motions whose scale is of the same order as the horizontal extent of the flow, usually the pipe or container diameter. In turn, these motions can have a substantial effect on the mean sedimentation velocity. Consequently, one might expect a pipe size effect that would manifest itself nondimensionally as a dependence on a parameter such as the ratio of the particle to pipe diameter, $2R/d$, or, perhaps, in a Froude number such as $(u_{SL})_0/(gd)^{\frac{1}{2}}$. Another source of discrepancy could be a dependence on the overall flow rate. Almost all of the data, including that of Mertes and Rhodes (1955), have been obtained from relatively quiescent sedimentation or fluidized bed experiments in which the overall flow rate is small and, therefore, the level of turbulence is limited to that produced by the relative motion between the particles and the suspending fluid. However, when the overall flow rate is increased so that even a single-phase flow of the suspending fluid would be turbulent, the mean sedimentation velocities may be significantly altered by the enhancement of the mixing and turbulent motions. Figure 2.9 presents data from some experiments by Bernier (1982) in which the relative velocity of bubbles of air in a vertical water flow were measured for various total volumetric fluxes, j. Small j values cause little deviation from the behavior at $j = 0$ and are consistent with the results of Figure 2.7. However, at larger j values for which a single-phase flow would be turbulent, the decrease in u_{GL} with increasing α almost completely disappears. Bernier surmised that this disappearance of the interaction effect is due to the increase in the turbulence level in the flow that essentially overwhelms any particle/particle or bubble/bubble interaction.

3

Bubble or Droplet Translation

3.1 Introduction

In Chapter 2 it was assumed that the particles were rigid and therefore not deformed, fissioned, or otherwise modified by the flow. However, there are many instances in which the particles that comprise the disperse phase are radically modified by the forces imposed by the continuous phase. Sometimes those modifications are radical enough to, in turn, affect the flow of the continuous phase. For example, the shear rates in the continuous phase may be sufficient to cause fission of the particles and this, in turn, may reduce the relative motion and therefore alter the global extent of phase separation in the flow.

The purpose of this chapter is to identify additional phenomena and issues that arise when the translating disperse phase consists of deformable *particles*, namely bubbles, droplets, or fissionable solid grains.

3.2 Deformation due to Translation

3.2.1 Dimensional Analysis

Because the fluid stresses due to translation may deform the bubbles, drops, or deformable solid particles that make up the disperse phase, we should consider not only the parameters governing the deformation but also the consequences in terms of the translation velocity and the shape. We concentrate here on bubbles and drops in which surface tension, S, acts as the force restraining deformation. However, the reader will realize that there would exist a similar analysis for deformable elastic particles. Furthermore, the discussion is limited to the case of *steady* translation, caused by gravity, g. Clearly the results could be extended to cover translation due to fluid acceleration by using an effective value of g as indicated in Section 2.4.2.

The characteristic force maintaining the sphericity of the bubble or drop is given by SR. Deformation will occur when the characteristic anisotropy in the fluid forces approaches SR; the magnitude of the anisotropic fluid force is given by $\mu_L W_\infty R$ for

3.2 Deformation due to Translation

$W_\infty R/\nu_L \ll 1$ or by $\rho_L W_\infty^2 R^2$ for $W_\infty R/\nu_L \gg 1$. Thus defining a Weber number, We $= 2\rho_L W_\infty^2 R/S$, deformation will occur when We/Re approaches unity for Re $\ll 1$ or when We approaches unity for Re $\gg 1$. But evaluation of these parameters requires knowledge of the terminal velocity, W_∞, and this may also be a function of the shape. Thus one must start by expanding the functional relation of Eq. (2.87), which determines W_∞ to include the Weber number:

$$F(\text{Re}, \text{We}, \text{Fr}) = 0 \tag{3.1}$$

This relation determines W_∞ where Fr is given by Eq. (2.85). Because all three dimensionless coefficients in this functional relation include both W_∞ and R, it is simpler to rearrange the arguments by defining another nondimensional parameter, the Haberman–Morton number (1953), Hm, that is a combination of We, Re, and Fr but does not involve W_∞. The Haberman–Morton number is defined as follows:

$$\text{Hm} = \frac{\text{We}^3}{\text{Fr}^2 \text{Re}^3} = \frac{g\mu_L^4}{\rho_L S^3}\left(1 - \frac{m_p}{\rho_L v}\right). \tag{3.2}$$

In the case of a bubble, $m_p \ll \rho_L v$ and therefore the factor in parenthesis is usually omitted. Then Hm becomes independent of the bubble size.

It follows that the terminal velocity of a bubble or drop can be represented by functional relation as follows:

$$F(\text{Re}, \text{Hm}, \text{Fr}) = 0 \quad \text{or} \quad F^*(\text{Re}, \text{Hm}, C_D) = 0 \tag{3.3}$$

and we confine the following discussion to the nature of this relation for bubbles ($m_p \ll \rho_L v$).

Some values for the Haberman–Morton number (with $m_p/\rho_L v = 0$) for various saturated liquids are shown in Figure 3.1; other values are listed in Table 3.1. Note that for all but the most viscous liquids, Hm is much less than unity. It is, of course, possible to have fluid accelerations much larger than g; however, this is unlikely to cause Hm values greater than unity in practical multiphase flows of most liquids.

3.2.2 Bubble Shapes and Terminal Velocities

Having introduced the Haberman–Morton number, we can now identify the conditions for departure from sphericity. For low Reynolds numbers (Re $\ll 1$) the terminal velocity is given by Re $\propto$ Fr^2. Then the shape deviates from spherical when We $\geq$ Re or, using Re $\propto$ Fr^2 and Hm $=$ We^{3}Fr^{-2}Re^{-4}, when

$$\text{Re} \geq \text{Hm}^{-\frac{1}{2}}. \tag{3.4}$$

Thus if Hm < 1 all bubbles for which Re $\ll 1$ remain spherical. However, there are some unusual circumstances in which Hm > 1 and then there is a range of Re, namely Hm$^{-\frac{1}{2}} < $ Re < 1, in which significant departure from sphericity might occur.

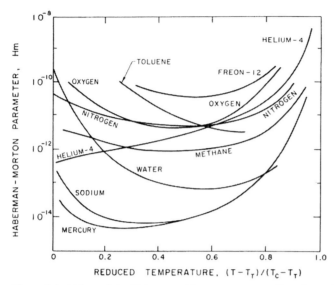

Figure 3.1. Values of the Haberman–Morton parameter, Hm, for various pure substances as a function of reduced temperature where T_T is the triple point temperature and T_C is the critical point temperature.

For high Reynolds numbers (Re $\gg$ 1) the terminal velocity is given by Fr $\approx O(1)$ and distortion occurs if We > 1. Using Fr = 1 and Hm = We^{3}Fr^{-2}Re^{-4} it follows that departure from sphericity occurs when

$$\text{Re} \gg \text{Hm}^{-\frac{1}{4}}. \tag{3.5}$$

Consequently, in the common circumstances in which Hm < 1, there exists a range of Reynolds numbers, Re < Hm$^{-\frac{1}{4}}$, in which sphericity is maintained; nonspherical shapes occur when Re > Hm$^{-\frac{1}{4}}$. For Hm > 1 departure from sphericity has already occurred at Re < 1, as discussed above.

Experimentally, it is observed that the initial departure from sphericity causes ellipsoidal bubbles that may oscillate in shape and have oscillatory trajectories (Hartunian and Sears 1957). As the bubble size is further increased to the point at which We $\approx$ 20, the bubble acquires a new asymptotic shape, known as a *spherical-cap bubble*. A photograph of a typical spherical-cap bubble is shown in Figure 3.2; the

Table 3.1. *Values of the Haberman–Morton numbers, $Hm = g\mu_L^4/\rho_L S^3$, for various liquids at normal temperatures*

Liquid	Hm	Liquid	Hm
Filtered water	0.25×10^{-10}	Turpentine	2.41×10^{-9}
Methyl alcohol	0.89×10^{-10}	Olive oil	7.16×10^{-3}
Mineral oil	1.45×10^{-2}	Syrup	0.92×10^{6}

3.2 Deformation due to Translation

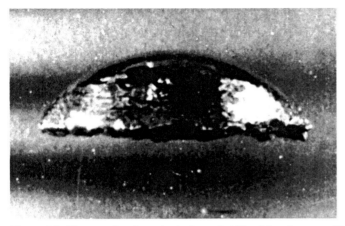

Figure 3.2. Photograph of a spherical cap bubble rising in water (from Davenport, Bradshaw, and Richardson 1967).

notation used to describe the approximate geometry of these bubbles is sketched in Figure 3.3. Spherical-cap bubbles were first investigated by Davies and Taylor (1950), who observed that the terminal velocity is simply related to the radius of curvature of the cap, R_C, or to the equivalent volumetric radius, R_B, by the following:

$$W_\infty = \frac{2}{3}(gR_C)^{\frac{1}{2}} = (gR_B)^{\frac{1}{2}}. \tag{3.6}$$

Assuming a typical laminar drag coefficient of $C_D = 0.5$, a spherical solid particle with the same volume would have the following terminal velocity:

$$W_\infty = (8gR_B/3C_D)^{\frac{1}{2}} = 2.3(gR_B)^{\frac{1}{2}}, \tag{3.7}$$

which is substantially higher than the spherical-cap bubble. From Eq. (3.6) it follows that the effective C_D for spherical-cap bubbles is 2.67 based on the area πR_B^2.

Wegener and Parlange (1973) have reviewed the literature on spherical-cap bubbles. Figure 3.4 is taken from their review and shows that the value of $W_\infty/(gR_B)^{\frac{1}{2}}$

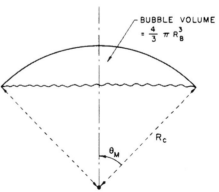

Figure 3.3. Notation used to describe the geometry of spherical cap bubbles.

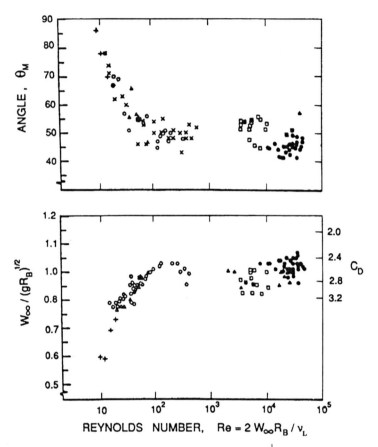

Figure 3.4. Data on the terminal velocity, $W_\infty/(gR_B)^{\frac{1}{2}}$, and the conical angle, θ_M, for spherical-cap bubbles studied by a number of different investigators (adapted from Wegener and Parlange 1973).

reaches a value of about 1 at a Reynolds number, $Re = 2W_\infty R_B/\nu_L$, of about 200 and, thereafter, remains fairly constant. Visualization of the flow reveals that, for Reynolds numbers less than about 360, the wake behind the bubble is laminar and takes the form of a toroidal vortex [similar to a Hill (1894) spherical vortex] shown in the left-hand photograph of Figure 3.5. The wake undergoes transition to turbulence about $Re = 360$, and bubbles at higher Re have turbulent wakes as illustrated in the right side of Figure 3.5. We should add that scuba divers have long observed that spherical-cap bubbles rising in the ocean seem to have a maximum size of the order of 30 cm in diameter. When they grow larger than this, they fission into two (or more) bubbles. However, the author has found no quantitative study of this fission process.

In closing, we note that the terminal velocities of the bubbles discussed here may be represented according to the functional relation of Eq. (3.3) as a family of $C_D(Re)$ curves for various Hm. Figure 3.6 has been extracted from the experimental data of Haberman and Morton (1953) and shows the dependence of $C_D(Re)$ on Hm at intermediate Re. The curves cover the spectrum from the low Re spherical bubbles

3.2 Deformation due to Translation

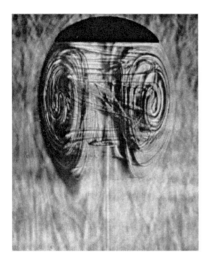

Figure 3.5. Flow visualizations of spherical-cap bubbles. On the left is a bubble with a laminar wake at Re ≈ 180 (from Wegener and Parlange 1973) and, on the right, a bubble with a turbulent wake at Re ≈ 17,000 (from Wegener, Sundell, and Parlange 1971, reproduced with permission of authors).

to the high Re spherical cap bubbles. The data demonstrate that, at higher values of Hm, the drag coefficient makes a relatively smooth transition from the low Reynolds number result to the spherical cap value of about 2.7. Lower values of Hm result in a deep minimum in the drag coefficient around a Reynolds number of about 200.

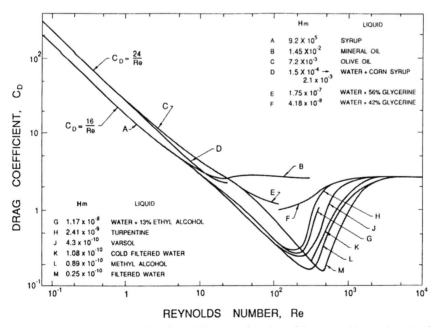

Figure 3.6. Drag coecients, C_D, for bubbles as a function of the Reynolds number, Re, for a range of Haberman–Morton numbers, Hm, as shown. Data from Haberman and Morton (1953).

Table 3.2. *Values of the temperature gradient of the surface tension,* $-dS/dT$, *for pure liquid/vapor interfaces (in kilograms per second squared per degree Kelvin)*

Interface	Temperature gradient	Interface	Temperature gradient
Water	2.02×10^{-4}	Methane	1.84×10^{-4}
Hydrogen	1.59×10^{-4}	Butane	1.06×10^{-4}
Helium-4	1.02×10^{-4}	Carbon dioxide	1.84×10^{-4}
Nitrogen	1.92×10^{-4}	Ammonia	1.85×10^{-4}
Oxygen	1.92×10^{-4}	Toluene	0.93×10^{-4}
Sodium	0.90×10^{-4}	Freon-12	1.18×10^{-4}
Mercury	3.85×10^{-4}	Uranium dioxide	1.11×10^{-4}

3.3 Marangoni Effects

Even if a bubble remains quite spherical, it can experience forces due to gradients in the surface tension, S, over the surface that modify the surface boundary conditions and therefore the translational velocity. These are called Marangoni effects. The gradients in the surface tension can be caused by a number of different factors. For example, gradients in the temperature, solvent concentration, or electric potential can create gradients in the surface tension. The *thermocapillary* effects due to temperature gradients have been explored by a number of investigators (for example, Young, Goldstein, and Block 1959) because of their importance in several technological contexts. For most of the range of temperatures, the surface tension decreases linearly with temperature, reaching zero at the critical point. Consequently, the controlling thermophysical property, dS/dT, is readily identified and more or less constant for any given fluid. Some typical data for dS/dT is presented in Table 3.2 and reveals a remarkably uniform value for this quantity for a wide range of liquids.

Surface tension gradients affect free surface flows because a gradient, dS/ds, in a direction, s, tangential to a surface clearly requires that a shear stress act in the negative s direction so that the surface is in equilibrium. Such a shear stress would then modify the boundary conditions (for example, the Hadamard–Rybczynski conditions used in Section 2.2.2), thus altering the flow and the forces acting on the bubble.

As an example of the Marangoni effect, we examine the steady motion of a spherical bubble in a viscous fluid when there exists a gradient of the temperature (or other controlling physical property), dT/dx_1, in the direction of motion (see Figure 2.1). We must first determine whether the temperature (or other controlling property) is affected by the flow. It is illustrative to consider two special cases from a spectrum of possibilities. The first and simplest special case, that is not so relevant to the thermocapillary phenomenon, is to assume that $T = (dT/dx_1)x_1$ throughout the flow

3.3 Marangoni Effects

field so that, on the surface of the bubble,

$$\left(\frac{1}{R}\frac{dS}{d\theta}\right)_{r=R} = -\sin\theta \left(\frac{dS}{dT}\right)\left(\frac{dT}{dx_1}\right). \tag{3.8}$$

Much more realistic is the assumption that thermal conduction dominates the heat transfer ($\nabla^2 T = 0$) and that there is no heat transfer through the surface of the bubble. Then it follows from the solution of Laplace's equation for the conductive heat transfer problem that

$$\left(\frac{1}{R}\frac{dS}{d\theta}\right)_{r=R} = -\frac{3}{2}\sin\theta \left(\frac{dS}{dT}\right)\left(\frac{dT}{dx_1}\right). \tag{3.9}$$

The latter is the solution presented by Young, Goldstein, and Block (1959), but it differs from Eq. (3.8) only in terms of the effective value of dS/dT. Here we employ Eq. (3.9) because we focus on thermocapillarity, but other possibilities such as Eq. (3.8) should be borne in mind.

For simplicity we continue to assume that the bubble remains spherical. This assumption implies that the surface tension differences are small compared with the absolute level of S and that the stresses normal to the surface are entirely dominated by the surface tension.

With these assumptions the tangential stress boundary condition for the spherical bubble becomes

$$\rho_L \nu_L \left(\frac{\partial u_\theta}{\partial r} - \frac{u_\theta}{r}\right)_{r=R} + \frac{1}{R}\left(\frac{dS}{d\theta}\right)_{r=R} = 0 \tag{3.10}$$

and this should replace the Hadamard–Rybczynski condition of zero shear stress that was used in Section 2.2.2. Applying the boundary condition given by Eqs. (3.10) and (3.9) (as well as the usual kinematic condition, $(u_r)_{r=R} = 0$) to the low Reynolds number solution given by Eqs. (2.11), (2.12), and (2.13) leads to the following:

$$A = -\frac{R^4}{4\rho_L \nu_L}\frac{dS}{dx_1}; \quad B = \frac{WR}{2} + \frac{R^2}{4\rho_L \nu_L}\frac{dS}{dx_1} \tag{3.11}$$

and consequently, from Eq. (2.14), the force acting on the bubble becomes the following:

$$F_1 = -4\pi\rho_L \nu_L WR - 2\pi R^2 \frac{dS}{dx_1}. \tag{3.12}$$

In addition to the normal Hadamard–Rybczynski drag (first term), we can identify a Marangoni force, $2\pi R^2 (dS/dx_1)$, acting on the bubble in the direction of *decreasing* surface tension. Thus, for example, the presence of a uniform temperature gradient, dT/dx_1, would lead to an additional force on the bubble of magnitude $2\pi R^2(-dS/dT)(dT/dx_1)$ in the direction of the warmer fluid because the surface tension decreases with temperature. Such thermocapillary effects have been observed and measured by Young, Goldstein, and Block (1959) and others.

Finally, we should comment on a related effect caused by surface contaminants that increase the surface tension. When a bubble is moving through liquid under the action, say, of gravity, convection may cause contaminants to accumulate on the downstream side of the bubble. This creates a positive $dS/d\theta$ gradient that, in turn, generates an effective shear stress acting in a direction opposite to the flow. Consequently, the contaminants tend to immobilize the surface. This causes the flow and the drag to change from the Hadamard–Rybczynski solution to the Stokes solution for zero tangential velocity. The effect is more pronounced for smaller bubbles because, for a given surface tension difference, the Marangoni force becomes larger relative to the buoyancy force as the bubble size decreases. Experimentally, this means that surface contamination usually results in Stokes drag for spherical bubbles smaller than a certain size and in Hadamard–Rybczynski drag for spherical bubbles larger than that size. Such a transition is observed in experiments measuring the rise velocity of bubbles and can be see in the data of Haberman and Morton (1953) included as Figure 3.6. Harper, Moore, and Pearson (1967) have analyzed the more complex hydrodynamic case of higher Reynolds numbers.

3.4 Bjerknes Forces

Another force that can be important for bubbles is that experienced by a bubble placed in an acoustic field. Termed the Bjerknes force, this nonlinear effect results from the the finite wavelength of the sound waves in the liquid. The frequency, wavenumber, and propagation speed of the stationary acoustic field are denoted by ω, κ, and c_L respectively where $\kappa = \omega/c_L$. The finite wavelength implies an instantaneous pressure gradient in the liquid and, therefore, a *buoyancy* force acting on the bubble.

To model this we express the instantaneous pressure, p, with the following:

$$p = p_0 + \text{Re}\{\tilde{p}^* \sin(\kappa x_i) e^{i\omega t}\}, \tag{3.13}$$

where p_0 is the mean pressure level, $\tilde{p}^*$ is the amplitude of the sound waves and x_i is the direction of wave propagation. Like any other pressure gradient, this produces an instantaneous force, F_i, on the bubble in the x_i direction given by the following:

$$F_i = -\frac{4}{3}\pi R^3 \left(\frac{dp}{dx_i}\right), \tag{3.14}$$

where R is the instantaneous radius of the spherical bubble. Because both R and dp/dx_i contain oscillating components, it follows that the combination of these in Eq. (3.14) leads to a nonlinear, time-averaged component in F_i, that we denote by $\bar{F}_i$. Expressing the oscillations in the volume or radius by the following:

$$R = R_e[1 + \text{Re}\{\varphi e^{i\omega t}\}] \tag{3.15}$$

3.5 Growing or Collapsing Bubbles

one can use the Rayleigh–Plesset equation (see Section 4.2.1) to relate the pressure and radius oscillations and thus obtain the following:

$$\text{Re}\{\varphi\} = \frac{\tilde{p}^*\left(\omega^2 - \omega_n^2\right)\sin(\kappa x_i)}{\rho_L R_e^2\left[\left(\omega^2 - \omega_n^2\right)^2 + \left(4\nu_L\omega/R_e^2\right)^2\right]}, \quad (3.16)$$

where ω_n is the natural frequency of volume oscillation of an individual bubble (see Section 4.4.1) and μ_L is the effective viscosity of the liquid in damping the volume oscillations. If ω is not too close to ω_n, a useful approximation is the following:

$$\text{Re}\{\varphi\} \approx \tilde{p}^*\sin(\kappa x_i)/\rho_L R_e^2(\omega^2 - \omega_n^2). \quad (3.17)$$

Finally, substituting Eqs. (3.13), (3.15), (3.16), and (3.17) into (3.14) one obtains the following:

$$\bar{F}_i = -2\pi R_e^3 \text{Re}\{\varphi\}\kappa\tilde{p}^*\cos(\kappa x_i) \approx -\frac{\pi\kappa R_e(\tilde{p}^*)^2 \sin(2\kappa x_i)}{\rho_L(\omega^2 - \omega_n^2)}. \quad (3.18)$$

This is known as the primary Bjerknes force because it follows from some of the effects discussed by that author (Bjerknes 1909). The effect was first properly identified by Blake (1949).

The form of the primary Bjerknes force produces some interesting bubble migration patterns in a stationary sound field. Note from Eq. (3.18) that if the excitation frequency, ω, is less than the bubble natural frequency, ω_n, then the primary Bjerknes force will cause migration of the bubbles away from the nodes in the pressure field and toward the antinodes (points of largest pressure amplitude). Conversely, if $\omega > \omega_n$ the bubbles tend to migrate from the antinodes to the nodes. A number of investigators (for example, Crum and Eller 1970) have observed the process by which small bubbles in a stationary sound field first migrate to the antinodes, where they grow by rectified diffusion (see Section 4.4.3) until they are larger than the resonant radius. They then migrate back to the nodes, where they may dissolve again when they experience only small pressure oscillations. Crum and Eller (1970) and have shown that the translational velocities of migrating bubbles are compatible with the Bjerknes force estimates given above.

3.5 Growing or Collapsing Bubbles

When the volume of a bubble changes significantly, that growth or collapse can also have a substantial effect on its translation. In this section we return to the discussion of high Re flow and specifically address the effects due to bubble growth or collapse. A bubble that grows or collapses close to a boundary may undergo translation due to the asymmetry induced by that boundary. A relatively simple example of the analysis of this class of flows is the case of the growth or collapse of a spherical bubble near a plane boundary, a problem first solved by Herring (1941) (see also Davies and Taylor 1942,

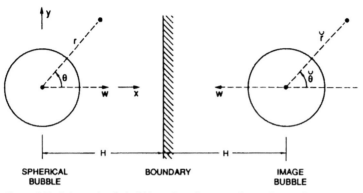

Figure 3.7. Schematic of a bubble undergoing growth or collapse close to a plane boundary. The associated translational velocity is denoted by W.

1943). Assuming that the only translational motion of the bubble is perpendicular to the plane boundary with velocity W, the geometry of the bubble and its image in the boundary are as shown in Figure 3.7. For convenience, we define additional polar coordinates, $(\breve{r}, \breve{\theta})$, with origin at the center of the image bubble. Assuming inviscid, irrotational flow, Herring (1941) and Davies and Taylor (1943) constructed the velocity potential, ϕ, near the bubble by considering an expansion in terms of R/H, where H is the distance of the bubble center from the boundary. Neglecting all terms that are of order R^3/H^3 or higher, the velocity potential can be obtained by superimposing the individual contributions from the bubble source/sink, the image source/sink, the bubble translation dipole, the image dipole, and one correction factor described below. This combination yields the following:

$$\phi = -\frac{R^2 \dot{R}}{r} - \frac{W R^3 \cos\theta}{2r^2} \pm \left\{ -\frac{R^2 \dot{R}}{\breve{r}} + \frac{W R^3 \cos\breve{\theta}}{2\breve{r}^2} - \frac{R^5 \dot{R} \cos\theta}{8H^2 r^2} \right\}. \quad (3.19)$$

The first and third terms are the source/sink contributions from the bubble and the image respectively. The second and fourth terms are the dipole contributions due to the translation of the bubble and the image. The last term arises because the source/sink in the bubble needs to be displaced from the bubble center by an amount $R^3/8H^2$ normal to the wall to satisfy the boundary condition on the surface of the bubble to order R^2/H^2. All other terms of order R^3/H^3 or higher are neglected in this analysis assuming that the bubble is sufficiently far from the boundary so that $H \gg R$. Finally, the sign choice on the last three terms of Eq. (3.19) is as follows: the upper, positive sign pertains to the case of a solid boundary and the lower, negative sign provides an approximate solution for a free surface boundary.

It remains to use this solution to determine the translational motion, $W(t)$, normal to the boundary. This is accomplished by invoking the condition that there is no net force on the bubble. Using the unsteady Bernoulli equation and the velocity potential and fluid velocities obtained from Eq. (3.19), Davies and Taylor (1943) evaluate the pressure at the bubble surface and thereby obtain an expression for the force, F_x, on

3.5 Growing or Collapsing Bubbles

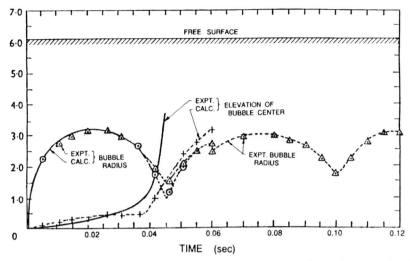

Figure 3.8. Data from Davies and Taylor (1943) on the mean radius and central elevation of a bubble in oil generated by a spark-initiated explosion of 1.32×10^6 ergs situated 6.05 cm below the free surface. The two measures of the bubble radius are one half of the horizontal span ($\triangle$) and one quarter of the sum of the horizontal and vertical spans ($\odot$). Theoretical calculations using Eq. (3.21) are indicated by the solid lines.

the bubble in the x direction:

$$F_x = -\frac{2\pi}{3} \left\{ \frac{d}{dt}(R^3 W) \pm \frac{3}{4} \frac{R^2}{H^2} \frac{d}{dt}\left(R^3 \frac{dR}{dt}\right) \right\}. \tag{3.20}$$

Adding the effect of buoyancy due to a component, g_x, of the gravitational acceleration in the x direction, Davies and Taylor then set the total force equal to zero and obtain the following equation of motion for $W(t)$:

$$\frac{d}{dt}(R^3 W) \pm \frac{3}{4} \frac{R^2}{H^2} \frac{d}{dt}\left(R^3 \frac{dR}{dt}\right) + \frac{4\pi R^3 g_x}{3} = 0. \tag{3.21}$$

In the absence of gravity this corresponds to the equation of motion first obtained by Herring (1941). Many of the studies of growing and collapsing bubbles near boundaries have been carried out in the context of underwater explosions (see Cole 1948). An example illustrating the solution of Eq. (3.21) and the comparison with experimental data is included in Figure 3.8 taken from Davies and Taylor (1943).

Another application of this analysis is to the translation of cavitation bubbles near walls. Here the motivation is to understand the development of impulsive loads on the solid surface. Therefore the primary focus is on bubbles close to the wall and the solution described above is of limited value because it requires $H \gg R$. However, considerable progress has been made in recent years in developing analytical methods for the solution of the inviscid free surface flows of bubbles near boundaries (Blake and Gibson 1987). One of the concepts that is particularly useful in determining the direction of bubble translation is based on a property of the flow first introduced by Kelvin (see Lamb 1932) and called the Kelvin impulse. This vector property applies to the flow generated by a finite particle or bubble in a fluid; it is denoted by I_{Ki} and defined by the following:

$$I_{Ki} = \rho_L \int_{S_B} \phi n_i \, dS, \tag{3.22}$$

where ϕ is the velocity potential of the irrotational flow, S_B is the surface of the bubble, and n_i is the outward normal at that surface (defined as positive into the bubble). If one visualizes a bubble in a fluid at rest, then the Kelvin impulse is the impulse that would have to be applied to the bubble to generate the motions of the fluid related to the bubble motion. Benjamin and Ellis (1966) were the first to demonstrate the value of this property in determining the interaction between a growing or collapsing bubble and a nearby boundary (see also Blake and Gibson 1987).

4

Bubble Growth and Collapse

4.1 Introduction

Unlike solid particles or liquid droplets, gas/vapor bubbles can grow or collapse in a flow and in doing so manifest a host of phenomena with technological importance. We devote this chapter to the fundamental dynamics of a growing or collapsing bubble in an infinite domain of liquid that is at rest far from the bubble. Although the assumption of spherical symmetry is violated in several important processes, it is necessary to first develop this baseline. The dynamics of clouds of bubbles or of bubbly flows are treated in later chapters.

4.2 Bubble Growth and Collapse

4.2.1 Rayleigh–Plesset Equation

Consider a spherical bubble of radius, $R(t)$ (where t is time), in an infinite domain of liquid whose temperature and pressure far from the bubble are T_∞ and $p_\infty(t)$ respectively. The temperature, T_∞, is assumed to be a simple constant because temperature gradients are not considered. Conversely, the pressure, $p_\infty(t)$, is assumed to be a known (and perhaps controlled) input that regulates the growth or collapse of the bubble.

Though compressibility of the liquid can be important in the context of bubble collapse, it will, for the present, be assumed that the liquid density, ρ_L, is a constant. Furthermore, the dynamic viscosity, μ_L, is assumed constant and uniform. It will also be assumed that the contents of the bubble are homogeneous and that the temperature, $T_B(t)$, and pressure, $p_B(t)$, within the bubble are always uniform. These assumptions may not be justified in circumstances that will be identified as the analysis proceeds.

The radius of the bubble, $R(t)$, will be one of the primary results of the analysis. As indicated in Figure 4.1, radial position within the liquid will be denoted by the distance, r, from the center of the bubble; the pressure, $p(r, t)$, radial outward velocity, $u(r, t)$, and temperature, $T(r, t)$, within the liquid will be so designated. Conservation

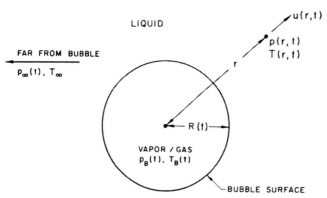

Figure 4.1. Schematic of a spherical bubble in an infinite liquid.

of mass requires that

$$u(r, t) = \frac{F(t)}{r^2}, \quad (4.1)$$

where $F(t)$ is related to $R(t)$ by a kinematic boundary condition at the bubble surface. In the idealized case of zero mass transport across this interface, it is clear that $u(R, t) = dR/dt$ and hence

$$F(t) = R^2 \frac{dR}{dt}. \quad (4.2)$$

This is often a good approximation even when evaporation or condensation is occurring at the interface (Brennen 1995) provided the vapor density is much smaller than the liquid density.

Assuming a Newtonian liquid, the Navier–Stokes equation for motion in the r direction,

$$-\frac{1}{\rho_L} \frac{\partial p}{\partial r} = \frac{\partial u}{\partial t} + u \frac{\partial u}{\partial r} - \nu_L \left\{ \frac{1}{r^2} \frac{\partial}{\partial r} \left(r^2 \frac{\partial u}{\partial r} \right) - \frac{2u}{r^2} \right\} \quad (4.3)$$

yields, after substituting for u from $u = F(t)/r^2$, the following:

$$-\frac{1}{\rho_L} \frac{\partial p}{\partial r} = \frac{1}{r^2} \frac{dF}{dt} - \frac{2F^2}{r^5}. \quad (4.4)$$

Note that the viscous terms vanish; indeed, the only viscous contribution to the Rayleigh–Plesset Eq. (4.8) comes from the dynamic boundary condition at the bubble surface. Equation (4.4) can be integrated to give the following:

$$\frac{p - p_\infty}{\rho_L} = \frac{1}{r} \frac{dF}{dt} - \frac{1}{2} \frac{F^2}{r^4} \quad (4.5)$$

after application of the condition $p \to p_\infty$ as $r \to \infty$.

To complete this part of the analysis, a dynamic boundary condition on the bubble surface must be constructed. For this purpose consider a control volume consisting of

4.2 Bubble Growth and Collapse

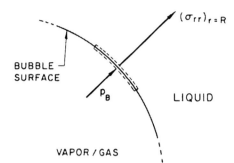

Figure 4.2. Portion of the spherical bubble surface.

a small, infinitely thin lamina containing a segment of interface (Figure 4.2). The net force on this lamina in the radially outward direction per unit area is as follows:

$$(\sigma_{rr})_{r=R} + p_B - \frac{2S}{R} \qquad (4.6)$$

or, because $\sigma_{rr} = -p + 2\mu_L \partial u/\partial r$, the force per unit area is as follows:

$$p_B - (p)_{r=R} - \frac{4\mu_L}{R}\frac{dR}{dt} - \frac{2S}{R}. \qquad (4.7)$$

In the absence of mass transport across the boundary (evaporation or condensation) this force must be zero, and substitution of the value for $(p)_{r=R}$ from Eq. (4.5) with $F = R^2 dR/dt$ yields the following generalized Rayleigh–Plesset equation for bubble dynamics:

$$\frac{p_B(t) - p_\infty(t)}{\rho_L} = R\frac{d^2R}{dt^2} + \frac{3}{2}\left(\frac{dR}{dt}\right)^2 + \frac{4v_L}{R}\frac{dR}{dt} + \frac{2S}{\rho_L R}. \qquad (4.8)$$

Given $p_\infty(t)$ this represents an equation that can be solved to find $R(t)$ provided $p_B(t)$ is known. In the absence of the surface tension and viscous terms, it was first derived and used by Rayleigh (1917). Plesset (1949) first applied the equation to the problem of traveling cavitation bubbles.

4.2.2 Bubble Contents

In addition to the Rayleigh–Plesset equation, considerations of the bubble contents are necessary. To be fairly general, it is assumed that the bubble contains some quantity of noncondensable gas whose partial pressure is p_{Go} at some reference size, R_o, and temperature, T_∞. Then, if there is no appreciable mass transfer of gas to or from the liquid, it follows that

$$p_B(t) = p_V(T_B) + p_{Go}\left(\frac{T_B}{T_\infty}\right)\left(\frac{R_o}{R}\right)^3. \qquad (4.9)$$

In some cases this last assumption is not justified, and it is necessary to solve a mass transport problem for the liquid in a manner similar to that used for heat diffusion (see Section 4.3.4).

It remains to determine $T_B(t)$. This is not always necessary because, under some conditions, the difference between the unknown T_B and the known T_∞ is negligible. But there are also circumstances in which the temperature difference, $(T_B(t) - T_\infty)$, is important and the effects caused by this difference dominate the bubble dynamics. Clearly the temperature difference, $(T_B(t) - T_\infty)$, leads to a different vapor pressure, $p_V(T_B)$, than would occur in the absence of such thermal effects, and this alters the growth or collapse rate of the bubble. It is therefore instructive to substitute Eq. (4.9) into Eq. (4.8) and thereby write the Rayleigh–Plesset equation in the following general form:

$$\overset{(1)}{\frac{p_V(T_\infty) - p_\infty(t)}{\rho_L}} + \overset{(2)}{\frac{p_V(T_B) - p_V(T_\infty)}{\rho_L}} + \overset{(3)}{\frac{p_{Go}}{\rho_L}\left(\frac{T_B}{T_\infty}\right)\left(\frac{R_o}{R}\right)^3}$$
$$= \underset{(4)}{R\frac{d^2R}{dt^2}} + \frac{3}{2}\left(\frac{dR}{dt}\right)^2 + \underset{(5)}{\frac{4v_L}{R}\frac{dR}{dt}} + \underset{(6)}{\frac{2S}{\rho_L R}}. \qquad (4.10)$$

The first term, (1), is the instantaneous tension or driving term determined by the conditions far from the bubble. The second term, (2), is referred to as the *thermal term*, and it will be seen that very different bubble dynamics can be expected depending on the magnitude of this term. When the temperature difference is small, it is convenient to use a Taylor expansion in which only the first derivative is retained to evaluate

$$\frac{p_V(T_B) - p_V(T_\infty)}{\rho_L} = A(T_B - T_\infty) \qquad (4.11)$$

where the quantity A may be evaluated from the following:

$$A = \frac{1}{\rho_L}\frac{dp_V}{dT} = \frac{\rho_V(T_\infty)\mathcal{L}(T_\infty)}{\rho_L T_\infty} \qquad (4.12)$$

using the Clausius–Clapeyron relation, $\mathcal{L}(T_\infty)$ being the latent heat of vaporization at the temperature T_∞. It is consistent with the Taylor expansion approximation to evaluate ρ_V and $\mathcal{L}$ at the known temperature T_∞. It follows that, for small temperature differences, term (2) in Eq. (4.10) is given by $A(T_B - T_\infty)$.

The degree to which the bubble temperature, T_B, departs from the remote liquid temperature, T_∞, can have a major effect on the bubble dynamics, and it is necessary to discuss how this departure might be evaluated. The determination of $(T_B - T_\infty)$ requires two steps. First, it requires the solution of the heat diffusion equation,

$$\frac{\partial T}{\partial t} + \frac{dR}{dt}\left(\frac{R}{r}\right)^2\frac{\partial T}{\partial r} = \frac{\mathcal{D}_L}{r^2}\frac{\partial}{\partial r}\left(r^2\frac{\partial T}{\partial r}\right), \qquad (4.13)$$

to determine the temperature distribution, $T(r, t)$, within the liquid ($\mathcal{D}_L$ is the thermal diffusivity of the liquid). Second, it requires an energy balance for the bubble. The

4.2 Bubble Growth and Collapse

heat supplied to the interface from the liquid is

$$4\pi R^2 k_L \left(\frac{\partial T}{\partial r}\right)_{r=R}, \qquad (4.14)$$

where k_L is the thermal conductivity of the liquid. Assuming that all of this is used for vaporization of the liquid (this neglects the heat used for heating or cooling the existing bubble contents, which is negligible in many cases), one can evaluate the mass rate of production of vapor and relate it to the known rate of increase of the volume of the bubble. This yields

$$\frac{dR}{dt} = \frac{k_L}{\rho_V \mathcal{L}} \left(\frac{\partial T}{\partial r}\right)_{r=R}, \qquad (4.15)$$

where k_L, ρ_V, and $\mathcal{L}$ should be evaluated at $T = T_B$. If, however, $T_B - T_\infty$ is small, it is consistent with the linear analysis described earlier to evaluate these properties at $T = T_\infty$.

The nature of the thermal effect problem is now clear. The thermal term in the Rayleigh–Plesset Eq. (4.10) requires a relation between $(T_B(t) - T_\infty)$ and $R(t)$. The energy balance Eq. (4.15) yields a relation between $(\partial T/\partial r)_{r=R}$ and $R(t)$. The final relation between $(\partial T/\partial r)_{r=R}$ and $(T_B(t) - T_\infty)$ requires the solution of the heat diffusion equation. It is this last step that causes considerable difficulty due to the evident nonlinearities in the heat diffusion equation; no exact analytic solution exists. However, the solution of Plesset and Zwick (1952) provides a useful approximation for many purposes. This solution is confined to cases in which the thickness of the thermal boundary layer, δ_T, surrounding the bubble is small compared with the radius of the bubble, a restriction that can be roughly represented by the identity

$$R \gg \delta_T \approx (T_\infty - T_B)/\left(\frac{\partial T}{\partial r}\right)_{r=R}. \qquad (4.16)$$

The Plesset–Zwick result is that

$$T_\infty - T_B(t) = \left(\frac{\mathcal{D}_L}{\pi}\right)^{\frac{1}{2}} \int_0^t \frac{[R(x)]^2 \left(\frac{\partial T}{\partial r}\right)_{r=R(x)} dx}{\left\{\int_x^t [R(y)]^4 dy\right\}^{\frac{1}{2}}}, \qquad (4.17)$$

where x and y are dummy time variables. Using Eq. (4.15) this can be written as follows:

$$T_\infty - T_B(t) = \frac{\mathcal{L}\rho_V}{\rho_L c_{PL} \mathcal{D}_L^{\frac{1}{2}}} \left(\frac{1}{\pi}\right)^{\frac{1}{2}} \int_0^t \frac{[R(x)]^2 \frac{dR}{dt} dx}{\left[\int_x^t R^4(y) dy\right]^{\frac{1}{2}}}. \qquad (4.18)$$

This can be directly substituted into the Rayleigh–Plesset equation to generate a complicated integro-differential equation for $R(t)$. However, for present purposes it is more instructive to confine our attention to regimes of bubble growth or collapse that can be approximated by the following relation:

$$R = R^* t^n, \qquad (4.19)$$

where R^* and n are constants. Then Eq. (4.18) reduces to the following:

$$T_\infty - T_B(t) = \frac{\mathcal{L}\rho_V}{\rho_L c_{PL} \mathcal{D}_L^{\frac{1}{2}}} R^* t^{n-\frac{1}{2}} C(n) \qquad (4.20)$$

where the constant

$$C(n) = n \left(\frac{4n+1}{\pi}\right)^{\frac{1}{2}} \int_0^1 \frac{z^{3n-1} dz}{(1-z^{4n+1})^{\frac{1}{2}}} \qquad (4.21)$$

and is of the order unity for most values of n of practical interest ($0 < n < 1$ in the case of bubble growth). Under these conditions the linearized form of the thermal term, (2), in the Rayleigh–Plesset Eq. (4.10) as given by Eqs. (4.11) and (4.12) becomes the following:

$$(T_B - T_\infty) \frac{\rho_V \mathcal{L}}{\rho_L T_\infty} = -\Sigma(T_\infty) C(n) R^* t^{n-\frac{1}{2}}, \qquad (4.22)$$

where the thermodynamic parameter is defined as follows:

$$\Sigma(T_\infty) = \frac{\mathcal{L}^2 \rho_V^2}{\rho_L^2 c_{PL} T_\infty \mathcal{D}_L^{\frac{1}{2}}}. \qquad (4.23)$$

In Section 4.3.1 it will be seen that this parameter, Σ, whose units are meters per second to the 3/2 power, is crucially important in determining the bubble dynamic behavior.

4.2.3 In the Absence of Thermal Effects; Bubble Growth

First we consider some of the characteristics of bubble dynamics in the absence of any significant thermal effects. This kind of bubble dynamic behavior is termed *inertially controlled* to distinguish it from the *thermally controlled* behavior discussed later. Under these circumstances the temperature in the liquid is assumed uniform and term (2) in the Rayleigh–Plesset Eq. (4.10) is zero.

For simplicity, it is assumed that the behavior of the gas in the bubble is polytropic so that

$$p_G = p_{Go} \left(\frac{R_o}{R}\right)^{3k}, \qquad (4.24)$$

where k is approximately constant. Clearly $k = 1$ implies a constant bubble temperature and $k = \gamma$ would model adiabatic behavior. It should be understood that accurate evaluation of the behavior of the gas in the bubble requires the solution of the mass,

4.2 Bubble Growth and Collapse

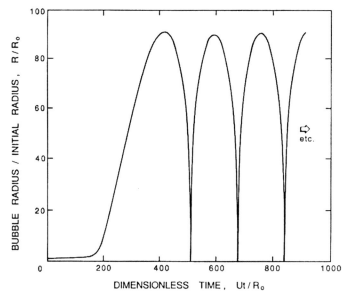

Figure 4.3. Typical solution of the Rayleigh–Plesset equation for a spherical bubble. The nucleus of radius, R_o, enters a low-pressure region at a dimensionless time of 0 and is convected back to the original pressure at a dimensionless time of 500. The low-pressure region is sinusoidal and symmetric about 250.

momentum, and energy equations for the bubble contents combined with appropriate boundary conditions that will include a thermal boundary condition at the bubble wall.

With these assumptions the Rayleigh–Plesset equation becomes

$$\frac{p_V(T_\infty) - p_\infty(t)}{\rho_L} + \frac{p_{Go}}{\rho_L}\left(\frac{R_o}{R}\right)^{3k} = R\frac{d^2R}{dt^2} + \frac{3}{2}\left(\frac{dR}{dt}\right)^2 + \frac{4\nu_L}{R}\frac{dR}{dt} + \frac{2S}{\rho_L R} \quad (4.25)$$

Equation (4.25) without the viscous term was first derived and used by Noltingk and Neppiras (1950, 1951); the viscous term was investigated first by Poritsky (1952).

Equation (4.25) can be readily integrated numerically to find $R(t)$ given the input $p_\infty(t)$, the temperature T_∞, and the other constants. Initial conditions are also required and, in the context of cavitating flows, it is appropriate to assume that the bubble begins as a microbubble of radius R_o in equilibrium at $t = 0$ at a pressure $p_\infty(0)$ so that

$$p_{Go} = p_\infty(0) - p_V(T_\infty) + \frac{2S}{R_o} \quad (4.26)$$

and that $dR/dt|_{t=0} = 0$. A typical solution for Eq. (4.25) under these conditions is shown in Figure 4.3; the bubble in this case experiences a pressure, $p_\infty(t)$, that first decreases below $p_\infty(0)$ and then recovers to its original value. The general features of this solution are characteristic of the response of a bubble as it passes through any low pressure region; they also reflect the strong nonlinearity of Eq. (4.25). The growth is fairly smooth and the maximum size occurs after the minimum pressure.

The collapse process is quite different. The bubble collapses catastrophically, and this is followed by successive rebounds and collapses. In the absence of dissipation mechanisms such as viscosity these rebounds would continue indefinitely without attenuation.

Analytic solutions to Eq. (4.25) are limited to the case of a step function change in p_∞. Nevertheless, these solutions reveal some of the characteristics of more general pressure histories, $p_\infty(t)$, and are therefore valuable to document. With a constant value of $p_\infty(t > 0) = p_\infty^*$, Eq. (4.25) is integrated by multiplying through by $2R^2 dR/dt$ and forming time derivatives. Only the viscous term cannot be integrated in this way, and what follows is confined to the inviscid case. After integration, application of the initial condition $(dR/dt)_{t=0} = 0$ yields the following:

$$\left(\frac{dR}{dt}\right)^2 = \frac{2(p_V - p_\infty^*)}{3\rho_L}\left\{1 - \frac{R_o^3}{R^3}\right\} + \frac{2p_{Go}}{3\rho_L(1-k)}\left\{\frac{R_o^{3k}}{R^{3k}} - \frac{R_o^3}{R^3}\right\} - \frac{2S}{\rho_L R}\left\{1 - \frac{R_o^2}{R^2}\right\} \tag{4.27}$$

where, in the case of isothermal gas behavior, the term involving p_{Go} becomes

$$2\frac{p_{Go}}{\rho_L}\frac{R_o^3}{R^3}\ln\left(\frac{R_o}{R}\right). \tag{4.28}$$

By rearranging Eq. (4.27) it follows that

$$t = R_o \int_1^{R/R_o} \left\{\frac{2(p_V - p_\infty^*)(1 - x^{-3})}{3\rho_L} + \frac{2p_{Go}(x^{-3k} - x^{-3})}{3(1-k)\rho_L} - \frac{2S(1 - x^{-2})}{\rho_L R_o x}\right\}^{-\frac{1}{2}} dx, \tag{4.29}$$

where, in the case $k = 1$, the gas term is replaced by the following:

$$\frac{2p_{Go}}{x^3}\ln x. \tag{4.30}$$

This integral can be evaluated numerically to find $R(t)$, albeit indirectly.

Consider first the characteristic behavior for bubble growth that this solution exhibits when $p_\infty^* < p_\infty(0)$. Equation (4.27) shows that the asymptotic growth rate for $R \gg R_o$ is given by the following:

$$\frac{dR}{dt} \to \left\{\frac{2(p_V - p_\infty^*)}{3\rho_L}\right\}^{\frac{1}{2}}. \tag{4.31}$$

Thus, following an initial period of acceleration, the velocity of the interface is relatively constant. It should be emphasized that Eq. (4.31) implies explosive growth of the bubble, in which the volume displacement is increasing like t^3.

4.2.4 In the Absence of Thermal Effects; Bubble Collapse

Now contrast the behavior of a bubble caused to collapse by an increase in p_∞ to p_∞^*. In this case when $R \ll R_o$ Eq. (4.27) yields the following:

$$\frac{dR}{dt} \to -\left(\frac{R_o}{R}\right)^{\frac{3}{2}} \left\{ \frac{2(p_\infty^* - p_V)}{3\rho_L} + \frac{2S}{\rho_L R_o} - \frac{2p_{Go}}{3(k-1)\rho_L}\left(\frac{R_o}{R}\right)^{3(k-1)} \right\}^{\frac{1}{2}}, \quad (4.32)$$

where, in the case of $k = 1$, the gas term is replaced by $2p_{Go}\ln(R_o/R)/\rho_L$. However, most bubble collapse motions become so rapid that the gas behavior is much closer to adiabatic than isothermal, and we will therefore assume $k \neq 1$.

For a bubble with a substantial gas content the asymptotic collapse velocity given by Eq. (4.32) will not be reached and the bubble will simply oscillate about a new, but smaller, equilibrium radius. Conversely, when the bubble contains very little gas, the inward velocity will continually increase (like $R^{-3/2}$) until the last term within the curly brackets reaches a magnitude comparable with the other terms. The collapse velocity will then decrease and a minimum size given by the following:

$$R_{\min} = R_o \left\{ \frac{1}{(k-1)} \frac{p_{Go}}{(p_\infty^* - p_V + 3S/R_o)} \right\}^{\frac{1}{3(k-1)}} \quad (4.33)$$

will be reached, following which the bubble will rebound. Note that, if p_{Go} is small, $R_{\min}$ could be very small indeed. The pressure and temperature of the gas in the bubble at the minimum radius are then given by p_m and T_m where

$$p_m = p_{Go} \left\{(k-1)\left(p_\infty^* - p_V + 3S/R_o\right)/p_{Go}\right\}^{k/(k-1)} \quad (4.34)$$

$$T_m = T_o \left\{(k-1)\left(p_\infty^* - p_V + 3S/R_o\right)/p_{Go}\right\}. \quad (4.35)$$

We comment later on the magnitudes of these temperatures and pressures (see Sections 5.2.2 and 5.3.3).

The case of zero gas content presents a special albeit somewhat hypothetical problem, because apparently the bubble will reach zero size and at that time have an infinite inward velocity. In the absence of both surface tension and gas content, Rayleigh (1917) was able to integrate Eq. (4.29) to obtain the time, t_{tc}, required for total collapse from $R = R_o$ to $R = 0$:

$$t_{tc} = 0.915 \left(\frac{\rho_L R_o^2}{p_\infty^* - p_V}\right)^{\frac{1}{2}}. \quad (4.36)$$

It is important at this point to emphasize that although the results for bubble growth in Section 4.2.3 are quite practical, the results for bubble collapse may be quite misleading. Apart from the neglect of thermal effects, the analysis was based on two other assumptions that may be violated during collapse. Later it is shown that the final stages of collapse may involve such high velocities (and pressures) that the assumption of liquid incompressibility is no longer appropriate. But, perhaps more important, it transpires (see Section 5.2.3) that a collapsing bubble loses its spherical symmetry in ways that can have important engineering consequences.

4.2.5 Stability of Vapor/Gas Bubbles

Apart from the characteristic bubble growth and collapse processes discussed in the last section, it is also important to recognize that the following equilibrium condition:

$$p_V - p_\infty + p_{Ge} - \frac{2S}{R_e} = 0 \qquad (4.37)$$

may not always represent a *stable* equilibrium state at $R = R_e$ with a partial pressure of gas p_{Ge}.

Consider a small perturbation in the size of the bubble from $R = R_e$ to $R = R_e(1 + \epsilon)$, $\epsilon \ll 1$, and the response resulting from the Rayleigh–Plesset equation. Care must be taken to distinguish two possible cases:

(1) The partial pressure of the gas remains the same at p_{Ge}.
(2) The mass of gas in the bubble and its temperature, T_B, remain the same.

From a practical point of view the Case (1) perturbation is generated over a length of time sufficient to allow adequate mass diffusion in the liquid so that the partial pressure of gas is maintained at the value appropriate to the concentration of gas dissolved in the liquid. Conversely, Case (2) is considered to take place too rapidly for significant gas diffusion. It follows that in Case (1) the gas term in the Rayleigh–Plesset Eq. (4.25) is p_{Ge}/ρ_L, whereas in Case (2) it is $p_{Ge} R_e^{3k}/\rho_L R^{3k}$. If n is defined as zero for Case (1) and $n = 1$ for Case (2) then substitution of $R = R_e(1 + \epsilon)$ into the Rayleigh–Plesset equation yields the following:

$$R\frac{d^2R}{dt^2} + \frac{3}{2}\left(\frac{dR}{dt}\right)^2 + \frac{4\nu_L}{R}\frac{dR}{dt} = \frac{\epsilon}{\rho_L}\left\{\frac{2S}{R_e} - 3nkp_{Ge}\right\}. \qquad (4.38)$$

Note that the right-hand side has the same sign as ϵ if

$$\frac{2S}{R_e} > 3nkp_{Ge} \qquad (4.39)$$

and a different sign if the reverse holds. Therefore, if the above inequality holds, the left-hand side of Eq. (4.38) implies that the velocity and/or acceleration of the bubble radius has the same sign as the perturbation, and hence the equilibrium is *unstable* because the resulting motion will cause the bubble to deviate further from $R = R_e$. Conversely, the equilibrium is stable if $np_{Ge} > 2S/3R_e$.

First consider Case (1) which must always be *unstable* because the inequality 4.39 always holds if $n = 0$. This is simply a restatement of the fact (discussed in Section 4.3.4) that, if one allows time for mass diffusion, then all bubbles will either grow or shrink indefinitely.

4.2 Bubble Growth and Collapse

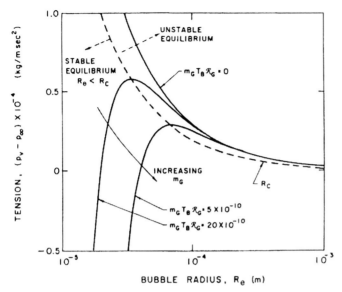

Figure 4.4. Stable and unstable bubble equilibrium radii as a function of the tension for various masses of gas in the bubble. Stable and unstable conditions are separated by the dotted line. Adapted from Daily and Johnson (1956).

Case (2) is more interesting because, in many of the practical engineering situations, pressure levels change over a period of time that is short compared with the time required for significant gas diffusion. In this case a bubble in stable equilibrium requires the following:

$$p_{Ge} = \frac{m_G T_B \mathcal{R}_G}{\frac{4}{3}\pi R_e^3} > \frac{2S}{3k R_e}. \tag{4.40}$$

where m_G is the mass of gas in the bubble and $\mathcal{R}_G$ is the gas constant. Indeed for a *given* mass of gas there exists a critical bubble size, R_c, where

$$R_c = \left\{ \frac{9k m_G T_B \mathcal{R}_G}{8\pi S} \right\}^{1/2}. \tag{4.41}$$

This critical radius was first identified by Blake (1949) and Neppiras and Noltingk (1951) and is often referred to as the Blake critical radius. All bubbles of radius $R_e < R_c$ can exist in stable equilibrium, whereas all bubbles of radius $R_e > R_c$ must be unstable. This critical size could be reached by decreasing the ambient pressure from p_∞ to the critical value, $p_{\infty c}$, where from Eqs. (4.41) and (4.37) it follows that

$$p_{\infty c} = p_V - \frac{4S}{3} \left\{ \frac{8\pi S}{9k m_G T_B \mathcal{R}_G} \right\}^{\frac{1}{2}}, \tag{4.42}$$

which is often called the Blake threshold pressure.

The isothermal case ($k = 1$) is presented graphically in Figure 4.4, where the solid lines represent equilibrium conditions for a bubble of size R_e plotted against the tension $(p_V - p_\infty)$ for various fixed masses of gas in the bubble and a fixed surface tension. The critical radius for any particular m_G corresponds to the maximum in each curve. The locus of the peaks is the graph of R_c values and is shown by the dashed line whose equation is $(p_V - p_\infty) = 4S/3R_e$. The region to the right of the dashed line represents unstable equilibrium conditions. This graphical representation was used by Daily and Johnson (1956) and is useful in visualizing the quasistatic response of a bubble when subjected to a decreasing pressure. Starting in the fourth quadrant under conditions in which the ambient pressure, $p_\infty > p_V$, and, assuming the mass of gas in the bubble is constant, the radius, R_e, will first increase as $(p_V - p_\infty)$ increases. The bubble will pass through a series of stable equilibrium states until the particular critical pressure corresponding to the maximum is reached. Any slight decrease in p_∞ below the value corresponding to this point will result in explosive cavitation growth regardless of whether p_∞ is further decreased. In the context of cavitation nucleation (Brennen 1995), it is recognized that a system consisting of small bubbles in a liquid can sustain a *tension* in the sense that it may be in equilibrium at liquid pressures below the vapor pressure. Due to surface tension, the maximum tension, $(p_V - p_\infty)$, that such a system could sustain would be $2S/R$. However, it is clear from the above analysis that stable equilibrium conditions do not exist in the range

$$\frac{4S}{3R} < (p_V - p_\infty) < \frac{2S}{R} \qquad (4.43)$$

and therefore the maximum tension should be given by $4S/3R$ rather than $2S/R$.

4.3 Thermal Effects

4.3.1 Thermal Effects on Growth

In Sections 4.2.3 through 4.2.5 some of the characteristics of bubble dynamics in the absence of thermal effects were explored. It is now necessary to examine the regime of validity of those analyses. First we evaluate the magnitude of the thermal term (2) in Eq. (4.10) [see also Eq. (4.22)] that was neglected to produce Eq. (4.25).

First examine the case of bubble growth. The asymptotic growth rate given by Eq. (4.31) is constant and hence in the characteristic case of a constant p_∞, terms (1), (3), (4), (5), and (6) in Eq. (4.10) are all either constant or diminishing in magnitude as time progresses. Note that a constant, asymptotic growth rate corresponds to the case

$$n = 1; \quad R^* = \{2(p_V - p_\infty^*)/3\rho_L\}^{\frac{1}{2}} \qquad (4.44)$$

in Eq. (4.19). Consequently, according to Eq. (4.22), the thermal term (2) in its linearized form for small $(T_\infty - T_B)$ is given by the following:

$$\text{term (2)} = \Sigma(T_\infty)C(1)R^* t^{\frac{1}{2}}. \qquad (4.45)$$

4.3 Thermal Effects

Under these conditions, even if the thermal term is initially negligible, it will gain in magnitude relative to all the other terms and will ultimately affect the growth in a major way. Parenthetically it should be added that the Plesset–Zwick assumption of a small thermal boundary layer thickness, δ_T, relative to R can be shown to hold throughout the inertially controlled growth period because δ_T increases like $(\mathcal{D}_L t)^{\frac{1}{2}}$, whereas R is increasing linearly with t. Only under circumstances of very slow growth might the assumption be violated.

Using the relation 4.45, one can therefore define a critical time, t_{c1} (called the first critical time), during growth when the order of magnitude of term (2) in Eq. (4.10) becomes equal to the order of magnitude of the retained terms, as represented by $(dR/dt)^2$. This first critical time is given by the following:

$$t_{c1} = \frac{(p_V - p_\infty^*)}{\rho_L} \cdot \frac{1}{\Sigma^2}, \qquad (4.46)$$

where the constants of the order of unity have been omitted for clarity. Thus t_{c1} depends not only on the tension $(p_V - p_\infty^*)/\rho_L$ but also on $\Sigma(T_\infty)$, a purely thermophysical quantity that is a function only of the liquid temperature. Recalling Eq. (4.23),

$$\Sigma(T) = \frac{\mathcal{L}^2 \rho_V^2}{\rho_L^2 c_{PL} T_\infty \mathcal{D}_L^{\frac{1}{2}}}, \qquad (4.47)$$

it can be anticipated that Σ^2 will change by many, many orders of magnitude in a given liquid as the temperature T_∞ is varied from the triple point to the critical point because Σ^2 is proportional to $(\rho_V/\rho_L)^4$. As a result the critical time, t_{c1}, will vary by many orders of magnitude. Some values of Σ for a number of liquids are plotted in Figure 4.5 as a function of the reduced temperature T/T_C. As an example, consider a typical cavitating flow experiment in a water tunnel with a tension of the order of 10^4 kg/m s^2. Because water at 20°C has a value of Σ of about 1 m/s$^{\frac{3}{2}}$, the first critical time is of the order of 10 s, which is very much longer than the time of growth of bubbles. Hence the bubble growth occurring in this case is unhindered by thermal effects; it is *inertially controlled* growth. Conversely, if the tunnel water were heated to 100°C or, equivalently, one observed bubble growth in a pot of boiling water at superheat of 2°K, then because $\Sigma \approx 10^3$ m/s$^{\frac{3}{2}}$ at 100°C the first critical time would be 10 μs. Thus virtually all the bubble growth observed would be *thermally controlled*.

4.3.2 Thermally Controlled Growth

When the first critical time is exceeded it is clear that the relative importance of the various terms in the Rayleigh–Plesset Eq. (4.10), will change. The most important terms become the driving term (1) and the thermal term (2), whose magnitude is much larger than that of the inertial terms (4). Hence if the tension $(p_V - p_\infty^*)$ remains constant, then the solution using the form of Eq. (4.22) for the thermal term must have

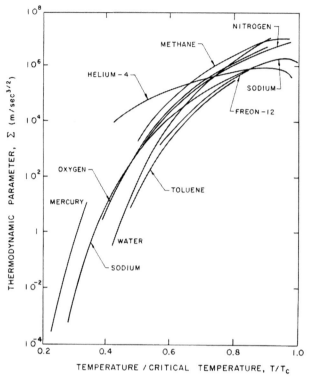

Figure 4.5. Values of the thermodynamic parameter, Σ, for various saturated liquids as a function of the reduced temperature, T/T_C.

$n = \frac{1}{2}$ and the asymptotic behavior is as follows:

$$R = \frac{(p_V - p_\infty^*) t^{\frac{1}{2}}}{\rho_L \Sigma(T_\infty) C(\frac{1}{2})} \quad or \quad n = \frac{1}{2}; \quad R^* = \frac{(p_V - p_\infty^*)}{\rho_L \Sigma(T_\infty) C(\frac{1}{2})}. \qquad (4.48)$$

Consequently, as time proceeds, the inertial, viscous, gaseous, and surface tension terms in the Rayleigh–Plesset equation all rapidly decline in importance. In terms of the superheat, ΔT, rather than the tension

$$R = \frac{1}{2C(\frac{1}{2})} \frac{\rho_L c_{PL} \Delta T}{\rho_V \mathcal{L}} (\mathcal{D}_L t)^{\frac{1}{2}}, \qquad (4.49)$$

where the group $\rho_L c_{PL} \Delta T / \rho_V \mathcal{L}$ is termed the Jakob number in the context of pool boiling and $\Delta T = T_w - T_\infty$, T_w being the wall temperature. We note here that this section addresses only the issues associated with bubble growth in the liquid bulk. The presence of a nearby wall (as is the case in most boiling) causes details and complications, the discussion of which is delayed until Chapter 6.

The result, Eq. (4.48), demonstrates that the rate of growth of the bubble decreases substantially after the first critical time, t_{c1}, is reached and that R subsequently increases like $t^{\frac{1}{2}}$ instead of t. Moreover, because the thermal boundary layer also increases like

4.3 Thermal Effects

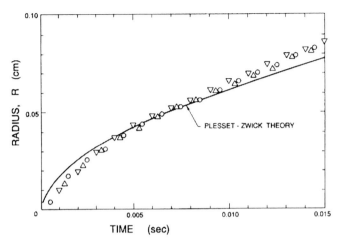

Figure 4.6. Experimental observations of the growth of three vapor bubbles (○, △, ▽) in superheated water at 103.1°C compared with the growth expected using the Plesset–Zwick theory (adapted from Dergarabedian 1953).

$(\mathcal{D}_L t)^{\frac{1}{2}}$, the Plesset–Zwick assumption remains valid indefinitely. An example of this thermally inhibited bubble growth is including in Figure 4.6, which is taken from Dergarabedian (1953). We observe that the experimental data and calculations using the Plesset–Zwick method agree quite well.

When bubble growth is caused by decompression so that $p_\infty(t)$ changes substantially with time during growth, the simple approximate solution of Eq. (4.48) no longer holds and the analysis of the unsteady thermal boundary layer surrounding the bubble becomes considerably more complex. One must then solve the diffusion Eq. (4.13), the energy equation [usually in the approximate form of Eq. (4.15)] and the Rayleigh–Plesset Eq. (4.10) simultaneously, though for the thermally controlled growth being considered here, most of the terms in Eq. (4.10) become negligible so that the simplification, $p_V(T_B) = p_\infty(t)$, is usually justified. When p_∞ is a constant this reduces to the problem treated by Plesset and Zwick (1952) and later addressed by Forster and Zuber (1954) and Scriven (1959). Several different approximate solutions to the general problem of thermally controlled bubble growth during liquid decompression have been put forward by Theofanous *et al.* (1969), Jones and Zuber (1978), and Cha and Henry (1981). All three analyses yield qualitatively similar results that also agree quite well with the experimental data of Hewitt and Parker (1968) for bubble growth in liquid nitrogen. Figure 4.7 presents a typical example of the data of Hewitt and Parker and a comparison with the three analytical treatments mentioned above.

Several other factors can complicate and alter the dynamics of thermally controlled growth. Nonequilibrium effects (Schrage 1953) can occur at very high evaporation rates where the liquid at the interface is no longer in thermal equilibrium with the vapor in the bubble and these have been explored by Theofanous *et al.* (1969) and

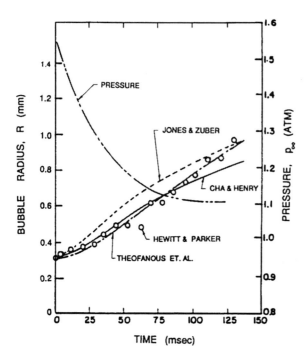

Figure 4.7. Data from Hewitt and Parker (1968) on the growth of a vapor bubble in liquid nitrogen (pressure/time history also shown) and comparison with the analytical treatments by Theofanous *et al.* (1969), Jones and Zuber (1978), and Cha and Henry (1981).

Plesset and Prosperetti (1977) among others. The consensus seems to be that this effect is insignificant except, perhaps, in some extreme circumstances. There is no clear indication in the experiments of any appreciable departure from equilibrium.

More important are the modifications to the heat transfer mechanisms at the bubble surface that may be caused by surface instabilities or by convective heat transfer. These are reviewed in Brennen (1995). Shepherd and Sturtevant (1982) and Frost and Sturtevant (1986) have examined rapidly growing nucleation bubbles near the limit of superheat and have found growth rates substantially larger than expected when the bubble was in the thermally controlled growth phase. Photographs (see Figure 4.8) reveal that the surfaces of those particular bubbles are rough and irregular. The enhancement of the heat transfer caused by this roughening is probably responsible

Figure 4.8. Typical photographs of a rapidly growing bubble in a droplet of superheated ether suspended in glycerine. The bubble is the dark, rough mass; the droplet is clear and transparent. The photographs, which are of different events, were taken 31, 44, and 58 μs after nucleation and the droplets are approximately 2 mm in diameter. Reproduced from Frost and Sturtevant (1986) with the permission of the authors.

4.3 Thermal Effects

for the larger than expected growth rates. Shepherd and Sturtevant (1982) attribute the roughness to the development of a baroclinic interfacial instability similar to the Landau–Darrieus instablity of flame fronts. In other circumstances, Rayleigh–Taylor instability of the interface could give rise to a similar effect (Reynolds and Berthoud 1981).

4.3.3 Cavitation and Boiling

The discussions of bubble dynamics in the last few sections lead, naturally, to two technologically important multiphase phenomena, namely cavitation and boiling. As we have delineated, the essential difference between cavitation and boiling is that bubble growth (and collapse) in boiling is inhibited by limitations on the heat transfer at the interface, whereas bubble growth (and collapse) in cavitation is limited not by heat transfer but only by inertial effects in the surrounding liquid. Cavitation is therefore an explosive (and implosive) process that is far more violent and damaging than the corresponding bubble dynamics of boiling. There are, however, many details that are relevant to these two processes and these are outlined in Chapters 5 and 6 respectively.

4.3.4 Bubble Growth by Mass Diffusion

In most of the circumstances considered in this chapter, it is assumed that the events occur too rapidly for significant mass transfer of contaminant gas to occur between the bubble and the liquid. Thus we assumed in Section 4.2.2 and elsewhere that the mass of contaminant gas in the bubble remained constant. It is convenient to reconsider this issue at this point, for the methods of analysis of mass diffusion will clearly be similar to those of thermal diffusion as described in Section 4.2.2 (see Scriven 1959). Moreover, there are some issues that require analysis of the rate of increase or decrease of the mass of gas in the bubble. One of the most basic issues is the fact that any and all of the gas-filled microbubbles that are present in a subsaturated liquid (and particularly in water) should dissolve away if the ambient pressure is sufficiently high. Henry's law states that the partial pressure of gas, p_{Ge}, in a bubble that is in equilibrium with a saturated concentration, c_∞, of gas dissolved in the liquid will be given by the following:

$$p_{Ge} = c_\infty \text{He}, \qquad (4.50)$$

where He is Henry's law constant for that gas and liquid combination (He decreases substantially with temperature). Consequently, if the ambient pressure, p_∞, is greater than $(c_\infty \text{He} + p_V - 2S/R)$, the bubble should dissolve away completely. Experience is contrary to this theory, and microbubbles persist even when the liquid is subjected to several atmospheres of pressure for an extended period; in most instances, this stabilization of nuclei is caused by surface contamination.

The process of mass transfer can be analyzed by noting that the concentration, $c(r, t)$, of gas in the liquid will be governed by a diffusion equation identical in form to Eq. (4.13) as follows:

$$\frac{\partial c}{\partial t} + \frac{dR}{dt}\left(\frac{R}{r}\right)^2 \frac{\partial c}{\partial r} = \frac{D}{r^2}\frac{\partial}{\partial r}\left(r^2 \frac{\partial c}{\partial r}\right), \qquad (4.51)$$

where D is the mass diffusivity, typically 2×10^{-5} cm^2/s for air in water at normal temperatures. As Plesset and Prosperetti (1977) demonstrate, the typical bubble growth rates due to mass diffusion are so slow that the convection term [the second term on the left-hand side of Eq. (4.51)] is negligible.

The simplest problem is that of a bubble of radius, R, in a liquid at a fixed ambient pressure, p_∞, and gas concentration, c_∞. In the absence of inertial effects the partial pressure of gas in the bubble will be p_{Ge} where

$$p_{Ge} = p_\infty - p_V + 2S/R \qquad (4.52)$$

and therefore the concentration of gas at the liquid interface is $c_s = p_{Ge}/\text{He}$. Epstein and Plesset (1950) found an approximate solution to the problem of a bubble in a liquid initially at uniform gas concentration, c_∞, at time $t = 0$ that takes the following form:

$$R\frac{dR}{dt} = \frac{D}{\rho_G}\frac{\{c_\infty - c_s(1 + 2S/Rp_\infty)\}}{(1 + 4S/3Rp_\infty)}\left\{1 + R(\pi Dt)^{-\frac{1}{2}}\right\}. \qquad (4.53)$$

where ρ_G is the density of gas in the bubble and c_s is the saturated concentration at the interface at the partial pressure given by Eq. (4.52) (the vapor pressure is neglected in their analysis). The last term in Eq. (4.53), $R(\pi Dt)^{-\frac{1}{2}}$, arises from a growing diffusion boundary layer in the liquid at the bubble surface. This layer grows like $(Dt)^{\frac{1}{2}}$. When t is large, the last term in Eq. (4.53) becomes small and the characteristic growth is given approximately by the following:

$$\{R(t)\}^2 - \{R(0)\}^2 \approx \frac{2D(c_\infty - c_s)t}{\rho_G}, \qquad (4.54)$$

where, for simplicity, we have neglected surface tension.

It is instructive to evaluate the typical duration of growth (or shrinkage). From Eq. (4.54) the time required for complete solution is t_{cs} where

$$t_{cs} \approx \frac{\rho_G\{R(0)\}^2}{2D(c_s - c_\infty)}. \qquad (4.55)$$

Typical values of $(c_s - c_\infty)/\rho_G$ are 0.01 (Plesset and Prosperetti 1977). Thus, in the absence of surface contaminant effects, a 10-μm bubble should completely dissolve in about 2.5 s.

Finally, we note that there is an important mass diffusion effect caused by ambient pressure oscillations in which nonlinearities can lead to bubble growth even in a subsaturated liquid. This is known as *rectified diffusion* and is discussed in Section 4.4.3.

4.4 Oscillating Bubbles

4.4.1 Bubble Natural Frequencies

In this and the sections that follow we consider the response of a bubble to oscillations in the prevailing pressure. We begin with an analysis of bubble natural frequencies in the absence of thermal effects and liquid compressibility effects. Consider the linearized dynamic solution of Eq. (4.25) when the pressure at infinity consists of a mean value, $\bar{p}_\infty$, upon which is superimposed a *small* oscillatory pressure of amplitude, $\tilde{p}$, and radian frequency, ω, so that

$$p_\infty = \bar{p}_\infty + \text{Re}\{\tilde{p} e^{j\omega t}\}. \qquad (4.56)$$

The linear dynamic response of the bubble is represented by the following:

$$R = R_e[1 + \text{Re}\{\varphi e^{j\omega t}\}], \qquad (4.57)$$

where R_e is the equilibrium size at the pressure $\bar{p}_\infty$ and the bubble radius response, φ, will in general be a complex number such that $R_e |\varphi|$ is the amplitude of the bubble radius oscillations. The phase of φ represents the phase difference between p_∞ and R.

For the present we assume that the mass of gas in the bubble, m_G, remains constant. Then substituting Eqs. (4.56) and (4.57) into Eq. (4.25), neglecting all terms of order $|\varphi|^2$ and using the equilibrium condition of Eq. (4.37), one finds the following:

$$\omega^2 - j\omega \frac{4\nu_L}{R_e^2} + \frac{1}{\rho_L R_e^2}\left\{\frac{2S}{R_e} - 3kp_{Ge}\right\} = \frac{\tilde{p}}{\rho_L R_e^2 \varphi}, \qquad (4.58)$$

where, as before,

$$p_{Ge} = \bar{p}_\infty - p_V + \frac{2S}{R_e} = \frac{3m_G T_B \mathcal{R}_G}{4\pi R_e^3}. \qquad (4.59)$$

It follows that for a given amplitude, $\tilde{p}$, the maximum or peak response amplitude occurs at a frequency, ω_p, given by the minimum value of the spectral radius of the left-hand side of Eq. (4.58):

$$\omega_p = \left\{\frac{(3kp_{Ge} - 2S/R_e)}{\rho_L R_e^2} - \frac{8\nu_L^2}{R_e^4}\right\}^{\frac{1}{2}} \qquad (4.60)$$

or in terms of $(\bar{p}_\infty - p_V)$ rather than p_{Ge}:

$$\omega_p = \left\{\frac{3k(\bar{p}_\infty - p_V)}{\rho_L R_e^2} + \frac{2(3k-1)S}{\rho_L R_e^3} - \frac{8\nu_L^2}{R_e^4}\right\}^{\frac{1}{2}}. \qquad (4.61)$$

At this peak frequency the amplitude of the response is, of course, inversely proportional to the damping as follows:

$$|\varphi|_{\omega=\omega_p} = \frac{\tilde{p}}{4\mu_L \{\omega_p^2 + \frac{4\nu_L^2}{R_e^4}\}^{\frac{1}{2}}}. \qquad (4.62)$$

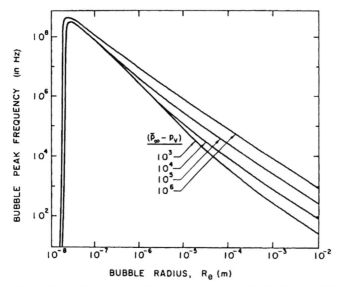

Figure 4.9. Bubble resonant frequency in water at 300°K ($S = 0.0717$, $\mu_L = 0.000863$, $\rho_L = 996.3$) as a function of the radius of the bubble for various values of $(\bar{p}_\infty - p_V)$ as indicated.

It is also convenient for future purposes to define the natural frequency, ω_n, of oscillation of the bubbles as the value of ω_p for zero damping:

$$\omega_n = \left\{ \frac{1}{\rho_L R_e^2} \left\{ 3k(\bar{p}_\infty - p_V) + 2(3k - 1)\frac{S}{R_e} \right\} \right\}^{\frac{1}{2}}. \quad (4.63)$$

The connection with the stability criterion of Section 4.2.5 is clear when one observes that no natural frequency exists for tensions $(p_V - \bar{p}_\infty) > 4S/3R_e$ (for isothermal gas behavior, $k = 1$); stable oscillations can only occur about a stable equilibrium.

Note from Eq. (4.61) that ω_p is a function only of $(\bar{p}_\infty - p_V)$, R_e, and the liquid properties. A typical graph for ω_p as a function of R_e for several $(\bar{p}_\infty - p_V)$ values is shown in Figure 4.9 for water at 300°K ($S = 0.0717$, $\mu_L = 0.000863$, $\rho_L = 996.3$). As is evident from Eq. (4.61), the second and third terms on the right-hand side dominate at very small R_e and the frequency is almost independent of $(\bar{p}_\infty - p_V)$. Indeed, no peak frequency exists below a size equal to about $2v_L^2 \rho_L/S$. For larger bubbles the viscous term becomes negligible and ω_p depends on $(\bar{p}_\infty - p_V)$. If the latter is positive, the natural frequency approaches zero like R_e^{-1}. In the case of tension, $p_V > \bar{p}_\infty$, the peak frequency does not exist above $R_e = R_c$.

For typical nuclei found in water (1 to 100 μm) the natural frequencies are of the order of 5 to 25 kHz. This has several important practical consequences. First, if one wishes to cause cavitation in water by means of an imposed acoustic pressure field, then the frequencies that will be most effective in producing a substantial concentration of large cavitation bubbles will be in this frequency range. This is also the frequency range employed in magnetostrictive devices used to oscillate solid material samples in water (or other liquid) to test the susceptibility of that material to cavitation damage

4.4 Oscillating Bubbles

(Knapp *et al.* 1970). Of course, the oscillation of the nuclei produced in this way will be highly nonlinear and therefore peak response frequencies will be significantly lower than those given above.

There are two important footnotes to this linear dynamic analysis of an oscillating bubble. First, the assumption that the gas in the bubble behaves polytropically is a dubious one. Prosperetti (1977) has analyzed the problem in detail with particular attention to heat transfer in the gas and has evaluated the effective polytropic exponent as a function of frequency. Not surprisingly the polytropic exponent increases from unity at very low frequencies to γ at intermediate frequencies. However, more unexpected behaviors develop at high frequencies. At the low and intermediate frequencies, the theory is largely in agreement with Crum's (1983) experimental measurements. Prosperetti, Crum, and Commander (1988) provide a useful summary of the issue.

A second, related concern is the damping of bubble oscillations. Chapman and Plesset (1971) presented a summary of the three primary contributions to the damping of bubble oscillations, namely that due to liquid viscosity, that due to liquid compressibility through acoustic radiation, and that due to thermal conductivity. It is particularly convenient to represent the three components of damping as three additive contributions to an effective liquid viscosity, μ_e, that can then be employed in the Rayleigh–Plesset equation in place of the actual liquid viscosity, μ_L:

$$\mu_e = \mu_L + \mu_t + \mu_a \tag{4.64}$$

where the *acoustic* viscosity, μ_a, is given by the following:

$$\mu_a = \frac{\rho_L \omega^2 R_e^3}{4 c_L}, \tag{4.65}$$

where c_L is the velocity of sound in the liquid. The *thermal* viscosity, μ_t, follows from the analysis by Prosperetti (1977) mentioned in the previous paragraph (see also Brennen 1995). The relative magnitudes of the three components of damping (or *effective* viscosity) can be quite different for different bubble sizes or radii, R_e. This is illustrated by the data for air bubbles in water at 20°C and atmospheric pressure that is taken from Chapman and Plesset (1971) and reproduced as Figure 4.10.

4.4.2 Nonlinear Effects

Due to the nonlinearities in the governing equations, particularly the Rayleigh–Plesset Eq. (4.10), the response of a bubble subjected to pressure oscillations will begin to exhibit important nonlinear effects as the amplitude of the oscillations is increased. In the last few sections of this chapter we briefly review some of these nonlinear effects. Much of the research appears in the context of acoustic cavitation, a subject with an extensive literature that is reviewed in detail elsewhere (Flynn 1964, Neppiras 1980; Plesset and Prosperetti 1977, Prosperetti 1982, 1984, Crum 1979, Young 1989). We include here a brief summary of the basic phenomena.

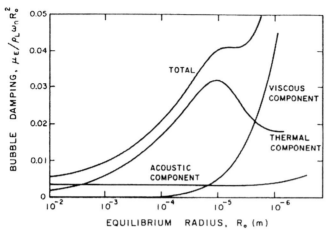

Figure 4.10. Bubble damping components and the total damping as a function of the equilibrium bubble radius, R_e, for water. Damping is plotted as an *effective* viscosity, μ_e, nondimensionalized as shown (from Chapman and Plesset 1971).

As the amplitude increases, the bubble *may* continue to oscillate stably. Such circumstances are referred to as *stable acoustic cavitation* to distinguish them from those of the *transient* regime described below. Several different nonlinear phenomena can affect stable acoustic cavitation in important ways. Among these are the production of subharmonics, the phenomenon of rectified diffusion (see Section 4.4.3) and the generation of Bjerknes forces (see Section 3.4). At larger amplitudes the change in bubble size during a single period of oscillation can become so large that the bubble undergoes a cycle of explosive cavitation growth and violent collapse similar to that described earlier in the chapter. Such a response is termed *transient acoustic cavitation* and is distinguished from stable acoustic cavitation by the fact that the bubble radius changes by several orders of magnitude during each cycle.

As Plesset and Prosperetti (1977) have detailed in their review of the subject, when a liquid that will inevitably contain microbubbles is irradiated with sound of a given frequency, ω, the nonlinear response results in harmonic dispersion that produces not only harmonics with frequencies that are integer multiples of ω (superharmonics) but, more unusually, subharmonics with frequencies less than ω of the form $m\omega/n$, where m and n are integers. Both the superharmonics and subharmonics become more prominent as the amplitude of excitation is increased. The production of subharmonics was first observed experimentally by Esche (1952), and possible origins of this nonlinear effect were explored in detail by Noltingk and Neppiras (1950, 1951), Flynn (1964), Borotnikova and Soloukin (1964), and Neppiras (1969), among others. Lauterborn (1976) examined numerical solutions for a large number of different excitation frequencies and was able to demonstrate the progressive development of the peak responses at subharmonic frequencies as the amplitude of the excitation is increased. Nonlinear effects not only create these subharmonic peaks but also cause the resonant peaks to be shifted to lower frequencies, creating discontinuities that correspond to bifurcations in the solutions. The weakly nonlinear

4.4 Oscillating Bubbles

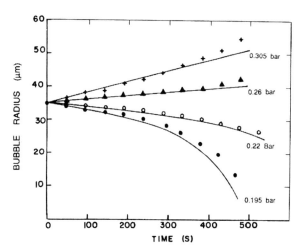

Figure 4.11. Examples from Crum (1980) of the growth (or shrinkage) of air bubbles in saturated water ($S = 68$ dynes/cm) due to rectified diffusion. Data is shown for four pressure amplitudes as shown. The lines are the corresponding theoretical predictions.

analysis of Brennen (1995) produces similar phenomena. In recent years, the modern methods of nonlinear dynamical systems analysis have been applied to this problem by Lauterborn and Suchla (1984), Smereka, Birnir, and Banerjee (1987), Parlitz *et al.* (1990), and others and have led to further understanding of the bifurcation diagrams and strange attractor maps that arise in the dynamics of single bubble oscillations.

Finally, we comment on the phenomenon of transient cavitation in which a phase of explosive cavitation growth and collapse occurs each cycle of the imposed pressure oscillation. We seek to establish the level of pressure oscillation at which this will occur, known as the threshold for transient cavitation (see Noltingk and Neppiras 1950, 1951, Flynn 1964, Young 1989). The answer depends on the relation between the radian frequency, ω, of the imposed oscillations and the natural frequency, ω_n, of the bubble. If $\omega \ll \omega_n$, then the liquid inertia is relatively unimportant in the bubble dynamics and the bubble will respond quasistatically. Under these circumstances the Blake criterion [see Section 4.2.5, Eq. (4.41)] will hold and the critical conditions will be reached when the minimum instantaneous pressure just reaches the critical Blake threshold pressure. On the other hand, if $\omega \gg \omega_n$, the issue will involve the dynamics of bubble growth since inertia will determine the size of the bubble perturbations. The details of this bubble dynamic problem have been addressed by Flynn (1964) and convenient guidelines are provided by Apfel (1981).

4.4.3 Rectified Mass Diffusion

When a bubble is placed in an oscillating pressure field, an important nonlinear effect can occur in the mass transfer of dissolved gas between the liquid and the bubble. This effect can cause a bubble to grow in response to the oscillating pressure when it would not otherwise do so. This effect is known as *rectified mass diffusion* (Blake 1949) and is important because it may cause nuclei to grow from a stable size to an unstable size and thus provide a supply of cavitation nuclei. Analytical models of the phenomenon were first put forward by Hsieh and Plesset (1961) and Eller and

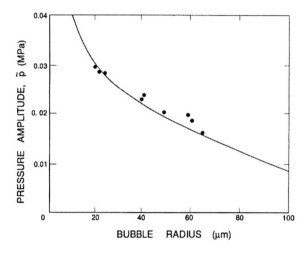

Figure 4.12. Data from Crum (1984) of the threshold pressure amplitude for rectified diffusion for bubbles in distilled water ($S = 68$ dynes/cm) saturated with air. The frequency of the sound is 22.1 kHz. The line is the theoretical prediction.

Flynn (1965), and reviews of the subject can be found in Crum (1980, 1984) and Young (1989).

Consider a gas bubble in a liquid with dissolved gas as described in Section 4.3.4. Now, however, we add an oscillation to the ambient pressure. Gas will tend to come out of solution into the bubble during that part of the oscillation cycle when the bubble is larger than the mean because the partial pressure of gas in the bubble is then depressed. Conversely, gas will redissolve during the other half of the cycle when the bubble is smaller than the mean. The linear contributions to the mass of gas in the bubble will, of course, balance so that the average gas content in the bubble will not be affected at this level. However, there are two nonlinear effects that tend to increase the mass of gas in the bubble. The first of these is due to the fact that release of gas by the liquid occurs during that part of the cycle when the surface area is larger, and therefore the influx during that part of the cycle is slightly larger than the efflux during the part of the cycle when the bubble is smaller. Consequently, there is a net flux of gas into the bubble that is quadratic in the perturbation amplitude. Second, the diffusion boundary layer in the liquid tends to be stretched thinner when the bubble is larger, and this also enhances the flux into the bubble during the part of the cycle when the bubble is larger. This effect contributes a second, quadratic term to the net flux of gas into the bubble.

Strasberg (1961) first explored the issue of the conditions under which a bubble would grow due to rectified diffusion. This and later analyses showed that, when an oscillating pressure is applied to a fluid consisting of a subsaturated or saturated liquid and seeded with microbubbles of radius R_e, there will exist a certain critical or threshold amplitude above which the microbubbles will begin to grow by rectified diffusion. The analytical expressions for the rate of growth and for the threshold pressure amplitudes agree quite well with the corresponding experimental measurements for distilled water saturated with air made by Crum (1980, 1984) (see Figures 4.11 and 4.12).

5

Cavitation

5.1 Introduction

Cavitation occurs in flowing liquid systems when the pressure falls sufficiently low in some region of the flow so that vapor bubbles are formed. Reynolds (1873) was among the first to attempt to explain the unusual behavior of ship propellers at higher rotational speeds by focusing on the possibility of the entrainment of air into the wakes of the propellor blades, a phenomenon we now term *ventilation*. He does not, however, seem to have envisaged the possibility of vapor-filled wakes, and it was left to Parsons (1906) to recognize the role played by vaporization. He also conducted the first experiments on *cavitation* and the phenomenon has been a subject of intensive research ever since because of the adverse effects it has on performance, because of the noise it creates and, most surprisingly, the damage it can do to nearby solid surfaces. In this chapter we examine various features and characteristics of cavitating flows.

5.2 Key Features of Bubble Cavitation

5.2.1 Cavitation Inception

It is conventional to characterize how close the pressure in the liquid flow is to the vapor pressure (and therefore the potential for cavitation) by means of the *cavitation number*, σ, defined by the following:

$$\sigma = \frac{p_\infty - p_V(T_\infty)}{\frac{1}{2}\rho_L U_\infty^2}, \tag{5.1}$$

where U_∞, p_∞, and T_∞ are respectively a reference velocity, pressure, and temperature in the flow (usually upstream quantities); ρ_L is the liquid density; and $p_V(T_\infty)$ is the saturated vapor pressure. In a particular flow as σ is reduced, cavitation will first be observed to occur at some particular value of σ called the incipient cavitation number and denoted by σ_i. Further reduction in σ below σ_i would cause an increase in the number and size of the vapor bubbles.

Suppose that prior to cavitation inception, the magnitude of the lowest pressure in the single phase flow is given by the minimum value of the coefficient of pressure, C_{pmin}. Note that C_{pmin} is a *negative* number and that its value could be estimated from either experiments on or calculations of the single phase flow. Then, *if cavitation inception were to occur when the minimum pressure reaches the vapor pressure* it would follow that the value of the critical inception number, σ_i, would be simply given by the following:

$$\sigma_i = -C_{pmin}. \tag{5.2}$$

Unfortunately, many factors can cause the actual values of σ_i to depart radically from $-C_{pmin}$ and much research has been conducted to explore these departures because of the importance of determining σ_i accurately. Among the important factors are the following:

1. The ability of the liquid to sustain a tension so that bubbles do not grow to observable size until the pressure falls a finite amount below the vapor pressure. The magnitude of this tension is a function of the contamination of the liquid and, in particular, the size and properties of the microscopic bubbles (*cavitation nuclei*)) that grow to produce the observable vapor bubbles (see, for example, Billet 1985).
2. Cavitation nuclei require a finite residence time in which to grow to observable size.
3. Measurements or calculations usually yield a minimum coefficient of pressure that is a time-averaged value. Conversely, many of the flows with which one must deal in practice are turbulent and, therefore, nuclei in the middle of turbulent eddies may experience pressures below the vapor pressure even when the mean pressure is greater than the vapor pressure.

Moreover, because water tunnel experiments designed to measure σ_i are often carried out at considerably reduced scale, it is also critical to know how to scale up these effects to accurately anticipate inception at the full scale. A detailed examination of these effects is beyond the scope of this text and the reader is referred to Knapp, Daily, and Hammitt (1970), Acosta and Parkin (1975), Arakeri (1979), and Brennen (1995) for further discussion.

The stability phenomenon described in Section 4.2.5 has important consequences in many cavitating flows. To recognize this, one must visualize a spectrum of sizes of cavitation nuclei being convected into a region of low pressure within the flow. Then the p_∞ in Eqs. (4.37) and (4.43) will be the local pressure in the liquid surrounding the bubble, and p_∞ must be less than p_V for explosive cavitation growth to occur. It is clear from the above analysis that all of the nuclei whose size, R, is greater than some critical value will become unstable, grow explosively, and cavitate, whereas those nuclei smaller than that critical size will react passively and will therefore not become visible to the eye. Though the actual response of the bubble is dynamic and p_∞ is changing continuously, we can nevertheless anticipate that the critical nuclei size will be given approximately by $4S/3(p_V - p_\infty)^*$, where $(p_V - p_\infty)^*$ is some

5.2 Key Features of Bubble Cavitation

representative measure of the tension in the low-pressure region. Note that the lower the pressure level, p_∞, the smaller the critical size and the larger the number of nuclei that are activated. This accounts for the increase in the number of bubbles observed in a cavitating flow as the pressure is reduced.

It will be useful to develop an estimate of the maximum size to which a cavitation bubble grows during its trajectory through a region where the pressure is below the vapor pressure. In a typical external flow around a body characterized by the dimension, ℓ, it follows from Eq. (4.31) that the rate of growth is roughly given by the following:

$$\frac{dR}{dt} = U_\infty(-\sigma - C_{\text{pmin}})^{\frac{1}{2}}. \tag{5.3}$$

It should be emphasized that Eq. (4.31) implies explosive growth of the bubble, in which the volume displacement is increasing like t^3. To obtain an estimate of the maximum size to which the cavitation bubble grows, R_m, a measure of the time it spends below vapor pressure is needed. Assuming that the pressure distribution near the minimum pressure point is roughly parabolic (see Brennen 1995) the length of the region below vapor pressure will be proportional to $\ell(-\sigma - C_{\text{pmin}})^{\frac{1}{2}}$ and therefore the time spent in that region will be the same quantity divided by U_∞. The result is that an estimate of maximum size, R_m, is as follows:

$$R_m \approx 2\ell(-\sigma - C_{\text{pmin}}) \tag{5.4}$$

where the factor 2 comes from the more detailed analysis of Brennen (1995). Note that, whatever their initial size, all activated nuclei grow to roughly the same maximum size because both the asymptotic growth rate [Eq. (4.31)] and the time available for growth are essentially independent of the size of the original nucleus. For this reason all of the bubbles in a bubbly cavitating flow grow to roughly the same size (Brennen 1995).

5.2.2 Cavitation Bubble Collapse

We now examine in more detail the mechanics of cavitation bubble collapse. As demonstrated in a preliminary way in Section 4.2.4, vapor or cavitation bubble collapse in the absence of thermal effects can lead to very large interface velocities and very high localized pressures. This violence has important technological consequences for it can damage nearby solid surfaces in critical ways. In this and the following few sections, we briefly review the fundamental processes associated with the phenomena of cavitation bubble collapse. For further details, the reader is referred to more specialized texts such as Knapp *et al.* (1975), Young (1989), and Brennen (1995).

The analysis of Section 4.2.4 allowed approximate evaluation of the magnitudes of the velocities, pressures, and temperatures generated by cavitation bubble collapse [Eqs. (4.32), (4.34), and (4.35)] under a number of assumptions, including that the bubble remains spherical. Though it will be shown in Section 5.2.3 that collapsing bubbles do not remain spherical, the spherical analysis provides a useful starting point.

When a cavitation bubble grows from a small nucleus to many times its original size, the collapse will begin at a maximum radius, R_m, with a partial pressure of gas, p_{Gm}, that is very small indeed. In a typical cavitating flow R_m is of the order of 100 times the original nuclei size, R_o. Consequently, if the original partial pressure of gas in the nucleus was about 1 bar the value of p_{Gm} at the start of collapse would be about 10^{-6} bar. If the typical pressure depression in the flow yields a value for $(p_\infty^* - p_\infty(0))$ of, say, 0.1 bar it would follow from Eq. (4.34) that the maximum pressure generated would be about 10^{10} bar and the maximum temperature would be 4×10^4 times the ambient temperature! Many factors, including the diffusion of gas from the liquid into the bubble and the effect of liquid compressibility, mitigate this result. Nevertheless, the calculation illustrates the potential for the generation of high pressures and temperatures during collapse and the potential for the generation of shock waves and noise.

Early work on collapse by Herring (1941), Gilmore (1952), and others focused on the inclusion of liquid compressibility to learn more about the production of shock waves in the liquid generated by bubble collapse. Modifications to the Rayleigh–Plesset equation that would allow for liquid compressibility were developed and these are reviewed by Prosperetti and Lezzi (1986). A commonly used variant is that proposed by Keller and Kolodner (1956); neglecting thermal, viscous, and surface tension effects this is as follows:

$$\left(1 - \frac{1}{c_L}\frac{dR}{dt}\right) R \frac{d^2R}{dt^2} + \frac{3}{2}\left(1 - \frac{1}{3c_L}\frac{dR}{dt}\right)\left(\frac{dR}{dt}\right)^2$$
$$= \left(1 + \frac{1}{c_L}\frac{dR}{dt}\right)\frac{1}{\rho_L}\{p_B - p_\infty - p_c(t + R/c_L)\} + \frac{R}{\rho_L c_L}\frac{dp_B}{dt}, \quad (5.5)$$

where c_L is the speed of sound in the liquid and $p_c(t)$ denotes the variable part of the pressure in the liquid at the location of the bubble center in the absence of the bubble.

However, as long as there is some noncondensable gas present in the bubble to decelerate the collapse, the primary importance of liquid compressibility is not the effect it has on the bubble dynamics (which is slight) but the role it plays in the formation of shock waves during the rebounding phase that follows collapse. Hickling and Plesset (1964) were the first to make use of numerical solutions of the compressible flow equations to explore the formation of pressure waves or shocks during the rebound phase. Figure 5.1 presents an example of their results for the pressure distributions in the liquid before (left) and after (right) the moment of minimum size. The graph on the right clearly shows the propagation of a pressure pulse or shock away from the bubble following the minimum size. As indicated in that figure, Hickling and Plesset concluded that the pressure pulse exhibits approximately geometric attenuation (like r^{-1}) as it propagates away from the bubble. Other numerical calculations have since been carried out by Ivany and Hammitt (1965), Tomita and Shima (1977), and Fujikawa and Akamatsu (1980), among others.

5.2 Key Features of Bubble Cavitation

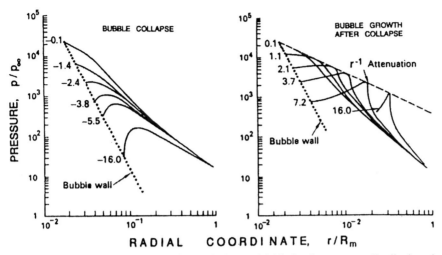

Figure 5.1. Typical results of Hickling and Plesset (1964) for the pressure distributions in the liquid before collapse (left) and after collapse (right) (without viscosity or surface tension). The parameters are $p_\infty = 1$ bar, $\gamma = 1.4$, and the initial pressure in the bubble was 10^{-3} bar. The values attached to each curve are proportional to the time before or after the minimum size.

Even if thermal effects are negligible for most of the collapse phase, they play a very important role in the final stage of collapse when the bubble contents are highly compressed by the inertia of the in-rushing liquid. The pressures and temperatures that are predicted to occur in the gas within the bubble during spherical collapse are very high indeed. Because the elapsed times are so small (of the order of microseconds), it would seem a reasonable approximation to assume that the noncondensable gas in the bubble behaves adiabatically. Typical of the adiabatic calculations is the work of Tomita and Shima (1977) who obtained maximum gas temperatures as high as 8800°K in the bubble center. But, despite the small elapsed times, Hickling (1963) demonstrated that heat transfer between the liquid and the gas is important because of the extremely high temperature gradients and the short distances involved. In later calculations Fujikawa and Akamatsu (1980) included heat transfer and, for a case similar to that of Tomita and Shima, found lower maximum temperatures and pressures of the order of 6700°K and 848 bar respectively at the bubble center. These temperatures and pressures only exist for a fraction of a microsecond.

All of these analyses assume spherical symmetry. We now focus attention on the stability of shape of a collapsing bubble before continuing discussion of the origins of cavitation damage.

5.2.3 Shape Distortion during Bubble Collapse

Like any other accelerating liquid–gas interface, the surface of a bubble is susceptible to Rayleigh–Taylor instability and is potentially unstable when the direction of the acceleration is from the less dense gas toward the denser liquid. Of course, the spherical

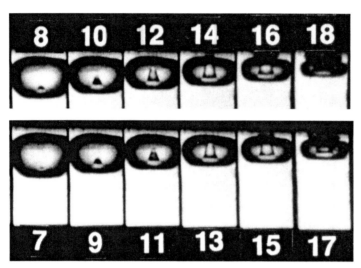

Figure 5.2. Series of photographs showing the development of the microjet in a bubble collapsing very close to a solid wall (at top of frame). The interval between the numbered frames is 2 μs and the frame width is 1.4 mm. From Tomita and Shima (1990), reproduced with permission of the authors.

geometry causes some minor quantitative departures from the behavior of a plane interface; these differences were explored by Birkhoff (1954) and Plesset and Mitchell (1956), who first analysed the Rayleigh–Taylor instability of bubbles. As expected a bubble is most unstable to nonspherical perturbations when it experiences the largest, positive values of d^2R/dt^2. During the growth and collapse cycle of a cavitation bubble, there is a brief and weakly unstable period during the initial phase of growth that can cause some minor roughening of the bubble surface (Reynolds and Berthoud 1981). But, much more important, is the rebound phase at the end of the collapse when compression of the bubble contents causes d^2R/dt^2 to switch from the small negative values of early collapse to very large positive values when the bubble is close to its minimum size.

This strong instability during the rebound phase appears to have several different consequences. When the bubble surroundings are strongly asymmetrical, for example, the bubble is close to a solid wall or a free surface, the dominant perturbation that develops is a reentrant jet. Of particular interest for cavitation damage is the fact that a nearby solid boundary can cause a reentrant microjet directed toward that boundary. The surface of the bubble furthest from the wall accelerates inward more rapidly than the side close to the wall and this results in a high-speed reentrant microjet that penetrates the bubble and can achieve very high speeds. Such microjets were first observed experimentally by Naude and Ellis (1961) and Benjamin and Ellis (1966). The series of photographs shown in Figure 5.2 represent a good example of the experimental observations of a developing reentrant jet. Figure 5.3 presents a comparison between the reentrant jet development in a bubble collapsing near a

5.2 Key Features of Bubble Cavitation

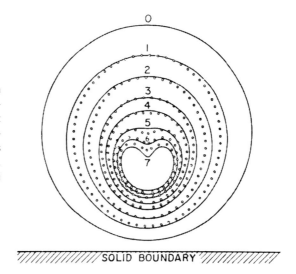

Figure 5.3. The collapse of a cavitation bubble close to a solid boundary in a quiescent liquid. The theoretical shapes of Plesset and Chapman (1971) (solid lines) are compared with the experimental observations of Lauterborn and Bolle (1975) (points). Figure adapted from Plesset and Prosperetti (1977).

solid wall as observed by Lauterborn and Bolle (1975) and as computed by Plesset and Chapman (1971). Note also that depth charges rely for their destructive power on a reentrant jet directed toward the submarine upon the collapse of the explosively generated bubble.

Other strong asymmetries can also cause the formation of a reentrant jet. A bubble collapsing near a free surface produces a reentrant jet directed *away* from the free surface (Chahine 1977). Indeed, there exists a critical flexibility for a nearby surface that separates the circumstances in which the reentrant jet is directed away from rather than toward the surface. Gibson and Blake (1982) demonstrated this experimentally and analytically and suggested flexible coatings or liners as a means of avoiding cavitation damage. Another possible asymmetry is the proximity of other, neighboring bubbles in a finite cloud of bubbles. Chahine and Duraiswami (1992) showed that the bubbles on the outer edge of such a cloud will tend to develop jets directed toward the center of the cloud.

When there is no strong asymmetry, the analysis of the Rayleigh–Taylor instability shows that the most unstable mode of shape distortion can be a much higher-order mode. These higher-order modes can dominate when a vapor bubble collapses far from boundaries. Thus observations of collapsing cavitation bubbles, although they may show a single vapor/gas volume prior to collapse, just after minimum size the *bubble* appears as a cloud of much smaller bubbles. An example of this is shown in Figure 5.4. Brennen (1995) shows how the most unstable mode depends on two parameters representing the effects of surface tension and noncondensable gas in the bubble. That most unstable mode number was later used in one of several analyses seeking to predict the number of fission fragments produced during collapse of a cavitating bubble (Brennen 2002).

Figure 5.4. Photographs of an ether bubble in glycerine before (left) and after (right) a collapse and rebound, both bubbles being about 5–6 mm across. Reproduced from Frost and Sturtevant (1986) with the permission of the authors.

5.2.4 Cavitation Damage

Perhaps the most ubiquitous engineering problem caused by cavitation is the material damage that cavitation bubbles can cause when they collapse in the vicinity of a solid surface. Consequently, this subject has been studied quite intensively for many years (see, for example, ASTM 1967; Thiruvengadam 1967, 1974; Knapp, Daily, and Hammitt 1970). The problem is a difficult one because it involves complicated unsteady flow phenomena combined with the reaction of the particular material of which the solid surface is made. Though there exist many empirical rules designed to help the engineer evaluate the potential cavitation damage rate in a given application, there remain a number of basic questions regarding the fundamental mechanisms involved. Cavitation bubble collapse is a violent process that generates highly localized, large-amplitude shock waves (Section 5.2.2) and microjets (Section 5.2.3). When this collapse occurs close to a solid surface, these intense disturbances generate highly localized and transient surface stresses. With softer material, individual pits caused by a single bubble collapse are often observed. But with the harder materials used in most applications it is the repetition of the loading due to repeated collapses that causes local surface fatigue failure and the subsequent detachment of pieces of material. Thus cavitation damage to metals usually has the crystalline appearance of fatigue failure. The damaged runner and pump impeller in Figures 5.5 and 5.6 are typical examples.

The issue of whether cavitation damage is caused by microjets or by shock waves generated when the remnant cloud of bubble reaches its minimum volume (or by both) has been debated for many years. In the 1940s and 1950s the focus was on the shock waves generated by spherical bubble collapse. When the phenomenon of the microjet was first observed, the focus shifted to studies of the impulsive pressures generated by microjets. First Shima *et al.* (1983) used high speed Schlieren photography to show that a spherical shock wave was indeed generated by the remnant cloud at the instant

5.2 Key Features of Bubble Cavitation

Figure 5.5. Major cavitation damage to the blades at the discharge from a Francis turbine.

of minimum volume. About the same time, Fujikawa and Akamatsu (1980) used a photoelastic material so that they could simultaneously observe the stresses in the solid and measure the acoustic pulses and were able to confirm that the impulsive stresses in the material were initiated at the same moment as the acoustic pulse. They also concluded that this corresponded to the instant of minimum volume and that the waves were not produced by the microjet. Later, however, Kimoto (1987) observed stress pulses that resulted both from microjet impingement and from the remnant cloud collapse shock.

The microjet phenomenon in a quiescent fluid has been extensively studied analytically as well as experimentally. Plesset and Chapman (1971) numerically calculated the distortion of an initially spherical bubble as it collapsed close to a solid boundary and, as Figure 5.3 demonstrates, their profiles are in good agreement with the experimental observations of Lauterborn and Bolle (1975). Blake and Gibson (1987) review the current state of knowledge, particularly the analytical methods for solving for bubbles collapsing near a solid or a flexible surface.

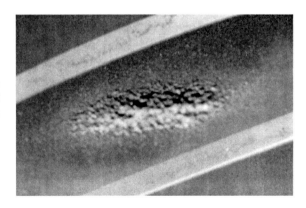

Figure 5.6. Photograph of localized cavitation damage on the blade of a mixed flow pump impeller made from an aluminum-based alloy.

Figure 5.7. Series of photographs of a hemispherical bubble collapsing against a wall showing the *pancaking* mode of collapse. From Benjamin and Ellis (1966) reproduced with permission of the first author.

It must also be noted that there are many circumstances in which it is difficult to discern a microjet. Some modes of bubble collapse near a wall involve a *pancaking* mode exemplified by the photographs in Figure 5.7 and in which no microjet is easily recognized.

Finally, it is important to emphasize that virtually all of the observations described above pertain to bubble collapse in an otherwise quiescent fluid. A bubble that grows and collapses in a flow is subject to other deformations that can significantly alter its collapse dynamics, modify or eliminate the microjet, and alter the noise and damage potential of the collapse process. In the next section some of these flow deformations are illustrated.

5.3 Cavitation Bubbles

5.3.1 Observations of Cavitating Bubbles

We end our brief survey of the dynamics of cavitating bubbles with some experimental observations of single bubbles (single cavitation events) in real flows for these reveal the complexity of the microfluid mechanics of individual bubbles. The focus here is on individual events springing from a single nucleus. The interactions between bubbles at higher nuclei concentrations are discussed later.

Pioneering observations of individual cavitation events were made by Knapp and his associates at the California Institute of Technology in the 1940s (see, for example, Knapp and Hollander 1948) using high-speed movie cameras capable of 20,000 frames/s. Shortly thereafter Plesset (1949), Parkin (1952), and others began to model these observations of the growth and collapse of traveling cavitation bubbles using modifications of Rayleigh's original equation of motion for a spherical bubble. However, observations of real flows demonstrate that even single cavitation bubbles are often highly distorted by the pressure gradients in the flow. Before describing some of the observations, it is valuable to consider the relative sizes of the cavitation bubbles and the viscous boundary layer. In the flow of a uniform stream of velocity, U, around an object such as a hydrofoil with typical dimension, ℓ, the thickness of the laminar boundary layer near the minimum pressure point will be given qualitatively by $\delta = (\nu_L \ell / U)^{\frac{1}{2}}$. Comparing this with the typical maximum bubble radius, R_m, given by Eq. (5.4), it follows that the ratio, δ / R_m, is roughly given by the

5.3 Cavitation Bubbles

Figure 5.8. A series of photographs illustrating, in profile, the growth and collapse of a traveling cavitation bubble in a flow around a 5.08-cm-diameter headform at $\sigma = 0.45$ and a speed of 9 m/s. the sequence is top left, top right, bottom left, bottom right; the flow is from right to left. The life-size width of each photograph is 0.73 cm. From Ceccio and Brennen (1991).

following:

$$\frac{\delta}{R_m} = \frac{1}{2(-\sigma - C_{pmin})} \left\{ \frac{\nu_L}{\ell U} \right\}^{\frac{1}{2}}. \tag{5.6}$$

Therefore, provided $(-\sigma - C_{pmin})$ is of the order of 0.1 or greater, it follows that for the high Reynolds numbers, $U\ell/\nu_L$, that are typical of most of the flows in which cavitation is a problem, the boundary layer is usually much thinner than the typical dimension of the bubble.

Recently, Ceccio and Brennen (1991) and Kuhn de Chizelle et al. (1992a,b) have made an extended series of observations of cavitation bubbles in the flow around axisymmetric bodies, including studies of the scaling of the phenomena. The observations at lower Reynolds numbers are exemplified by the photographs of bubble profiles in Figure 5.8. In all cases the shape during the initial growth phase is that of a spherical cap, the bubble being separated from the wall by a thin layer of liquid of the same order of magnitude as the boundary layer thickness. Later developments depend on the geometry of the headform and the Reynolds number. In some cases as the bubble enters the region of adverse pressure gradient, the exterior frontal surface is pushed inward, causing the profile of the bubble to appear wedgelike. Thus the collapse is initiated on the exterior frontal surface of the bubble, and this often leads to the bubble fissioning into forward and aft bubbles as seen in Figure 5.8. At the same time, the bubble acquires significant spanwise vorticity through its interactions with the boundary layer during the growth phase. Consequently, as the collapse proceeds, this vorticity is concentrated and the bubble evolves into one (or two or possibly more)

Figure 5.9. Examples of bubble fission (upper left), the instability of the liquid layer under a traveling cavitation bubble (upper right) and the attached tails (lower). From Ceccio and Brennen (1991) experiments with a 5.08-cm diameter ITTC headform at $\sigma = 0.45$ and a speed of 8.7 m/s. The flow is from right to left. The life-size widths of the photographs are 0.63, 0.80, and 1.64 cm respectively.

short cavitating vortices with spanwise axes. These vortex bubbles proceed to collapse and seem to rebound as a cloud of much smaller bubbles. Ceccio and Brennen (1991) (see also Kumar and Brennen 1993) conclude that the flow-induced fission prior to collapse can have a substantial effect on the noise produced.

Two additional phenomena were observed. In some cases the layer of liquid underneath the bubble would become disrupted by some instability, creating a bubbly layer of fluid that subsequently gets left behind the main bubble (see Figure 5.9). Second, it sometimes happened that when a bubble passed a point of laminar separation, it triggered the formation of local *attached cavitation* streaks at the lateral or spanwise extremities of the bubble, as seen in Figure 5.9. Then, as the main bubble proceeds downstream, these *streaks* or *tails* of attached cavitation are stretched out behind the main bubble, the trailing ends of the tails being attached to the solid surface. Tests at much higher Reynolds numbers (Kuhn de Chizelle *et al.* 1992a,b) revealed that these *events with tails* occured more frequently and would initiate attached cavities over the entire wake of the bubble as seen in Figure 5.10. Moreover, the attached cavitation would tend to remain for a longer period after the main bubble had disappeared. Eventually, at the highest Reynolds numbers tested, it appeared that the passage of a single bubble was sufficient to trigger a *patch* of attached cavitation (Figure 5.10, bottom) that would persist for an extended period after the bubble had long disappeared.

In summary, cavitation bubbles are substantially deformed and their dynamics and acoustics altered by the flow fields in which they occur. This necessarily changes the noise and damage produced by those cavitation events.

5.3.2 Cavitation Noise

The violent and catastrophic collapse of cavitation bubbles results in the production of noise that is a consequence of the momentary large pressures that are generated when the contents of the bubble are highly compressed. Consider the flow in the

5.3 Cavitation Bubbles

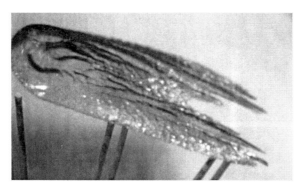

Figure 5.10. Typical cavitation events from the scaling experiments of Kuhn de Chizelle *et al.* (1992b) showing transient bubble-induced patches, the upper one occurring on a 50.8-cm-diameter Schiebe headform at $\sigma = 0.605$ and a speed of 15 m/s and the lower one on a 25.4-cm headform at $\sigma = 0.53$ and a speed of 15 m/s. The flow is from right to left. The life-size widths of the photographs are 6.3 cm (top) and 7.6 cm (bottom).

liquid caused by the volume displacement of a growing or collapsing cavity. In the far field the flow will approach that of a simple source, and it is clear that Eq. (4.5) for the pressure will be dominated by the first term on the right-hand side (the unsteady inertial term) because it decays more slowly with radius r than the second term. If we denote the time-varying volume of the cavity by $V(t)$ and substitute using Eq. (4.2), it follows that the time-varying component of the pressure in the far field is given by the following:

$$p_a = \frac{\rho_L}{4\pi \mathcal{R}} \frac{d^2 V}{dt^2}, \qquad (5.7)$$

where p_a is the radiated acoustic pressure and we denote the distance, r, from the cavity center to the point of measurement by $\mathcal{R}$ (for a more thorough treatment see Dowling and Ffowcs Williams 1983 and Blake 1986b). Because the noise is directly proportional to the second derivative of the volume with respect to time, it is clear that the noise pulse generated at bubble collapse occurs because of the very large and positive values of d^2V/dt^2 when the bubble is close to its minimum size. It is conventional (see, for example, Blake 1986b) to present the sound level using a root

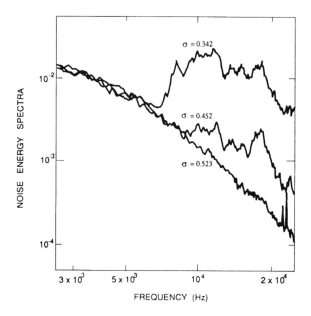

Figure 5.11. Acoustic power spectra from a model spool valve operating under noncavitating ($\sigma = 0.523$) and cavitating ($\sigma = 0.452$ and 0.342) conditions (from the investigation of Martin et al. 1981).

mean square pressure or *acoustic* pressure, p_s, defined by the following:

$$p_s^2 = \overline{p_a^2} = \int_0^\infty \mathcal{G}(f) df \qquad (5.8)$$

and to represent the distribution over the frequency range, f, by the spectral density function, $\mathcal{G}(f)$.

To the researcher or engineer, the crackling noise that accompanies cavitation is one of the most evident characteristics of the phenomenon. The onset of cavitation is often detected first by this noise rather than by visual observation of the bubbles. Moreover, for the practical engineer it is often the primary means of detecting cavitation in devices such as pumps and valves. Indeed, several empirical methods have been suggested that estimate the rate of material damage by measuring the noise generated (for example, Lush and Angell 1984).

The noise due to cavitation in the orifice of a hydraulic control valve is typical, and spectra from such an experiment are presented in Figure 5.11. The lowest curve at $\sigma = 0.523$ represents the turbulent noise from the noncavitating flow. Below the incipient cavitation number (about 0.523 in this case) there is a dramatic increase in the noise level at frequencies of about 5 kHz and above. The spectral peak between 5 and 10 kHz corresponds closely to the expected natural frequencies of the nuclei present in the flow (see Section 4.4.1).

Most of the analytical approaches to cavitation noise build on knowledge of the dynamics of collapse of a single bubble. Fourier analyses of the radiated acoustic pressure due to a single bubble were first visualized by Rayleigh (1917) and implemented by Mellen (1954) and Fitzpatrick and Strasberg (1956). In considering such Fourier analyses, it is convenient to nondimensionalize the frequency by the typical time span

5.3 Cavitation Bubbles

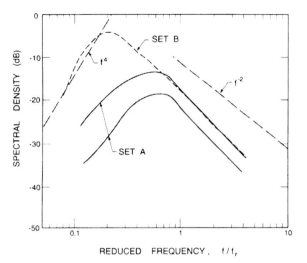

Figure 5.12. Acoustic power spectra of the noise from a cavitating jet. Shown are mean lines through two sets of data constructed by Blake and Sevik (1982) from the data by Jorgensen (1961). Typical asymptotic behaviors are also indicated. The reference frequency, f_r, is $(p_\infty/\rho_L d^2)^{\frac{1}{2}}$, where d is the jet diameter.

of the whole event or, equivalently, by the collapse time, t_{tc}, given by Eq. (4.36). Now consider the frequency content of $\mathcal{G}(f)$ using the dimensionless frequency, ft_{tc}. Because the volume of the bubble increases from zero to a finite value and then returns to zero, it follows that for $ft_{tc} < 1$ the Fourier transform of the volume is independent of frequency. Consequently d^2V/dt^2 will be proportional to f^2 and therefore $\mathcal{G}(f) \propto f^4$ (see Fitzpatrick and Strasberg 1956). This is the origin of the left-hand asymptote in Figure 5.12.

The behavior at intermediate frequencies for which $ft_{tc} > 1$ has been the subject of more speculation and debate. Mellen (1954) and others considered the typical equations governing the collapse of a spherical bubble in the absence of thermal effects and noncondensable gas [Eq. (4.32)] and concluded that, because the velocity $dR/dt \propto R^{-\frac{3}{2}}$, it follows that $R \propto t^{\frac{2}{5}}$. Therefore the Fourier transform of d^2V/dt^2 leads to the asymptotic behavior $\mathcal{G}(f) \propto f^{-\frac{2}{5}}$. The error in this analysis is the neglect of the noncondensable gas. When this is included and when the collapse is sufficiently advanced, the last term in the square brackets of Eq. (4.32) becomes comparable with the previous terms. Then the behavior is quite different from $R \propto t^{-\frac{2}{5}}$. Moreover, the values of d^2V/dt^2 are much larger during this rebound phase, and therefore the frequency content of the rebound phase will dominate the spectrum. It is therefore not surprising that the $f^{-\frac{2}{5}}$ is not observed in practice. Rather, most of the experimental results seem to exhibit an intermediate frequency behavior like f^{-1} or f^{-2}. Jorgensen (1961) measured the noise from submerged, cavitating jets and found a behavior like f^{-2} at the higher frequencies (see Figure 5.12). However, most of the experimental data for cavitating bodies or hydrofoils exhibit a weaker decay. The data by Arakeri and Shangumanathan (1985) from cavitating headform experiments show a very consistent f^{-1} trend over almost the entire frequency range, and very similar results have been obtained by Ceccio and Brennen (1991).

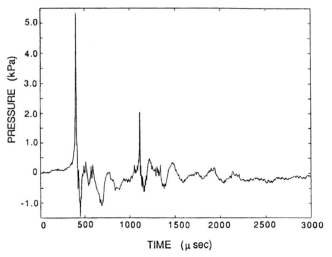

Figure 5.13. A typical acoustic signal from a single collapsing bubble. From Ceccio and Brennen (1991).

Ceccio and Brennen (1991) recorded the noise from individual cavitation bubbles in a flow; a typical acoustic signal from their experiments is reproduced in Figure 5.13. The large positive pulse at about 450 μs corresponds to the first collapse of the bubble. This first pulse in Figure 5.13 is followed by some facility-dependent oscillations and by a second pulse at about 1100 μs. This corresponds to the second collapse that follows the rebound from the first collapse.

A good measure of the magnitude of the collapse pulse is the acoustic impulse, I, defined as the area under the pulse or

$$I = \int_{t_1}^{t_2} p_a dt, \tag{5.9}$$

where t_1 and t_2 are times before and after the pulse at which p_a is zero. For later purposes we also define a dimensionless impulse, I^*, as follows:

$$I^* = 4\pi I \mathcal{R}/\rho_L U \ell^2, \tag{5.10}$$

where U and ℓ are the reference velocity and length in the flow. The average acoustic impulses for individual bubble collapses on two axisymmetric headforms (ITTC and Schiebe headforms) are compared in Figure 5.14 with impulses predicted from integration of the Rayleigh–Plesset equation. Because these theoretical calculations assume that the bubble remains spherical, the discrepancy between the theory and the experiments is not too surprising. Indeed one interpretation of Figure 5.14 is that the theory can provide an order of magnitude estimate and an upper bound on the noise produced by a single bubble. In actuality, the departure from sphericity produces a less focused collapse and therefore less noise.

The next step is to consider the synthesis of cavitation noise from the noise produced by individual cavitation bubbles or events. If the impulse produced by each event is

5.3 Cavitation Bubbles

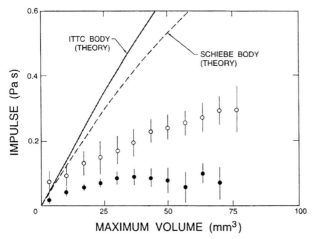

Figure 5.14. Comparison of the acoustic impulse, I, produced by the collapse of a single cavitation bubble on two axisymmetric headforms as a function of the maximum volume prior to collapse. (Open symbols) Average data for Schiebe headform; (closed symbols) ITTC headform; vertical lines indicate 1 standard deviation. Also shown are the corresponding results from the solution of the Rayleigh–Plesset equation. From Ceccio and Brennen (1991).

denoted by I and the number of events per unit time is denoted by $\dot{n}$, the sound pressure level, p_s, is given by the following:

$$p_s = I\dot{n}. \tag{5.11}$$

Consider the scaling of cavitation noise that is implicit in this construct. Both the experimental results and the analysis based on the Rayleigh–Plesset equation indicate that the nondimensional impulse produced by a single cavitation event is strongly correlated with the maximum volume of the bubble prior to collapse and is almost independent of the other flow parameters. It follows from Eqs. (5.7) and (5.9) that

$$I^* = \frac{1}{U\ell^2}\left\{\left(\frac{dV}{dt}\right)_{t_2} - \left(\frac{dV}{dt}\right)_{t_1}\right\} \tag{5.12}$$

and the values of dV/dt at the moments $t = t_1, t_2$ when $d^2V/dt^2 = 0$ may be obtained from the Rayleigh–Plesset equation. If the bubble radius at the time t_1 is denoted by R_x and the coefficient of pressure in the liquid at that moment is denoted by C_{px}, then

$$I^* \approx 8\pi\left(\frac{R_x}{\ell}\right)^2 (C_{px} - \sigma)^{\frac{1}{2}}. \tag{5.13}$$

Numerical integrations of the Rayleigh–Plesset equation for a range of typical circumstances yield $R_x/R_m \approx 0.62$ where R_m is the maximum volumetric radius and that $(C_{px} - \sigma) \propto R_m/\ell$ (in these calculations ℓ was the headform radius) so that

$$I^* \approx \beta\left(\frac{R_m}{\ell}\right)^{\frac{5}{2}}. \tag{5.14}$$

The aforementioned integrations of the Rayleigh–Plesset equation yield a factor of proportionality, β, of about 35. Moreover, the upper envelope of the experimental data of which Figure 5.14 is a sample appears to correspond to a value of $\beta \approx 4$. We note that a quite similar relation between I^* and R_m/ℓ emerges from the analysis by Esipov and Naugol'nykh (1973) of the compressive sound wave generated by the collapse of a gas bubble in a compressible liquid.

From the above relations, it follows that

$$I \approx \frac{\beta}{12}\rho_L U R_m^{\frac{5}{2}}/\mathcal{R}\ell^{\frac{1}{2}}. \tag{5.15}$$

Consequently, the evaluation of the impulse from a single event is completed by an estimate of R_m such as that of Eq. (5.4). Because that estimate has R_m independent of U for a given cavitation number, it follows that I is linear with U.

The event rate, $\dot{n}$, can be considerably more complicated to evaluate than might at first be thought. If all the nuclei flowing through a certain, known streamtube (say with a cross-sectional area, A_n, in the upstream flow) were to cavitate similarly, then the result would be as follows:

$$\dot{n} = nA_n U, \tag{5.16}$$

where n is the nuclei concentration (number/unit volume) in the incoming flow. Then it follows that the acoustic pressure level resulting from substituting Eqs. (5.16) and (5.15) into Eq. (5.11) and using Eq. (5.4) becomes the following:

$$p_s \approx \frac{\beta}{3}\rho_L U^2 A_n n \ell^2 (-\sigma - C_{pmin})^{\frac{5}{2}}/\mathcal{R}, \tag{5.17}$$

where we have omitted some of the constants of order unity. For the relatively simple flows considered here, Eq. (5.17) yields a sound pressure level that scales with U^2 and with ℓ^4 because $A_n \propto \ell^2$. This scaling with velocity does correspond roughly to that which has been observed in some experiments on traveling bubble cavitation, for example, those of Blake, Wolpert, and Geib (1977) and Arakeri and Shangumanathan (1985). The former observe that $p_s \propto U^m$ where $m = 1.5$ to 2.

Different scaling laws will apply when the cavitation is generated by turbulent fluctuations such as in a turbulent jet (see, for example, Ooi 1985 and Franklin and McMillan 1984). Then the typical tension experienced by a nucleus as it moves along a disturbed path in a turbulent flow is very much more difficult to estimate. Consequently, the models for the sound pressure due to cavitation in a turbulent flow and the scaling of that sound with velocity are less well understood.

5.3.3 Cavitation Luminescence

Though highly localized both temporally and spatially, the extremely high temperatures and pressures that can occur in the noncondensable gas during collapse are believed to be responsible for the phenomenon known as luminescence, the emission

5.3 Cavitation Bubbles

of light that is observed during cavitation bubble collapse. The phenomenon was first observed by Marinesco and Trillat (1933), and a number of different explanations were advanced to explain the emissions. The fact that the light was being emitted at collapse was first demonstrated by Meyer and Kuttruff (1959). They observed cavitation on the face of a rod oscillating magnetostrictively and correlated the light with the collapse point in the growth-and-collapse cycle. The balance of evidence now seems to confirm the suggestion by Noltingk and Neppiras (1950) that the phenomenon is caused by the compression and adiabatic heating of the noncondensable gas in the collapsing bubble. As we discussed previously in Sections 4.2.4 and 5.2.2, temperatures of the order of $6000°K$ can be anticipated on the basis of uniform compression of the noncondensable gas; the same calculations suggest that these high temperatures will last for only a fraction of a microsecond. Such conditions would explain the emission of light. Indeed, the measurements of the spectrum of sonoluminescence by Taylor and Jarman (1970), Flint and Suslick (1991), and others suggest a temperature of about $5000°K$. However, some recent experiments by Barber and Putterman (1991) indicate much higher temperatures and even shorter emission durations of the order of picoseconds. Speculations on the explanation for these observations have centered on the suggestion by Jarman (1960) that the collapsing bubble forms a spherical, inward-propagating shock in the gas contents of the bubble and that the focusing of the shock at the center of the bubble is an important reason for the extremely high apparent *temperatures* associated with the sonoluminescence radiation. It is, however, important to observe that spherical symmetry is essential for this mechanism to have any significant consequences. One would therefore expect that the distortions caused by a flow would not allow significant shock focusing and would even reduce the effectiveness of the basic compression mechanism.

6

Boiling and Condensation

6.1 Introduction

The fundamentals of bubble growth or collapse during boiling or condensation were described in Chapter 4 and particularly in the sections dealing with thermally inhibited growth or collapse. This chapter deals with a number of additional features of these processes. In many industrial contexts in which boiling or condensation occurs, the presence of a nearby solid surface is necessary for the rapid supply or removal of the latent heat inherent in the phase change. The presence of this wall modifies the flow patterns and other characteristics of these multiphase flows and this chapter addresses those additional phenomena.

In all cases the heat flux per unit area through the solid surface is denoted by $\dot{q}$; the wall temperature is denoted by T_w and the bulk liquid temperature by T_b (or T_L). The temperature difference, $\Delta T = T_w - T_b$, is a ubiquitous feature of all these problems. Moreover, in almost all cases the pressure differences within the flow are sufficiently small that the saturated liquid/vapor temperature, T_e, can be assumed uniform. Then, to a first approximation, boiling at the wall occurs when $T_w > T_e$ and $T_b \leq T_e$. When $T_b < T_e$ and the liquid must be heated to T_e before bubbles occur, the situation is referred to as subcooled boiling. Conversely, condensation at the wall occurs when $T_w < T_e$ and $T_b \geq T_e$. When $T_b > T_e$ and the vapor must be cooled to T_e before liquid appears, the situation is referred to as super-heated condensation.

The solid surface may be a plane vertical or horizontal containing surface or it may be the interior or exterior of a circular pipe. Another factor influencing the phenomena is whether there is a substantial fluid flow (convection) parallel to the solid surface. For some of the differences between these various geometries and imposed flow conditions the reader is referred to texts such as Collier and Thome (1994), Hsu and Graham (1976), or Whalley (1987). In the next section we review the phenomena associated with a plane horizontal boundary with no convection. Later sections deal with vertical surfaces.

6.2 Horizontal Surfaces

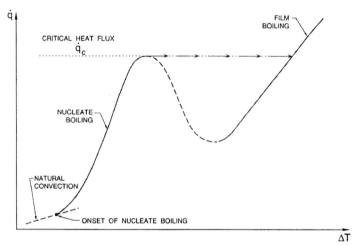

Figure 6.1. Pool boiling characteristics.

6.2 Horizontal Surfaces

6.2.1 Pool Boiling

Perhaps the most common configuration, known as *pool boiling* is when a pool of liquid is heated from below through a horizontal surface. For present purposes we assume that the heat flux, $\dot{q}$, is uniform. A uniform bulk temperature far from the wall is maintained because the mixing motions generated by natural convection (and, in boiling, by the motions of the bubbles) mean that most of the liquid is at a fairly uniform temperature. In other words, the temperature difference ΔT occurs within a thin layer next to the wall.

In pool boiling the relation between the heat flux, $\dot{q}$, and ΔT is as sketched in Figure 6.1 and events develop with increasing ΔT as follows. When the pool as a whole has been heated to a temperature close to T_e, the onset of nucleate boiling occurs. Bubbles form at nucleation sites on the wall and grow to a size at which the buoyancy force overcomes the surface tension forces acting at the line of attachment of the bubble to the wall. The bubbles then break away and rise through the liquid.

In a steady-state process, the vertically upward heat flux, $\dot{q}$, should be the same at all elevations above the wall. Close to the wall the situation is complex for several mechanisms increase the heat flux above that for pure conduction through the liquid. First the upward flux of vapor away from the wall must be balanced by an equal downward mass flux of liquid and this brings cooler liquid into closer proximity to the wall. Second, the formation and movement of the bubbles enhances mixing in the liquid near the wall and thus increases heat transfer from the wall to the liquid. Third, the flux of heat to provide the latent heat of vaporization that supplies vapor to the bubbles increases the total heat flux. While a bubble is still attached to the wall, vapor may be formed at the surface of the bubble closest to the wall and then condense on

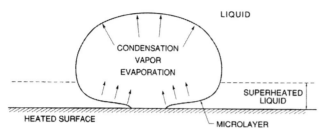

Figure 6.2. Sketch of nucleate boiling bubble with microlayer.

the surface furthest from the wall, thus creating a heat pipe effect. This last mode of heat transfer is sketched in Figure 6.2 and requires the presence of a thin layer of liquid under the bubble known as the *microlayer*.

At distances further from the wall (Figure 6.3) the dominant component of $\dot{q}$ is simply the enthalpy flux difference between the upward flux of vapor and the downward flux of liquid. Assuming this enthalpy difference is given approximately by the latent heat, $\mathcal{L}$, it follows that the upward volume flux of vapor, j_V, is given by $\dot{q}/\rho_V \mathcal{L}$, where ρ_V is the saturated vapor density at the prevailing pressure. Because mass must be conserved the downward mass flux of liquid must be equal to the upward mass flux of vapor and it follows that the downward liquid volume flux should be $\dot{q}/\rho_L \mathcal{L}$, where ρ_L is the saturated liquid density at the prevailing pressure.

To complete the analysis, estimates are needed for the number of nucleation sites per unit area of the wall ($N^* m^{-2}$), the frequency (f) with which bubbles leave each site and the equivalent volumetric radius (R) upon departure. Given the upward velocity of the bubbles (u_V) this allows evaluation of the volume fraction and volume flux of vapor bubbles from the following:

$$\alpha = \frac{4\pi R^3 N^* f}{3 u_V}; \quad j_V = \frac{4}{3}\pi R^3 N^* f \qquad (6.1)$$

and it then follows that

$$\dot{q} = \frac{4}{3}\pi R^3 N^* f \rho_V \mathcal{L}. \qquad (6.2)$$

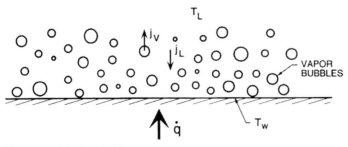

Figure 6.3. Nucleate boiling.

6.2 Horizontal Surfaces

As ΔT is increased both the site density N^* and the bubble frequency f increase until, at a certain critical heat flux, $\dot{q}_c$, a complete film of vapor blankets the wall. This is termed *boiling crisis*. Normally one is concerned with systems in which the heat flux rather than the wall temperature is controlled, and, because the vapor film provides a substantial barrier to heat transfer, such systems experience a large increase in the wall temperature when the boiling crisis occurs. This development is sketched in Figure 6.1. The increase in wall temperature can be very hazardous and it is therefore important to be able to predict the boiling crisis and the heat flux at which this occurs. There are a number of detailed analyses of the boiling crisis and for such detail the reader is referred to Zuber *et al.* (1959, 1961), Rohsenow and Hartnett (1973), Hsu and Graham (1976), Whalley (1987), or Collier and Thome (1994). This important fundamental process is discussed in Chapter 14 as a classic example of the flooding phenomenon in multiphase flows.

6.2.2 Nucleate Boiling

As Eq. (6.2) illustrates, quantitative understanding and prediction of nucleate boiling requires detailed information on the quantities N^*, f, R, and u_V and thus knowledge not only of the number of nucleation sites per unit area but also of the cyclic sequence of events as each bubble grows and detaches from a particular site. Though detailed discussion of the nucleation sites is beyond the scope of this book, it is well established that increasing ΔT activates increasingly smaller (and therefore more numerous) sites (Griffith and Wallis 1960) so that N^* increases rapidly with ΔT. The cycle of events at each nucleation site as bubbles are created, grow, and detach is termed the *ebullition cycle* and consists of the following:

1. A period of bubble growth during which the bubble growth rate is directly related to the rate of heat supply to each site, $\dot{q}/N^*$. In the absence of inertial effects and assuming that all this heat is used for evaporation (in a more precise analysis some fraction is used to heat the liquid), the bubble growth rate is then given by $dR/dt = C\dot{q}/4\pi R^2 \rho_V \mathcal{L} N^*$ where C is some constant that will be influenced by complicating factors such as the geometry of the bubble attachment to the wall and the magnitude of the temperature gradient in the liquid normal to the wall (see, for example, Hsu and Graham 1976).
2. The moment of detachment when the upward buoyancy forces exceed the surface tension forces at the bubble/wall contact line. This leads to a bubble size, R_d, upon detachment given qualitatively by

$$R_d = C \left[\frac{S}{g(\rho_L - \rho_V)} \right]^{\frac{1}{2}}, \qquad (6.3)$$

where the constant C will depend on surface properties such as the contact angle but is of the order of 0.005 (Fritz 1935). With the growth rate from the growth phase analysis this fixes the time for growth.

3. The waiting period during which the local cooling of the wall in the vicinity of the nucleation site is diminished by conduction within the wall surface and after which the growth of another bubble is initiated.

Obviously the sum of the growth time and the waiting period leads to the bubble frequency, f.

In addition, the rate of rise of the bubbles can be estimated using the methods of Chapters 2 and 3. As discussed later in Section 14.3.3, the downward flow of liquid must also be taken into account in evaluating u_V.

These are the basic elements involved in characterizing nucleate boiling though there are many details for which the reader is referred to the texts by Rohsenow and Hartnett (1973), Hsu and Graham (1976), Whalley (1987), or Collier and Thome (1994). Note that the concepts involved in the analysis of nucleate boiling on an inclined or vertical surface do not differ greatly. The addition of an imposed flow velocity parallel to the wall will alter some details because, for example, the analysis of the conditions governing bubble detachment must include consideration of the resulting drag on the bubble.

6.2.3 Film Boiling

At or near boiling crisis a film of vapor is formed that coats the surface and substantially impedes heat transfer. This vapor layer presents the primary resistance to heat transfer because the heat must be conducted through the layer. It follows that the thickness of the layer, δ, is given approximately by the following:

$$\delta = \frac{\Delta T k_V}{\dot{q}}. \tag{6.4}$$

However, these flows are usually quite unsteady because the vapor/liquid interface is unstable to Rayleigh–Taylor instability (see Sections 7.5.1 and 14.3.3). The result of this unsteadiness of the interface is that vapor bubbles are introduced into the liquid and travel upwards while liquid droplets are also formed and fall down through the vapor toward the hot surface. These droplets are evaporated near the surface producing an upward flow of vapor. Equation (6.4) then needs modification to account for the heat transfer across the thin layer under the droplet.

The droplets do not normally touch the hot surface because the vapor created on the droplet surface nearest the wall creates a lubrication layer that suspends the droplet. This is known as the Leidenfrost effect. It is readily observed in the kitchen when a drop of water is placed on a hot plate. Note, however, that the thermal resistance takes a similar form to that in Eq. (6.4) though the temperature difference in the vicinity of the droplet now occurs across the much thinner layer under the droplet rather than across the film thickness, δ.

6.2 Horizontal Surfaces

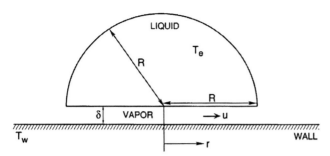

Figure 6.4. Hemispherical model of liquid drop for the Leidenfrost analysis.

6.2.4 Leidenfrost Effect

To analyze the Leidenfrost effect, we assume the simple geometry shown in Figure 6.4 in which a thin, uniform layer of vapor of thickness δ separates the hemispherical droplet (radius, R) from the wall. The droplet is assumed to have been heated to the saturation temperature T_e and the temperature difference $T_w - T_e$ is denoted by ΔT. Then the heat flux per unit surface area across the vapor layer is given by $k_V \Delta T/\delta$ and this causes a mass rate of evaporation of liquid at the droplet surface of $k_V \Delta T/\delta \mathcal{L}$. The outward radial velocity of vapor at a radius of r from the center of the vapor layer, $u(r)$ (see Figure 6.4), must match the total rate of volume production of vapor inside this radius, $\pi r^2 k_V \Delta T/\rho_V \delta \mathcal{L}$. Assuming that we use mean values of the quantities k_V, ρ_V, and $\mathcal{L}$ or that these do not vary greatly within the flow, this implies that the value of u averaged over the layer thickness must be given by the following:

$$u(r) = \frac{k_V \Delta T}{2\rho_V \mathcal{L}} \frac{r}{\delta^2}. \tag{6.5}$$

This connects the velocity $u(r)$ of the vapor to the thickness δ of the vapor layer. A second relation between these quantities is obtained by considering the equation of motion for the viscous outward radial flow of vapor (assuming the liquid velocities are negligible). This is simply a radial Poiseuille flow in which the mean velocity across the gap, $u(r)$, must be given by the following:

$$u(r) = -\frac{\delta^2}{12\mu_V} \frac{dp}{dr}, \tag{6.6}$$

where $p(r)$ is the pressure distribution in the vapor layer. Substituting for $u(r)$ from Eq. (6.5) and integrating we obtain the pressure distribution in the vapor layer as follows:

$$p(r) = p_a + \frac{3k_V \mu_V \Delta T}{\rho_V \mathcal{L}} \frac{(R^2 - r^2)}{2\delta^4}, \tag{6.7}$$

where p_a is the surrounding atmospheric pressure. Integrating the pressure difference, $p(r) - p_a$, to find the total upward force on the droplet and equating this to the difference between the weight of the droplet and the buoyancy force, $2\pi(\rho_L - \rho_V)R^3/3$, yields the following expression for the thickness, δ, of the vapor layer:

$$\frac{\delta}{R} = \left[\frac{9k_V \mu_V \Delta T}{8\rho_V(\rho_L - \rho_V)g\mathcal{L}R^3}\right]^{\frac{1}{4}}. \tag{6.8}$$

Substituting this result back into the expression for the velocity and then evaluating the mass flow rate of vapor and consequently the rate of loss of mass of the droplet, one can find the following expression for the lifetime, t_t, of a droplet of initial radius, R_o, as follows:

$$t_t = 4\left[\frac{2\mu_V}{9\rho_V g}\right]^{\frac{1}{4}}\left[\frac{(\rho_L - \rho_V)\mathcal{L}R_o}{k_V \Delta T}\right]^{\frac{3}{4}}. \tag{6.9}$$

As a numerical example, a water droplet with a radius of 2 mm at a saturated temperature of about 400 K near a wall with a temperature of 500 K will have a film thickness of just 40 μm but a lifetime of just over 1 hr. Note that as ΔT, k_V, or g go up the lifetime goes down as expected; conversely, increasing R_o or μ_V has the opposite effect.

6.3 Vertical Surfaces

Boiling on a heated vertical surface is qualitatively similar to that on a horizontal surface except for the upward liquid and vapor velocities caused by natural convection. Often this results in a cooler liquid and a lower surface temperature at lower elevations and a progression through various types of boiling as the flow proceeds upward. Figure 6.5 provides an illustrative example. Boiling begins near the bottom of the heated rod and the bubbles increase in size as they are convected upward. At a well-defined elevation, boiling crisis (Section 14.3.3 and Figure 6.1) occurs and marks the transition to film boiling at a point about 5/8 of the way up the rod in the photograph. At this point, the material of the rod or pipe experiences an abrupt and substantial rise in surface temperature as described in Section 14.3.3.

The nucleate boiling regime was described earlier. The film boiling regime is a little different than that described in Section 6.2.3 and is addressed in the following section.

6.3.1 Film Boiling

The first analysis of film boiling on a vertical surface was due to Bromley (1950) and proceeds as follows. Consider a small element of the vapor layer of length dy and thickness $\delta(y)$ as shown in Figure 6.6. The temperature difference between the wall and the vapor/liquid interface is ΔT. Therefore the mass rate of conduction of heat from the wall and through the vapor to the vapor/liquid interface per unit surface area of the wall is given approximately by $k_V \Delta T/\delta$, where k_V is the thermal conductivity of the vapor. In general some of this heat flux will be used to evaporate liquid at the interface

6.3 Vertical Surfaces

Figure 6.5. The evolution of convective boiling around a heated rod, reproduced from Sherman and Sabersky (1981) with permission.

Figure 6.6. Sketch for the film boiling analysis.

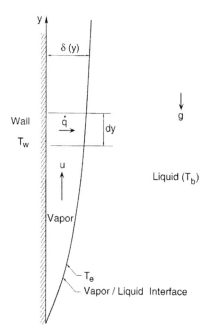

and some will be used to heat the liquid outside the layer from its bulk temperature, T_b to the saturated vapor/liquid temperature of the interface, T_e. If the subcooling is small, the latter heat sink is small compared with the former and, for simplicity in this analysis, it is assumed that this is the case. Then the mass rate of evaporation at the interface (per unit area of that interface) is $k_V \Delta T / \delta \mathcal{L}$. Denoting the mean velocity of the vapor in the layer by $u(y)$, continuity of vapor mass within the layer requires that

$$\frac{d(\rho_V u \delta)}{dy} = \frac{k_V \Delta T}{\delta \mathcal{L}}. \tag{6.10}$$

Assuming that we use mean values for ρ_V, k_V, and $\mathcal{L}$, this is a differential relation between $u(y)$ and $\delta(y)$. A second relation between these two quantities can be obtained by considering the equation of motion for the vapor in the element dy. That vapor mass will experience a pressure denoted by $p(y)$ that must be equal to the pressure in the liquid if surface tension is neglected. Moreover, if the liquid motions are neglected so that the pressure variation in the liquid is hydrostatic, it follows that the net force acting on the vapor element as a result of these pressure variations will be $\rho_L g \delta dy$ per unit depth normal to the sketch. Other forces per unit depth acting on the vapor element will be its weight, $\rho_V g \delta dy$, and the shear stress at the wall that is estimated roughly by $\mu_V u / \delta$. Then if the vapor momentum fluxes are neglected the balance of forces on the vapor element yields the following:

$$u = \frac{(\rho_L - \rho_V) g \delta^2}{\mu_V}. \tag{6.11}$$

Substituting this expression for u into Eq. (6.10) and solving for $\delta(y)$, assuming that the origin of y is chosen to be the origin or virtual origin of the vapor layer where $\delta = 0$, we obtain the following expression for $\delta(y)$:

$$\delta(y) = \left[\frac{4 k_V \Delta T \mu_V}{3 \rho_V (\rho_L - \rho_V) g \mathcal{L}}\right]^{\frac{1}{4}} y^{\frac{1}{4}}. \tag{6.12}$$

This defines the geometry of the film.

We can then evaluate the heat flux $\dot{q}(y)$ per unit surface area of the plate; the local heat transfer coefficient, $\dot{q}/\Delta T$, becomes the following:

$$\frac{\dot{q}(y)}{\Delta T} = \left[\frac{3 \rho_V (\rho_L - \rho_V) g \mathcal{L} k_V^3}{4 \Delta T \mu_V}\right]^{\frac{1}{4}} y^{-\frac{1}{4}}. \tag{6.13}$$

Note that this is singular at $y = 0$. It also follows by integration that the overall heat transfer coefficient for a plate extending from $y = 0$ to $y = \ell$ is as follows:

$$\left(\frac{4}{3}\right)^{\frac{3}{4}} \left[\frac{\rho_V (\rho_L - \rho_V) g \mathcal{L} k_V^3}{\Delta T \mu_V \ell}\right]^{\frac{1}{4}}. \tag{6.14}$$

This characterizes the film boiling heat transfer coefficients in the upper right of Figure 6.1. Although many features of the flow have been neglected, this relation

6.4 Condensation

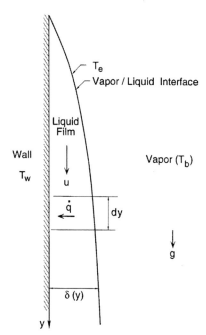

Figure 6.7. Sketch for the film condensation analysis.

gives good agreement with the experimental observations (Westwater 1958). Other geometrical arrangements such as heated circular pipes on which film boiling is occurring will have a similar functional dependence on the properties of the vapor and liquid (Collier and Thome 1994, Whalley 1987).

6.4 Condensation

The spectrum of flow processes associated with condensation on a solid surface are almost a mirror image of those involved in boiling. Thus drop condensation on the underside of a cooled horizontal plate or on a vertical surface is very analogous to nucleate boiling. The phenomenon is most apparent as the misting up of windows or mirrors. When the population of droplets becomes large they run together to form condensation films, the dominant form of condensation in most industrial contexts. Because of the close parallels with boiling flows, it would be superfluous to repeated the analyses for condensation flows. However, in the next section we include the specifics of one example, namely film condensation on a vertical surface. For more detail on condensation flows the reader is referred to the reviews by Butterworth (1977).

6.4.1 Film Condensation

The circumstance of film condensation on a vertical plate as sketched in Figure 6.7 allows an analysis that is precisely parallel to that for film boiling detailed in section 6.3.1. The obvious result is a film thickness, $\delta(y)$, (where y is now measured vertically downward) given by the following:

$$\delta(y) = \left[\frac{4k_L(-\Delta T)\mu_L}{3\rho_L(\rho_L - \rho_V)g\mathcal{L}}\right]^{\frac{1}{4}} y^{\frac{1}{4}} \qquad (6.15)$$

a local heat transfer coefficient given by the following:

$$\frac{\dot{q}(y)}{\Delta T} = \left[\frac{3\rho_L(\rho_L - \rho_V)g\mathcal{L}k_L^3}{4(-\Delta T)\mu_L}\right]^{\frac{1}{4}} y^{-\frac{1}{4}} \qquad (6.16)$$

and the following overall heat transfer coefficient for a plate of length ℓ:

$$\left(\frac{4}{3}\right)^{\frac{3}{4}} \left[\frac{\rho_L(\rho_L - \rho_V)g\mathcal{L}k_L^3}{(-\Delta T)\mu_L\ell}\right]^{\frac{1}{4}}. \qquad (6.17)$$

Clearly the details of film condensation will be different for different geometric configurations of the solid surface (inclined walls, horizontal tubes, etc.) and for laminar or turbulent liquid films. For such details, the reader is referred to the valuable review by Collier and Thome (1994).

7

Flow Patterns

7.1 Introduction

From a practical engineering point of view one of the major design difficulties in dealing with multiphase flow is that the mass, momentum, and energy transfer rates and processes can be quite sensitive to the geometric distribution or topology of the components within the flow. For example, the geometry may strongly effect the interfacial area available for mass, momentum, or energy exchange between the phases. Moreover, the flow within each phase or component will clearly depend on that geometric distribution. Thus we recognize that there is a complicated two-way coupling between the flow in each of the phases or components and the geometry of the flow (as well as the rates of change of that geometry). The complexity of this two-way coupling presents a major challenge in the study of multiphase flows and there is much that remains to be done before even a superficial understanding is achieved.

An appropriate starting point is a phenomenological description of the geometric distributions or *flow patterns* that are observed in common multiphase flows. This chapter describes the flow patterns observed in horizontal and vertical pipes and identifies a number of the instabilities that lead to transition from one flow pattern to another.

7.2 Topologies of Multiphase Flow

7.2.1 Multiphase Flow Patterns

A particular type of geometric distribution of the components is called a flow pattern or flow regime and many of the names given to these flow patterns (such as annular flow or bubbly flow) are now quite standard. Usually the flow patterns are recognized by visual inspection, although other means such as analysis of the spectral content of the unsteady pressures or the fluctuations in the volume fraction have been devised for those circumstances in which visual information is difficult to obtain (Jones and Zuber 1974).

For some of the simpler flows, such as those in vertical or horizontal pipes, a substantial number of investigations have been conducted to determine the dependence of the flow pattern on component volume fluxes, (j_A, j_B), on volume fraction and on the fluid properties such as density, viscosity, and surface tension. The results are often displayed in the form of a *flow regime map* that identifies the flow patterns occurring in various parts of a parameter space defined by the component flow rates. The flow rates used may be the volume fluxes, mass fluxes, momentum fluxes, or other similar quantities depending on the author. Perhaps the most widely used of these flow pattern maps is that for horizontal gas/liquid flow constructed by Baker (1954). Summaries of these flow pattern studies and the various empirical laws extracted from them are a common feature in reviews of multiphase flow (see, for example, Wallis 1969 or Weisman 1983).

The boundaries between the various flow patterns in a flow pattern map occur because a regime becomes unstable as the boundary is approached and growth of this instability causes transition to another flow pattern. Like the laminar-to-turbulent transition in single phase flow, these multiphase transitions can be rather unpredictable because they may depend on otherwise minor features of the flow, such as the roughness of the walls or the entrance conditions. Hence, the flow pattern boundaries are not distinctive lines but more poorly defined transition zones.

But there are other serious difficulties with most of the existing literature on flow pattern maps. One of the basic fluid mechanical problems is that these maps are often dimensional and therefore apply only to the specific pipe sizes and fluids employed by the investigator. A number of investigators (for example Baker 1954, Schicht 1969, or Weisman and Kang 1981) have attempted to find generalized coordinates that would allow the map to cover different fluids and pipes of different sizes. However, such generalizations can only have limited value because several transitions are represented in most flow pattern maps and the corresponding instabilities are governed by different sets of fluid properties. For example, one transition might occur at a critical Weber number, whereas another boundary may be characterized by a particular Reynolds number. Hence, even for the simplest duct geometries, there exist no universal, dimensionless flow pattern maps that incorporate the full, parametric dependence of the boundaries on the fluid characteristics.

Beyond these difficulties there are a number of other troublesome questions. In single-phase flow it is well established that an entrance length of 30 to 50 diameters is necessary to establish fully developed turbulent pipe flow. The corresponding entrance lengths for multiphase flow patterns are less well established and it is quite possible that some of the reported experimental observations are for temporary or developing flow patterns. Moreover, the implicit assumption is often made that there exists a unique flow pattern for given fluids with given flow rates. It is by no means certain that this is the case. Indeed, in Chapter 16, it is shown that even very simple models of multiphase flow can lead to conjugate states. Consequently, there may be several possible flow patterns whose occurence may depend on the initial conditions, specifically on the manner in which the multiphase flow is generated.

7.2 Topologies of Multiphase Flow

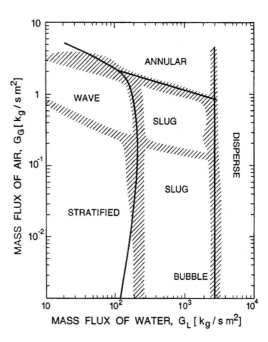

Figure 7.1. Flow regime map for the horizontal flow of an air/water mixture in a 5.1-cm-diameter pipe with flow regimes as defined in Figure 7.2. Hatched regions are observed regime boundaries, lines are theoretical predictions. Adapted from Weisman (1983).

In summary, there remain many challenges associated with a fundamental understanding of flow patterns in multiphase flow and considerable research is necessary before reliable design tools become available. In this chapter we concentrate on some of the qualitative features of the boundaries between flow patterns and on the underlying instabilities that give rise to those transitions.

7.2.2 Examples of Flow Regime Maps

Despite the issues and reservations discussed in the preceding section it is useful to provide some examples of flow regime maps along with the definitions that help distinguish the various regimes. We choose to select the first examples from the flows of mixtures of gas and liquid in horizontal and vertical tubes, mostly because these flows are of considerable industrial interest. However, many other types of flow regime maps could be used as examples and some appear elsewhere in this book; examples are the flow regimes described in the next section and those for granular flows indicated in Figure 13.5.

We begin with gas/liquid flows in horizontal pipes (see, for example, Hubbard and Dukler 1966, Wallis 1969, Weisman 1983). Figure 7.1 shows the occurence of different flow regimes for the flow of an air/water mixture in a horizontal, 5.1-cm diameter pipe where the regimes are distinguished visually using the definitions in Figure 7.2. The experimentally observed transition regions are shown by the hatched areas in Figure 7.1. The solid lines represent theoretical predictions, some of which are discussed later in this chapter. Note that in a mass flux map like this the ratio

130   Flow Patterns

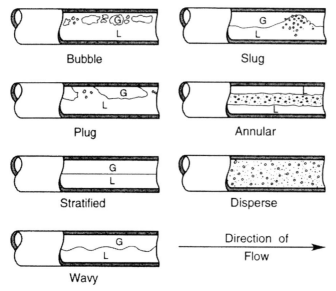

Figure 7.2. Sketches of flow regimes for flow of air/water mixtures in a horizontal, 5.1-cm-diameter pipe. Adapted from Weisman (1983).

of the ordinate to the abscissa is $\mathcal{X}/(1-\mathcal{X})$ and therefore the mass quality, $\mathcal{X}$, is known at every point in the map.

Other examples of flow regime maps for horizontal air/water flow (by different investigators) are shown in Figures 7.3 and 7.4. These maps plot the volumetric fluxes rather than the mass fluxes but because the densities of the liquid and gas in these experiments are relatively constant, there is a rough equivalence. Note that in a volumetric

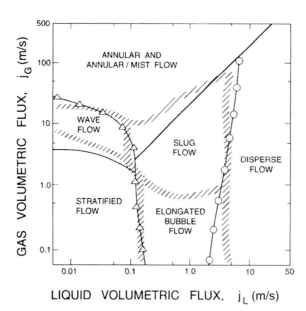

Figure 7.3. A flow regime map for the flow of an air/water mixture in a horizontal, 2.5-cm-diameter pipe at 25°C and 1 bar. Solid lines and points are experimental observations of the transition conditions while the hatched zones represent theoretical predictions. From Mandhane et al. (1974).

7.2 Topologies of Multiphase Flow

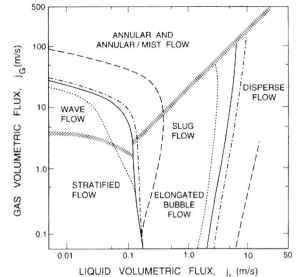

Figure 7.4. Same as Figure 7.3 but showing changes in the flow regime boundaries for various pipe diameters: 1.25 cm (dotted lines), 2.5 cm (solid lines), 5 cm (dash-dot lines), and 30 cm (dashed lines). From Mandhane *et al.* (1974).

flux map the ratio of the ordinate to the abscissa is $\beta/(1-\beta)$ and therefore the volumetric quality, β, is known at every point in the map. There are many industrial processes in which the mass quality is a key flow parameter and therefore mass flux maps are often preferred.

Figure 7.4 shows how the boundaries were observed to change with pipe diameter. Moreover, Figures 7.1 and 7.4 appear to correspond fairly closely. Note that both show well-mixed regimes occuring above some critical liquid flux and above some critical gas flux; we expand further on this in Section 7.3.1.

7.2.3 Slurry Flow Regimes

As a further example, consider the flow regimes manifest by slurry (solid/liquid mixture) flow in a horizontal pipeline. When the particles are small so that their settling velocity is much less than the turbulent mixing velocities in the fluid and when the volume fraction of solids is low or moderate, the flow will be well mixed. This is termed the *homogeneous* flow regime (Figure 7.5) and typically occurs only in practical slurry

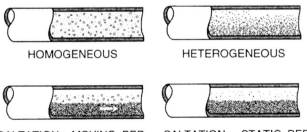

Figure 7.5. Flow regimes for slurry flow in a horizontal pipeline.

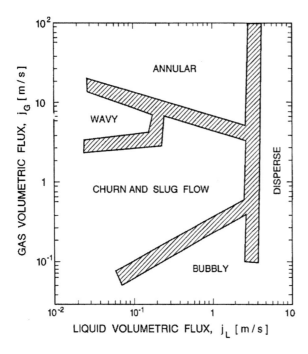

Figure 7.6. A flow regime map for the flow of an air/water mixture in a vertical, 2.5-cm-diameter pipe showing the experimentally observed transition regions hatched; the flow regimes are sketched in Figure 7.7. Adapted from Weisman (1983).

pipelines when all the particle sizes are of the order of tens of microns or less. When somewhat larger particles are present, vertical gradients will occur in the concentration and the regime is termed *heterogeneous*; moreover the larger particles will tend to sediment faster and so a vertical size gradient will also occur. The limit of this heterogeneous flow regime occurs when the particles form a packed bed in the bottom of the pipe. When a packed bed develops, the flow regime is known as a *saltation* flow. In a saltation flow, solid material may be transported in two ways, either because the bed moves en masse or because material in suspension above the bed is carried along by the suspending fluid. Further analyses of these flow regimes, their transitions, and their pressure gradients are included in Sections 8.2.1, 8.2.2, and 8.2.3. For further detail, the reader is referred to Shook and Roco (1991), Zandi and Govatos (1967), and Zandi (1971).

7.2.4 Vertical Pipe Flow

When the pipe is oriented vertically, the regimes of gas/liquid flow are a little different as illustrated in Figures 7.6 and 7.7 (see, for example, Hewitt and Hall Taylor 1970, Butterworth and Hewitt 1977, Hewitt 1982, Whalley 1987). Another vertical flow regime map is shown in Figure 7.8, this one using momentum flux axes rather than volumetric or mass fluxes. Note the wide range of flow rates in Hewitt and Roberts (1969) flow regime map and the fact that they correlated both air/water data at atmospheric pressure and steam/water flow at high pressure.

Typical photographs of vertical gas/liquid flow regimes are shown in Figure 7.9. At low gas volume fractions of the order of a few percentage, the flow is an

7.2 Topologies of Multiphase Flow

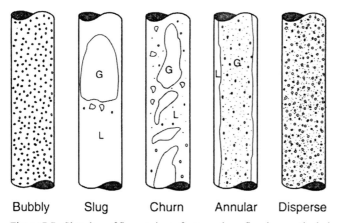

Figure 7.7. Sketches of flow regimes for two-phase flow in a vertical pipe. Adapted from Weisman (1983).

amalgam of individual ascending bubbles (left photograph). Note that the visual appearance is deceptive; most people would judge the volume fraction to be significantly larger than 1%. As the volume fraction is increased (the middle photograph has $\alpha = 4.5$%), the flow becomes unstable at some critical volume fraction, which in the case illustrated is ~ 15%. This instability produces large-scale mixing motions that dominate the flow and have a scale comparable to the pipe diameter. At still larger volume fractions, large unsteady gas volumes accumulate within these mixing motions and produce the flow regime known as churn-turbulent flow (right photograph).

It should be added that flow regime information such as that presented in Fig. 7.8 appears to be valid both for flows that are not evolving with axial distance along the pipe and for flows, such as those in boiler tubes, in which the volume fraction is

Figure 7.8. The vertical flow regime map of Hewitt and Roberts (1969) for flow in a 3.2-cm-diameter tube, validated for both air/water flow at atmospheric pressure and steam/water flow at high pressure.

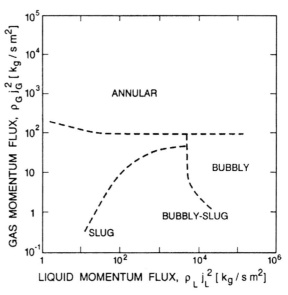

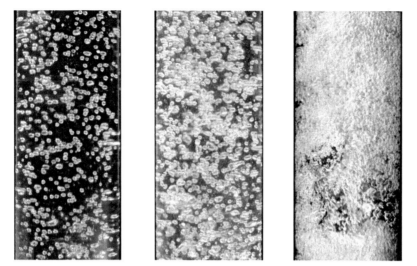

Figure 7.9. Photographs of air/water flow in a 10.2-cm-diameter vertical pipe (Kytömaa 1987). (Left) 1% air; (middle) 4.5% air; (right) >15% air.

increasing with axial position. Figure 7.10 provides a sketch of the kind of evolution one might expect in a vertical boiler tube based on the flow regime maps given above. It is interesting to compare and contrast this flow pattern evolution with the inverted case of convective boiling surrounding a heated rod in Figure 6.4.

7.2.5 Flow Pattern Classifications

One of the most fundamental characteristics of a multiphase flow pattern is the extent to which it involves global separation of the phases or components. At the two ends of the spectrum of separation characteristics are those flow patterns that are termed *disperse* and those that are termed *separated*. A disperse flow pattern is one in which one phase or component is widely distributed as drops, bubbles, or particles in the other *continuous* phase. Conversely, a *separated* flow consists of separate, parallel streams of the two (or more) phases. Even within each of these limiting states there are various degrees of component separation. The asymptotic limit of a disperse flow in which the disperse phase is distributed as an infinite number of infinitesimally small particles, bubbles, or drops is termed a *homogeneous* multiphase flow. As discussed in Sections 2.4.2 and 9.2 this limit implies zero relative motion between the phases. However, there are many practical disperse flows, such as bubbly or mist flow in a pipe, in which the flow is quite disperse in that the particle size is much smaller than the pipe dimensions but in which the relative motion between the phases is significant.

Within separated flows there are similar gradations or degrees of phase separation. The low-velocity flow of gas and liquid in a pipe that consists of two single-phase

7.2 Topologies of Multiphase Flow

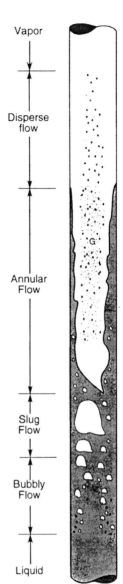

Figure 7.10. The evolution of the steam/water flow in a vertical boiler tube.

streams can be designated a *fully separated* flow. Conversely, most annular flows in a vertical pipe consist of a film of liquid on the walls and a central core of gas that contains a significant number of liquid droplets. These droplets are an important feature of annular flow and therefore the flow can only be regarded as partially separated.

To summarize: one of the basic characteristics of a flow pattern is the degree of separation of the phases into streamtubes of different concentrations. The degree of separation will, in turn, be determined by (a) some balance between the fluid mechanical processes enhancing dispersion and those causing segregation, (b) the initial conditions or mechanism of generation of the multiphase flow, or (c) some mix of both effects. In Section 7.3.1 we discuss the fluid mechanical processes referred to in (a).

A second basic characteristic that is useful in classifying flow patterns is the level of intermittency in the volume fraction. Examples of intermittent flow patterns are slug flows in both vertical and horizontal pipe flows and the occurrence of interfacial waves in horizontal separated flow. The first separation characteristic was the degree of separation of the phases between streamtubes; this second, intermittency, characteristic can be viewed as the degree of periodic separation in the streamwise direction. The slugs or waves are kinematic or concentration waves (sometimes called continuity waves) and a general discussion of the structure and characteristics of such waves is contained in Chapter 16. Intermittency is the result of an instability in which kinematic waves grow in an otherwise nominally steady flow to create significant streamwise separation of the phases.

In the rest of this chapter we describe how these ideas of cross-streamline separation and intermittency can lead to an understanding of the limits of specific multiphase flow regimes. The mechanics of limits on disperse flow regimes are discussed first in Sections 7.3 and 7.4. Limits on separated flow regimes are outlined in Section 7.5.

7.3 Limits of Disperse Flow Regimes

7.3.1 Disperse Phase Separation and Dispersion

To determine the limits of a disperse phase flow regime, it is necessary to identify the dominant processes enhancing separation and those causing dispersion. By far the most common process causing phase separation is due to the difference in the densities of the phases and the mechanisms are therefore functions of the ratio of the density of the disperse phase to that of the continuous phase, ρ_D/ρ_C. Then the buoyancy forces caused either by gravity or, in a nonuniform or turbulent flow, by the Lagrangian fluid accelerations will create a relative velocity between the phases whose magnitude is denoted by W_p. Using the analysis of Section 2.4.2, we can conclude that the ratio W_p/U (where U is a typical velocity of the mean flow) is a function only of the Reynolds number, $Re = 2UR/\nu_C$, and the parameters X and Y are defined by Eqs. (2.91) and (2.92). The particle size, R, and the streamwise extent of the flow, ℓ, both occur in the dimensionless parameters Re, X, and Y. For low velocity flows in which $U^2/\ell \ll g$, ℓ is replaced by g/U^2 and hence a Froude number, gR/U^2, rather than R/ℓ appears in the parameter X. This then establishes a velocity, W_p, that characterizes the relative motion and therefore the phase separation due to density differences.

As an aside we note that there are some fluid mechanical phenomena that can cause phase separation even in the absence of a density difference. For example, Ho and Leal (1974) explored the migration of neutrally buoyant particles in shear flows at low Reynolds numbers. These effects are usually suffciently weak compared with those due to density differences that they can be neglected in many applications.

7.3 Limits of Disperse Flow Regimes

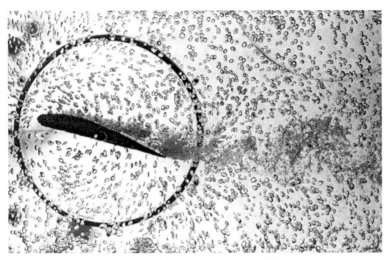

Figure 7.11. Bubbly flow around a NACA 4412 hydrofoil (10-cm chord) at an angle of attack; flow is from left to right. From the work of Ohashi *et al.* (1990), reproduced with the author's permission.

In a quiescent multiphase mixture the primary mechanism of phase separation is sedimentation (see Chapter 16) though more localized separation can also occur as a result of the inhomogeneity instability described in Section 7.4. In flowing mixtures the mechanisms are more complex and, in most applications, are controlled by a balance between the buoyancy/gravity forces and the hydrodynamic forces. In high-Reynolds-number, turbulent flows, the turbulence can cause either dispersion or segregation. Segregation can occur when the relaxation time for the particle or bubble is comparable with the typical time of the turbulent fluid motions. When $\rho_D/\rho_C \gg 1$ as, for example, with solid particles suspended in a gas, the particles are centrifuged out of the more intense turbulent eddies and collect in the shear zones in between (see, for example, Squires and Eaton 1990, Elghobashi and Truesdell 1993). Conversely, when $\rho_D/\rho_C \ll 1$ as, for example, with bubbles in a liquid, the bubbles tend to collect in regions of low pressure such as in the wake of a body or in the centers of vortices (see, for example, Pan and Banerjee 1997). We previously included a photograph (Figure 1.6) showing heavier particles centrifuged out of vortices in a turbulent channel flow. Here, as a counterpoint, we include the photograph, Figure 7.11, from Ohashi *et al.* (1990) showing the flow of a bubbly mixture around a hydrofoil. Note the region of higher void fraction (more than four times the upstream void fraction according to the measurements) in the wake on the suction side of the foil. This accumulation of bubbles on the suction side of a foil or pump blade has importance consequences for performance as discussed in Section 7.3.3.

Counteracting the above separation processes are dispersion processes. In many engineering contexts the principal dispersion is caused by the turbulent or other unsteady motions in the continuous phase. Figure 7.11 also illustrates this process for the concentrated regions of high void fraction in the wake are dispersed as they are carried downstream. The shear created by unsteady velocities can also cause either fission or fusion of the disperse phase bubbles, drops, or particles, but we delay discussion of this additional complexity until the next section. For the present it is necessary only to characterize the mixing motions in the continuous phase by a typical velocity, W_t. Then the degree of separation of the phases will clearly be influenced by the relative magnitudes of W_p and W_t or, specifically, by the ratio W_p/W_t. Disperse flow will occur when $W_p/W_t \ll 1$ and separated flow when $W_p/W_t \gg 1$. The corresponding flow pattern boundary should be given by some value of W_p/W_t of order unity. For example, in slurry flows in a horizontal pipeline, Thomas (1962) suggested a value of W_p/W_t of 0.2 based on his data.

7.3.2 Example: Horizontal Pipe Flow

As a quantitative example, we pursue the case of the flow of a two-component mixture in a long horizontal pipe. The separation velocity, W_p, due to gravity, g, would then be given qualitatively by Eq. (2.74) or (2.83), namely

$$W_p = \frac{2R^2 g}{9\nu_C} \left(\frac{\Delta \rho}{\rho_C} \right) \quad \text{if} \quad 2W_p R / \nu_C \ll 1 \tag{7.1}$$

or

$$W_p = \left\{ \frac{2}{3} \frac{Rg}{C_D} \frac{\Delta \rho}{\rho_C} \right\}^{\frac{1}{2}} \quad \text{if} \quad 2W_p R / \nu_C \gg 1, \tag{7.2}$$

where R is the particle, droplet, or bubble radius; ν_C and ρ_C are the kinematic viscosity and density of the continuous fluid; and $\Delta \rho$ is the density difference between the components. Furthermore, the typical turbulent velocity will be some function of the friction velocity, $(\tau_w/\rho_C)^{\frac{1}{2}}$, and the volume fraction, α, of the disperse phase. The effect of α is less readily quantified so, for the present, we concentrate on dilute systems ($\alpha \ll 1$) in which

$$W_t \approx \left(\frac{\tau_w}{\rho_C} \right)^{\frac{1}{2}} = \left\{ \frac{d}{4\rho_C} \left(-\frac{dp}{ds} \right) \right\}^{\frac{1}{2}}, \tag{7.3}$$

where d is the pipe diameter and dp/ds is the pressure gradient. Then the transition condition, $W_p/W_t = K$ (where K is some number of order unity), can be rewritten as follows:

$$\left(-\frac{dp}{ds} \right) \approx \frac{4\rho_C}{K^2 d} W_p^2 \tag{7.4}$$

$$\approx \frac{16}{81 K^2} \frac{\rho_C R^4 g^2}{\nu_C^2 d} \left(\frac{\Delta \rho}{\rho_C} \right)^2 \quad \text{for} \quad 2W_p R/\nu_C \ll 1 \tag{7.5}$$

$$\approx \frac{32}{3 K^2} \frac{\rho_C Rg}{C_D d} \left(\frac{\Delta \rho}{\rho_C} \right) \quad \text{for} \quad 2W_p R/\nu_C \gg 1. \tag{7.6}$$

7.3 Limits of Disperse Flow Regimes

In summary, the expression on the right-hand side of Eq. (7.5) [or (7.6)] yields the pressure drop at which W_p/W_t exceeds the critical value of K and the particles will be maintained in suspension by the turbulence. At lower values of the pressure drop the particles will settle out and the flow will become separated and stratified.

This criterion on the pressure gradient may be converted to a criterion on the flow rate by using some version of the turbulent pipe flow relation between the pressure gradient and the volume flow rate, j. For example, one could conceive of using, as a first approximation, a typical value of the turbulent friction factor, $f = \tau_w/\frac{1}{2}\rho_C j^2$ (where j is the total volumetric flux). In the case of $2W_p R/\nu_C \gg 1$, this leads to a critical volume flow rate, $j = j_c$, given by the following:

$$j_c = \left\{ \frac{8}{3K^2 f} \frac{gD}{C_D} \frac{\Delta\rho}{\rho_C} \right\}^{\frac{1}{2}}. \tag{7.7}$$

With $8/3K^2 f$ replaced by an empirical constant, this is the general form of the critical flow rate suggested by Newitt et al. (1955) for horizontal slurry pipeline flow; for $j > j_c$ the flow regime changes from saltation flow to heterogeneous flow (see Figure 7.5). Alternatively, one could write this nondimensionally using a Froude number defined as $Fr = j_c/(gd)^{\frac{1}{2}}$. Then the criterion yields a critical Froude number given by the following:

$$Fr^2 = \frac{8}{3K^2 f C_D} \frac{\Delta\rho}{\rho_C} \tag{7.8}$$

If the common expression for the turbulent friction factor, namely $f = 0.31/(jd/\nu_C)^{\frac{1}{4}}$ is used in Eq. (7.7) that expression becomes the following:

$$j_c = \left\{ \frac{17.2}{K^2 C_D} \frac{gRd^{\frac{1}{4}}}{\nu_C^{\frac{1}{4}}} \frac{\Delta\rho}{\rho_C} \right\}^{\frac{4}{7}}. \tag{7.9}$$

A numerical example will help relate this criterion to the boundary of the disperse phase regime in the flow regime maps. For the case of Figure 7.3 and using for simplicity, $K = 1$ and $C_D = 1$, then with a drop or bubble size, $R = 3$ mm, Eq. (7.9) gives a value of j_c of 3 m/s when the continuous phase is liquid (bubbly flow) and a value of 40 m/s when the continuous phase is air (mist flow). These values are in good agreement with the total volumetric flux at the boundary of the disperse flow regime in Figure 7.3, which, at low j_G, is about 3 m/s and at higher j_G (volumetric qualities above 0.5) is about 30–40 m/s.

Another approach to the issue of the critical velocity in slurry pipeline flow is to consider the velocity required to fluidize a packed bed in the bottom of the pipe (see, for example, Durand and Condolios 1952 or Zandi and Govatos 1967). This is described further in Section 8.2.3.

7.3.3 Particle Size and Particle Fission

In the preceding sections, the transition criteria determining the limits of the disperse flow regime included the particle, bubble, or drop size or, more specifically, the dimensionless parameter $2R/d$ as illustrated by the criteria of Eqs. (7.5), (7.6), and (7.9). However, these criteria require knowledge of the size of the particles, $2R$, and this is not always accessible particularly in bubbly flow. Even when there may be some knowledge of the particle or bubble size in one region or at one time, the various processes of fission and fusion need to be considered in determining the appropriate $2R$ for use in these criteria. One of the serious complications is that the size of the particles, bubbles, or drops is often determined by the flow itself because the flow shear tends to cause fission and therefore limit the maximum size of the surviving particles. Then the flow regime may depend on the particle size that in turn depends on the flow and this two-way interaction can be difficult to unravel. Figure 7.11 illustrates this problem because one can observe many smaller bubbles in the flow near the suction surface and in the wake that clearly result from fission in the highly sheared flow near the suction surface. Another example from the flow in pumps is described in the next section.

When the particles are very small, a variety of forces may play a role in determining the effective particle size and some comments on these are included later in Section 7.3.7. But often the bubbles or drops are sufficiently large that the dominant force resisting fission is due to surface tension while the dominant force promoting fission is the shear in the flow. We confine the present discussion to these circumstances. Typical regions of high shear occur in boundary layers, in vortices or in turbulence. Frequently, the larger drops or bubbles are fissioned when they encounter regions of high shear and do not subsequently coalesce to any significant degree. Then, the characteristic force resisting fission would be given by SR, whereas the typical shear force causing fission might be estimated in several ways. For example, in the case of pipe flow the typical shear force could be characterized by $\tau_w R^2$. Then, assuming that the flow is initiated with larger particles that are then fissioned by the flow, we would estimate that $R = S/\tau_w$. This is used in the next section to estimate the limits of the bubbly or mist flow regime in pipe flows.

In other circumstances, the shearing force in the flow might be described by $\rho_C(\dot\gamma R)^2 R^2$ where $\dot\gamma$ is the typical shear rate and ρ_C is the density of the continuous phase. This expression for the fission force assumes a high Reynolds number in the flow around the particle or explicitly that $\rho_C \dot\gamma R^2/\mu_C \gg 1$ where μ_C is the dynamic viscosity of the continuous phase. But if $\rho_C \dot\gamma R^2/\mu_C \ll 1$, then a more appropriate estimate of the fission force would be $\mu_C \dot\gamma R^2$. Consequently, the maximum particle size, R_m, one would expect to see in the flow in these two regimes would be as follows:

$$R_m = \left\{\frac{S}{\mu_C \dot\gamma}\right\} \quad \text{for} \quad \rho_C \dot\gamma R^2/\mu_C \ll 1$$

$$or \left\{\frac{S}{\rho_C \dot\gamma^2}\right\}^{\frac{1}{3}} \quad \text{for} \quad \rho_C \dot\gamma R^2/\mu_C \gg 1, \quad (7.10)$$

7.3 Limits of Disperse Flow Regimes

Figure 7.12. A bubbly air/water mixture (volume fraction about 4%) entering an axial flow impeller (a 10.2-cm-diameter scale model of the SSME low-pressure liquid oxygen impeller) from the right. The inlet plane is roughly in the center of the photograph and the tips of the blades can be seen to the left of the inlet plane.

respectively. Note that in both instances the maximum size decreases with increasing shear rate.

7.3.4 Examples of Flow-Determined Bubble Size

An example of the use of the above relations can be found in the important area of two-phase pump flows and we quote here data from studies of the pumping of bubbly liquids. The issue here is the determination of the volume fraction at which the pump performance is seriously degraded by the presence of the bubbles. It transpires that, in most practical pumping situations, the turbulence and shear at inlet and around the leading edges of the blades of the pump (or other turbomachine) tend to fission the bubbles and thus determine the size of the bubbles in the blade passages. An illustration is included in Figure 7.12 which shows an air/water mixture progressing through an axial flow impeller; the bubble size downstream of the inlet plane is much smaller that that approaching the impeller.

The size of the bubbles within the blade passages is important because it is the migration and coalescence of these bubbles that appear to cause degradation in the performance. Because the velocity of the relative motion depends on the bubble size, it follows that the larger the bubbles the more likely it is that large voids will form within the blade passage due to migration of the bubbles toward regions of lower pressure (Furuya 1985, Furuya and Maekawa 1985). As Patel and Runstadler (1978) observed during experiments on centrifugal pumps and rotating passages, regions of low pressure occur not only on the suction sides of the blades but also under the shroud of a centrifugal pump. These large voids or gas-filled wakes can cause

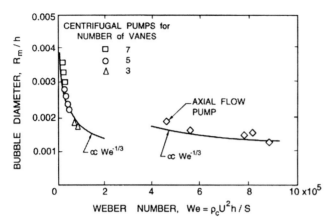

Figure 7.13. The bubble sizes, R_m, observed in the blade passages of centrifugal and axial flow pumps as a function of Weber number, where h is the blade spacing (adapted from Murakami and Minemura 1978).

substantial changes in the deviation angle of the flow leaving the impeller and hence lead to substantial degradation in the pump performance.

The key is therefore the size of the bubbles in the blade passages and some valuable data on this has been compiled by Murakami and Minemura (1977, 1978) for both axial and centrifugal pumps. This is summarized in Figure 7.4, where the ratio of the observed bubble size, R_m, to the blade spacing, h, is plotted against the Weber number, We $= \rho_C U^2 h/S$ (U is the blade tip velocity). Rearranging the first version of Eq. (7.10), estimating that the inlet shear is proportional to U/h and adding a proportionality constant, C, because the analysis is qualitative, we would expect that $R_m = C/\text{We}^{\frac{1}{3}}$. The dashed lines in Figure 7.13 are examples of this prediction and exhibit behavior very similar to the experimental data. In the case of the axial pumps, the effective value of the coefficient, $C = 0.15$.

A different example is provided by cavitating flows in which the highest shear rates occur during the collapse of the cavitation bubbles. As discussed in Section 5.2.3, these high shear rates cause individual cavitation bubbles to fission into many smaller fragments so that the bubble size emerging from the region of cavitation bubble collapse is much smaller than the size of the bubbles entering that region. The phenomenon is exemplified by Figure 7.14, which shows the growth of the cavitating bubbles on the suction surface of the foil, the collapse region near the trailing edge and the much smaller bubbles emerging from the collapse region. Some analysis of the fission due to cavitation bubble collapse is contained in Brennen (2002).

7.3.5 Bubbly or Mist Flow Limits

Returning now to the issue of determining the boundaries of the bubbly (or mist flow) regime in pipe flows, and using the expression $R = S/\tau_w$ for the bubble size

7.3 Limits of Disperse Flow Regimes

Figure 7.14. Traveling bubble cavitation on the surface of a NACA 4412 hydrofoil at zero incidence angle, a speed of 13.7 m/s, and a cavitation number of 0.3. The flow is from left to right, the leading edge of the foil is just to the left of the white glare patch on the surface, and the chord is 7.6 cm (Kermeen 1956).

in Eq. (7.6), the transition between bubbly disperse flow and separated (or partially separated flow) is described by the following relation:

$$\left\{ \frac{-\frac{dp}{ds}}{g\Delta\rho} \right\}^{\frac{1}{2}} \left\{ \frac{S}{gd^2\Delta\rho} \right\}^{-\frac{1}{4}} = \left\{ \frac{64}{3K^2 C_D} \right\}^{\frac{1}{4}} = \text{constant}. \tag{7.11}$$

This is the analytical form of the flow regime boundary suggested by Taitel and Dukler (1976) for the transition from disperse bubbly flow to a more separated state. Taitel and Dukler also demonstrate that when the constant in Eq. (7.11) is of the order of unity, the boundary agrees well with that observed experimentally by Mandhane et al. (1974). This agreement is shown in Figure 7.3. The same figure serves to remind us that there are other transitions that Taitel and Dukler were also able to model with qualitative arguments. They also demonstrate, as mentioned earlier, that each of these transitions typically scale differently with the various nondimensional parameters governing the characteristics of the flow and the fluids.

7.3.6 Other Bubbly Flow Limits

As the volume fraction of gas or vapor is increased, a bubbly flow usually transitions to a mist flow, a metamorphosis that involves a switch in the continuous and disperse phases. However, there are several additional comments on this metamorphosis that need to be noted.

First, at very low flow rates, there are circumstances in which this transition does not occur at all and the bubbly flow becomes a foam. Though the precise conditions necessary for this development are not clear, foams and their rheology have been the subject of considerable study. The mechanics of foams are beyond the scope of this book; the reader is referred to the review of Kraynik (1988) and the book of Weaire and Hutzler (2001).

Second, though it is rarely mentioned, the reverse transition from mist flow to bubbly flow as the volume fraction decreases involves energy dissipation and an increase in pressure. This transition has been called a *mixing shock* (Witte 1969) and typically occurs when a droplet flow with significant relative motion transitions to a bubbly flow with negligible relative motion. Witte (1969) has analyzed these mixing shocks and obtains expressions for the compression ratio across the mixing shock as a function of the upstream slip and Euler number.

7.3.7 Other Particle Size Effects

In Sections 7.3.3 and 7.3.5 we outlined one class of circumstances in which bubble fission is an important facet of the disperse phase dynamics. It is, however, important to add, even if briefly, that there are many other mechanisms for particle fission and fusion that may be important in a disperse phase flow. When the particles are submicron or micron sized, intermolecular and electromagnetic forces can become critically important in determining particle aggregation in the flow. These phenomena are beyond the scope of this book and the reader is referred to texts such as Friedlander (1977) or Flagan and Seinfeld (1988) for information on the effects these forces have on flows involving particles and drops. It is, however, valuable to add that gas/solid suspension flows with larger particles can also exhibit important effects as a result of electrical charge separation and the forces that those charges create between particles or between the particles and the walls of the flow. The process of electrification or charge separation is often a very important feature of such flows (Boothroyd 1971). Pneumatically driven flows in grain elevators or other devices can generate huge electropotential differences (as large as hundreds of kilovolts) that can, in turn, cause spark discharges and consequently dust explosions. In other devices, particularly electrophotographic copiers, the charge separation generated in a flowing toner/carrier mixture is a key feature of such devices. Electromagnetic and intermolecular forces can also play a role in determining the bubble or droplet size in gas/liquid flows (or flows of immiscible liquid mixtures).

7.4 Inhomogeneity Instability

In Section 7.3.1 we presented a qualitative evaluation of phase separation processes driven by the combination of a density difference and a fluid acceleration. Such a combination does not necessarily imply separation within a homogeneous quiescent mixture (except through sedimentation). However, it transpires that local phase separation may also occur through the development of an inhomogeneity instability whose origin and consequences we describe in the next two sections.

7.4.1 Stability of Disperse Mixtures

It transpires that a homogeneous, quiescent multiphase mixture may be internally unstable as a result of gravitationally induced relative motion. This instability was first described for fluidized beds by Jackson (1963). It results in horizontally oriented, vertically propagating volume fraction waves or layers of the disperse phase. To evaluate the stability of a uniformly dispersed two component mixture with uniform relative velocity induced by gravity and a density difference, Jackson constructed a model consisting of the following system of equations:

1. The number continuity equation [Eq. (1.30)] for the particles (density, ρ_D, and volume fraction, $\alpha_D = \alpha$):

$$\frac{\partial \alpha}{\partial t} + \frac{\partial (\alpha u_D)}{\partial y} = 0, \tag{7.12}$$

   where all velocities are in the vertically upward direction.

2. Volume continuity for the suspending fluid (assuming constant density, ρ_C, and zero mass interaction, $\mathcal{I}_N = 0$):

$$\frac{\partial \alpha}{\partial t} - \frac{\partial ((1-\alpha)u_C)}{\partial y} = 0. \tag{7.13}$$

3. Individual phase momentum equations [Eq. (1.42)] for both the particles and the fluid assuming constant densities and no deviatoric stress:

$$\rho_D \alpha \left\{ \frac{\partial u_D}{\partial t} + u_D \frac{\partial u_D}{\partial y} \right\} = -\alpha \rho_D g + \mathcal{F}_D \tag{7.14}$$

$$\rho_C (1-\alpha) \left\{ \frac{\partial u_C}{\partial t} + u_C \frac{\partial u_C}{\partial y} \right\} = -(1-\alpha)\rho_C g - \frac{\partial p}{\partial y} - \mathcal{F}_D. \tag{7.15}$$

4. A force interaction term of the form given by Eq. (1.44). Jackson constructs a component, $\mathcal{F}'_{Dk}$, due to the relative motion of the form

$$\mathcal{F}'_D = q(\alpha)(1-\alpha)(u_C - u_D), \tag{7.16}$$

   where q is assumed to be some function of α. Note that this is consistent with a low Reynolds number flow.

Jackson then considered solutions of these equations that involve small, linear perturbations or waves in an otherwise homogeneous mixture. Thus the flow was decomposed into the following:

1. A uniform, homogeneous fluidized bed in which the mean values of u_D and u_C are respectively zero and some adjustable constant. To maintain generality, we will characterize the relative motion by the drift flux, $j_{CD} = \alpha(1-\alpha)u_C$.
2. An unsteady linear perturbation in the velocities, pressure, and volume fraction of the form $\exp\{i\kappa y + (\zeta - i\omega)t\}$ that models waves of wavenumber, κ, and

frequency, ω, traveling in the y direction with velocity ω/κ and increasing in amplitude at a rate given by ζ.

Substituting this decomposition into the system of equations described above yields the following expression for $(\zeta - i\omega)$:

$$(\zeta - i\omega)\frac{j_{CD}}{g} = \pm K_2 \left\{1 + 4iK_3 + 4K_1 K_3^2 - 4iK_3(1+K_1)K_4\right\}^{\frac{1}{2}} - K_2(1 + 2iK_3) \quad (7.17)$$

where the constants K_1 through K_3 are given by the following:

$$K_1 = \frac{\rho_D}{\rho_C}\frac{(1-\alpha)}{\alpha}; \quad K_2 = \frac{(\rho_D - \rho_C)\alpha(1-\alpha)}{2\{\rho_D(1-\alpha) + \rho_C\alpha\}}$$

$$K_3 = \frac{\kappa j_{CD}^2}{g\alpha(1-\alpha)^2\{\rho_D/\rho_C - 1\}} \quad (7.18)$$

and K_4 is given by the following:

$$K_4 = 2\alpha - 1 + \frac{\alpha(1-\alpha)}{q}\frac{dq}{d\alpha}. \quad (7.19)$$

It transpires that K_4 is a critical parameter in determining the stability and it, in turn, depends on how q, the factor of proportionality in Eq. (7.16), varies with α. Here we examine two possible functions, $q(\alpha)$. The Carman–Kozeny Eq. (2.96) for the pressure drop through a packed bed is appropriate for slow viscous flow and leads to $q \propto \alpha^2/(1-\alpha)^2$; from Eq. (7.19) this yields $K_4 = 2\alpha + 1$ and is an example of low-Reynolds-number flow. As a representative example of higher Reynolds number flow we take the relation 2.100 due to Wallis (1969) and this leads to $q \propto \alpha/(1-\alpha)^{b-1}$ (recall Wallis suggests $b = 3$); this yields $K_4 = b\alpha$. We examine both of these examples of the form of $q(\alpha)$.

Note that the solution [Eq. (7.17)] yields the nondimensional frequency and growth rate of waves with wavenumber, κ, as functions of just three dimensionless variables, the volume fraction, α, the density ratio, ρ_D/ρ_C, and the relative motion parameter, $j_{CD}/(g/\kappa)^{\frac{1}{2}}$, similar to a Froude number. Note also that Eq. (7.17) yields two roots for the dimensionless frequency, $\omega j_{CD}/g$, and growth rate, $\zeta j_{CD}/g$. Jackson demonstrates that the negative sign choice is an attenuated wave; consequently we focus exclusively on the positive sign choice that represents a wave that propagates in the direction of the drift flux, j_{CD}, and grows exponentially with time. It is also easy to see that the growth rate tends to infinity as $\kappa \to \infty$. However, it is meaningless to consider wavelengths less than the interparticle distance and therefore the focus should be on waves of this order because they will predominate. Therefore, in the discussion below, it is assumed that the κ^{-1} values of primary interest are of the order of the typical interparticle distance.

Figure 7.15 presents typical dimensionless growth rates for various values of the parameters α, ρ_D/ρ_C, and $j_{CD}/(g/k)^{\frac{1}{2}}$ for both the Carman–Kozeny and Wallis

7.4 Inhomogeneity Instability

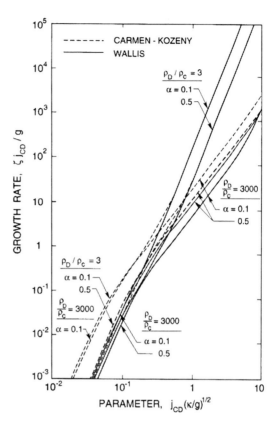

Figure 7.15. The dimensionless growth rate $\zeta j_{CD}/g$ plotted against the parameter $j_{CD}/(g/\kappa)^{\frac{1}{2}}$ for various values of α and ρ_D/ρ_C and for both $K_4 = 2\alpha + 1$ and $K_4 = 3\alpha$.

expressions for K_4. In all cases the growth rate increases with the wavenumber κ, confirming the fact that the fastest growing wavelength is the smallest that is relevant. We note, however, that a more complete linear analysis by Anderson and Jackson (1968) (see also Homsy et al. 1980, Jackson 1985, Kytömaa 1987) that includes viscous effects yields a wavelength that has a maximum growth rate. Figure 7.15 also demonstrates that the effect of void fraction is modest; although the lines for $\alpha = 0.5$ lie below those for $\alpha = 0.1$ this must be weighed in conjunction with the fact that the interparticle distance is greater in the latter case. Gas and liquid fluidized beds are typified by ρ_D/ρ_C values of 3000 and 3 respectively; because the lines for these two cases are not far apart, the primary difference is the much larger values of j_{CD} in gas-fluidized beds. Everything else being equal, increasing j_{CD} means following a line of slope 1 in Figure 7.15 and this implies much larger values of the growth rate in gas-fluidized beds. This is in accord with the experimental observations.

As a postscript, it must be noted that the above analysis leaves out many effects that may be consequential. As previously mentioned, the inclusion of viscous effects is important at least for lower Reynolds number flows. At higher particle Reynolds numbers, even more complex interactions can occur as particles encounter the wakes of other particles. For example, Fortes et al. (1987) demonstrated the complexity of particle/particle interactions under those circumstances and Joseph (1993) provides a summary of how the inhomogeneities or volume fraction waves evolve with such interactions. General analyses of kinematic waves are contained in Chapter 16 and the reader is referred to that chapter for details.

7.4.2 Inhomogeneity Instability in Vertical Flows

In vertical flows, the inhomogeneity instability described in the previous section will mean the development of intermittency in the volume fraction. The short-term result of this instability is the appearance of vertically propagating, horizontally oriented kinematic waves (see Chapter 16) in otherwise nominally steady flows. They have been most extensively researched in fluidized beds but have also be observed experimentally in vertical bubbly flows by Bernier (1982), Boure and Mercadier (1982), Kytomaa and Brennen (1990) (who also examined solid/liquid mixtures at large Reynolds numbers) and analyzed by Biesheuvel and Gorissen (1990). (Some further comment on these bubbly flow measurements is contained in Section 16.2.3.)

As they grow in amplitude these wavelike volume fraction perturbations seem to evolve in several ways depending on the type of flow and the manner in which it is initiated. In turbulent gas/liquid flows they result in large gas volumes or slugs with a size close to the diameter of the pipe. In some solid/liquid flows they produce a series of periodic vortices, again with a dimension comparable with that of the pipe diameter. But the long-term consequences of the inhomogeneity instability have been most carefully studied in the context of fluidized beds. Following the work of Jackson (1963), El-Kaissy and Homsy (1976) studied the evolution of the kinematic waves experimentally and observed how they eventually lead, in fluidized beds, to three-dimensional structures known as *bubbles*. These are *not* gas bubbles but three-dimensional, bubblelike zones of low particle concentration that propagate upward through the bed while their structure changes relatively slowly. They are particularly evident in wide fluidized beds where the lateral dimension is much larger than the typical interparticle distance. Sometimes bubbles are directly produced by the sparger or injector that creates the multiphase flow. This tends to be the case in gas-fluidized beds where, as illustrated in the preceding section, the rate of growth of the inhomogeneity is much greater than in liquid fluidized beds and thus bubbles are instantly formed.

Because of their ubiquity in industrial processes, the details of the three-dimensional flows associated with fluidized-bed bubbles have been extensively studied both experimentally (see, for example, Davidson and Harrison 1963, Davidson *et al.* 1985) and analytically (Jackson 1963, Homsy *et al.* 1980). Roughly spherical or spherical cap in shape, these zones of low solids volume fraction always rise in a fluidized bed (see Figure 7.16). When the density of bubbles is low, single bubbles are observed to rise with a velocity, W_B, given empirically by Davidson and Harrison (1963) as

$$W_B = 0.71 g^{\frac{1}{2}} V_B^{\frac{1}{6}}, \tag{7.20}$$

7.4 Inhomogeneity Instability 149

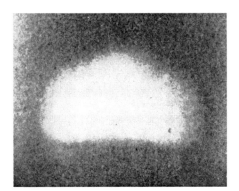

Figure 7.16. (Left) X-ray image of fluidized bed bubble (about 5 cm in diameter) in a bed of glass beads (courtesy of P.T. Rowe). (Right) View from above of bubbles breaking the surface of a sand/air fluidized bed (courtesy of J.F. Davidson).

where V_B is the volume of the bubble. Both the shape and rise velocity have many similarities to the spherical-cap bubbles discussed in Section 3.2.2. The rise velocity, W_B may be either faster or slower than the upward velocity of the suspending fluid, u_C, and this implies two types of bubbles that Catipovic *et al.* (1978) call *fast* and *slow* bubbles respectively. Figure 7.17 qualitatively depicts the nature of the streamlines of the flow relative to the bubbles for fast and slow bubbles. The same article provides a flow regime map, Figure 7.18 indicating the domains of

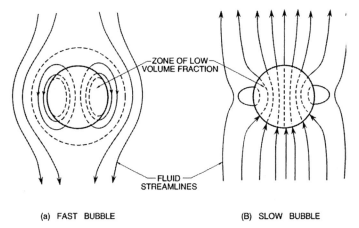

Figure 7.17. Sketches of the fluid streamlines relative to a fluidized bed *bubble* of low volume fraction for a *fast* bubble (left) and a *slow* bubble. Adapted from Catipovic *et al.* (1978).

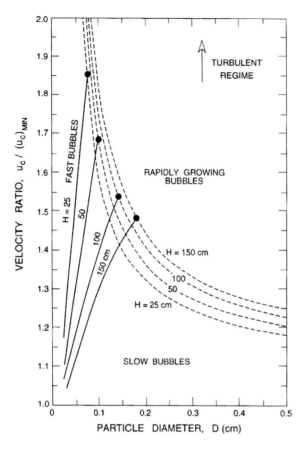

Figure 7.18. Flow regime map for fluidized beds with large particles (diameter, D), where $(u_C)_{min}$ is the minimum fluidization velocity and H is the height of the bed. Adapted from Catipovic *et al.* (1978).

fast bubbles, slow bubbles, and rapidly growing bubbles. When the particles are smaller other forces become important, particularly those that cause particles to stick together. In gas-fluidized beds the flow regime map of Geldart (1973), reproduced as Figure 7.19, is widely used to determine the flow regime. With very small particles (Group C) the cohesive effects dominate and the bed behaves like a plug, although the suspending fluid may create holes in the plug. With somewhat larger particles (Group A), the bed exhibits considerable expansion before bubbling begins. Group B particles exhibit bubbles as soon as fluidization begins (fast bubbles) and, with even larger particles (Group D), the bubbles become slow bubbles.

Aspects of the flow regime maps in Figures 7.18 and 7.19 qualitatively reflect the results of the instability analysis of the last section. Larger particles and larger fluid velocities imply larger j_{CD} values and therefore, according to instability analysis, larger growth rates. Thus, in the upper right side of both figures we find rapidly growing bubbles. Moreover, in the instability analysis it transpires that the ratio of the wave speed, ω/κ (analogous to the bubble velocity) to the typical fluid velocity, j_{CD}, is a continuously decreasing function of the parameter, $j_{CD}/(g/\kappa)^{\frac{1}{2}}$. Indeed, $\omega/j_{CD}\kappa$

7.5 Limits on Separated Flow

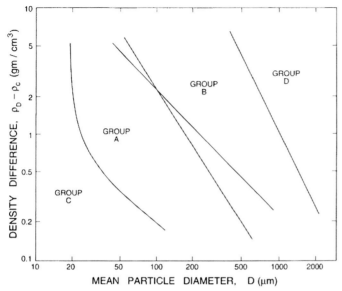

Figure 7.19. Flow regime map for fluidized beds with small particles (diameter, D). Adapted from Geldart (1973).

decreases from values greater than unity to values less than unity as $j_{CD}/(g/\kappa)^{\frac{1}{2}}$ increases. This is entirely consistent with the progression from fast bubbles for small particles (small j_{CD}) to slow bubbles for larger particles.

For further details on bubbles in fluidized beds the reader is referred to the extensive literature, including the books of Zenz and Othmer (1960), Cheremisinoff and Cheremisinoff (1984), Davidson *et al.* (1985), and Gibilaro (2001).

7.5 Limits on Separated Flow

We now leave disperse flow limits and turn to the mechanisms that limit separated flow regimes.

7.5.1 Kelvin–Helmoltz Instability

Separated flow regimes such as stratified horizontal flow or vertical annular flow can become unstable when waves form on the interface between the two fluid streams (subscripts 1 and 2). As indicated in Figure 7.20, the densities of the fluids will be denoted by ρ_1 and ρ_2 and the velocities by u_1 and u_2. If these waves continue to grow in amplitude they will cause a transition to another flow regime, typically one with greater intermittency and involving plugs or slugs. Therefore, to determine this particular boundary of the separated flow regime, it is necessary to investigate the potential growth of the interfacial waves, whose wavelength is denoted by λ (wavenumber, $\kappa = 2\pi/\lambda$). Studies of such waves have a long history originating with the work of Kelvin and

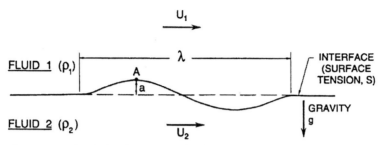

Figure 7.20. Sketch showing the notation for Kelvin–Helmholtz instability.

Helmholtz and the phenomena they revealed have come to be called Kelvin–Helmholtz instabilities (see, for example, Yih 1965). In general this class of instabilities involves the interplay between at least two of the following three types of forces:

- a buoyancy force due to gravity and proportional to the difference in the densities of the two fluids. This can be characterized by $g\ell^3 \Delta\rho$, where $\Delta\rho = \rho_1 - \rho_2$, g is the acceleration due to gravity, and ℓ is a typical dimension of the waves. This force may be stabilizing or destabilizing depending on the orientation of gravity, g, relative to the two fluid streams. In a horizontal flow in which the upper fluid is lighter than the lower fluid the force is stabilizing. When the reverse is true the buoyancy force is destabilizing and this causes Rayleigh–Taylor instabilities. When the streams are vertical as in vertical annular flow the role played by the buoyancy force is less clear.
- a surface tension force characterized by $S\ell$ that is always stabilizing.
- a Bernoulli effect that implies a change in the pressure acting on the interface caused by a change in velocity resulting from the displacement, a, of that surface. For example, if the upward displacement of the point A in Figure 7.21 were to cause an increase in the local velocity of fluid 1 and a decrease in the local velocity of fluid 2, this would imply an induced pressure difference at the point A that would increase the amplitude of the distortion, a. Such Bernoulli forces depend on the difference in the velocity of the two streams, $\Delta u = u_1 - u_2$, and are characterized by $\rho(\Delta u)^2 \ell^2$, where ρ and ℓ are a characteristic density and dimension of the flow.

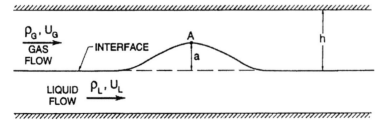

Figure 7.21. Sketch showing the notation for stratified flow instability.

7.5 Limits on Separated Flow

The interplay between these forces is most readily illustrated by a simple example. Neglecting viscous effects, one can readily construct the planar, incompressible potential flow solution for two semi-infinite horizontal streams separated by a plane horizontal interface (as in Figure 7.20) on which small-amplitude waves have formed. Then it is readily shown (Lamb 1879, Yih 1965) that Kelvin–Helmholtz instability will occur when

$$\frac{g\Delta\rho}{\kappa} + S\kappa - \frac{\rho_1\rho_2(\Delta u)^2}{\rho_1 + \rho_2} < 0. \tag{7.21}$$

The contributions from the three previously mentioned forces are self-evident. Note that the surface tension effect is stabilizing because that term is always positive, the buoyancy effect may be stabilizing or destabilizing depending on the sign of $\Delta\rho$, and the Bernoulli effect is always destabilizing. Clearly, one subset of this class of Kelvin–Helmholtz instabilities are the Rayleigh–Taylor instabilities that occur in the absence of flow ($\Delta u = 0$) when $\Delta\rho$ is negative. In that static case, the above relation shows that the interface is unstable to all wave numbers less than the critical value, $\kappa = \kappa_c$, where

$$\kappa_c = \left(\frac{g(-\Delta\rho)}{S}\right)^{\frac{1}{2}}. \tag{7.22}$$

In the next two sections we focus on the instabilities induced by the destabilizing Bernoulli effect for these can often cause instability of a separated flow regime.

7.5.2 Stratified Flow Instability

As a first example, consider the stability of the horizontal stratified flow depicted in Figure 7.21, where the destabilizing Bernoulli effect is primarily opposed by a stabilizing buoyancy force. An approximate instability condition is readily derived by observing that the formation of a wave (such as that depicted in Figure 7.21) will lead to a reduced pressure, p_A, in the gas in the orifice formed by that wave. The reduction below the mean gas pressure, $\bar{p}_G$, is given by Bernoulli's equation as follows:

$$p_A - \bar{p}_G = -\rho_G u_G^2 a/h, \tag{7.23}$$

provided $a \ll h$. The restraining pressure is given by the buoyancy effect of the elevated interface, namely $(\rho_L - \rho_G)ga$. It follows that the flow will become unstable when

$$u_G^2 > gh\Delta\rho/\rho_G. \tag{7.24}$$

In this case the liquid velocity has been neglected because it is normally small compared with the gas velocity. Consequently, the instability criterion provides an upper limit on the gas velocity that is, in effect, the velocity difference. Taitel and Dukler (1976) compared this prediction for the boundary of the stratified flow regime in a horizontal pipe of diameter, d, with the experimental observations of Mandhane *et al.* (1974) and found substantial agreement. This can be demonstrated by observing that, from Eq. (7.24),

$$j_G = \alpha u_G = C(\alpha)\alpha(gd\Delta\rho/\rho_G)^{\frac{1}{2}}, \qquad (7.25)$$

where $C(\alpha) = (h/d)^{\frac{1}{2}}$ is some simple monotonically increasing function of α that depends on the pipe cross section. For example, for the 2.5-cm pipe of Figure 7.3 the factor $(gd\Delta\rho/\rho_G)^{\frac{1}{2}}$ in Eq. (7.25) will have a value of approximately 15 m/s. As shown in Figure 7.3, this is in close agreement with the value of j_G at which the flow at low j_L departs from the stratified regime and begins to become wavy and then annular. Moreover, the factor $C(\alpha)\alpha$ should decrease as j_L increases and, in Figure 7.3, the boundary between stratified flow and wavy flow also exhibits this decrease.

7.5.3 Annular Flow Instability

As a second example consider vertical annular flow that becomes unstable when the Bernoulli force overcomes the stabilizing surface tension force. From Eq. (7.21), this implies that disturbances with wavelengths greater than a critical value, λ_c, will be unstable and that

$$\lambda_c = 2\pi S(\rho_1 + \rho_2)/\rho_1\rho_2(\Delta u)^2. \qquad (7.26)$$

For a liquid stream and a gas stream (as is normally the case in annular flow) and with $\rho_L \ll \rho_G$ this becomes the following:

$$\lambda_c = 2\pi S/\rho_G(\Delta u)^2. \qquad (7.27)$$

Now consider the application of this criterion to the flow regime maps for vertical pipe flow included in Figures 7.6 and 7.8. We examine the stability of a well-developed annular flow at a high gas volume fraction where $\Delta u \approx j_G$. Then, for a water/air mixture, Eq. (7.27) predicts critical wavelengths of 0.4 and 40 cm for $j_G = 10$ m/s and $j_G = 1$ m/s respectively. In ther words, at low values of j_G only larger wavelengths are unstable and this seems to be in accord with the breakup of the flow into large slugs. Conversely, at higher j_G flow rates, even quite small wavelengths are unstable and the liquid gets torn apart into the small droplets carried in the core gas flow.

8

Internal Flow Energy Conversion

8.1 Introduction

One of the most common requirements of a multiphase flow analysis is the prediction of the energy gains and losses as the flow proceeds through the pipes, valves, pumps, and other components that make up an internal flow system. In this chapter we attempt to provide a few insights into the physical processes that influence these energy conversion processes in a multiphase flow. The literature contains a plethora of engineering correlations for pipe friction and some data for other components such as pumps. This chapter provides an overview and some references to illustrative material but does not pretend to survey these empirical methodologies.

As might be expected, frictional losses in straight uniform pipe flows have been the most widely studied of these energy conversion processes and so we begin with a discussion of that subject, focusing first on disperse or nearly disperse flows and then on separated flows. In the last part of the chapter, we consider multiphase flows in pumps, in part because of the ubiquity of these devices and in part because they provide a second example of the multiphase flow effects in internal flows.

8.2 Frictional Loss in Disperse Flow

8.2.1 Horizontal Flow

We begin with a discussion of disperse horizontal flow. There exists a substantial body of data relating to the frictional losses or pressure gradient, $(-dp/ds)$, in a straight pipe of circular cross section (the coordinate s is measured along the axis of the pipe). Clearly $(-dp/ds)$ is a critical factor in the design of many systems, for example, slurry pipelines. Therefore a substantial database exists for the flows of mixtures of solids and water in horizontal pipes. The hydraulic gradient is usually nondimensionalized using the pipe diameter, d, the density of the suspending phase (ρ_L if liquid), and either

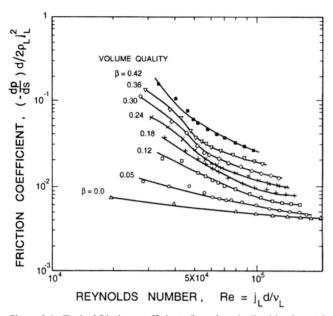

Figure 8.1. Typical friction coefficients (based on the liquid volumetric flux and the liquid density) plotted against Reynolds number (based on the liquid volumetric flux and the liquid viscosity) for the horizontal pipeline flow ($d = 5.2$ cm) of sand ($D = 0.018$ cm) and water at $21°C$ (Lazarus and Neilson 1978).

the total volumetric flux, j, or the volumetric flux of the suspending fluid (j_L if liquid). Thus, commonly used friction coefficients are as follows:

$$C_f = \frac{d}{2\rho_L j_L^2}\left(-\frac{dp}{ds}\right) \quad or \quad C_f = \frac{d}{2\rho_L j^2}\left(-\frac{dp}{ds}\right) \tag{8.1}$$

and, in parallel with the traditional Moody diagram for single-phase flow, these friction coefficients are usually presented as functions of a Reynolds number for various mixture ratios as characterized by the volume fraction, α, or the volume quality, β, of the suspended phase. Commonly used Reynolds numbers are based on the pipe diameter, the viscosity of the suspending phase (v_L if liquid) and either the total volumetric flux, j, or the volumetric flux of the suspending fluid.

For a more complete review of slurry pipeline data the reader is referred to Shook and Roco (1991) and Lazarus and Neilsen (1978). For the solids/gas flows associated with the pneumatic conveying of solids, Soo (1983) provides a good summary. For boiling flows or for gas/liquid flows, the reader is referred to the reviews of Hsu and Graham (1976) and Collier and Thome (1994).

The typical form of the friction coefficient data is illustrated in Figures 8.1 and 8.2 taken from Lazarus and Neilson (1978). Typically the friction coefficient increases markedly with increasing concentration and this increase is more significant the lower the Reynolds number. Note that the measured increases in the friction coefficient can exceed an order of magnitude. For a given particle size and density, the flow in a given pipe becomes increasingly homogeneous as the flow rate is increased because,

8.2 Frictional Loss in Disperse Flow

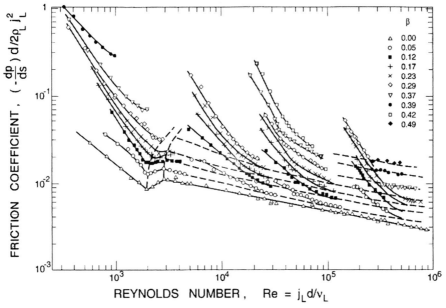

Figure 8.2. Typical friction coefficients (based on the liquid volumetric flux and the liquid density) plotted against Reynolds number (based on the liquid volumetric flux and the liquid viscosity) for the horizontal pipeline flow of four different solid/liquid mixtures (Lazarus and Neilson 1978).

as discussed in Section 7.3.1, the typical mixing velocity is increasing, whereas the typical segregation velocity remains relatively constant. The friction coefficient is usually increased by segregation effects, so, for a given pipe and particles, part of the decrease in the friction coefficient with increasing flow rate is due to the normal decrease with Reynolds number and part is due to the increasing homogeneity of the flow. Figure 8.2, taken from Lazarus and Neilson, shows how the friction coefficient curves for a variety of solid/liquid flows, tend to asymptote at higher Reynolds numbers to a family of curves (shown by the dashed lines) on which the friction coefficient is a function only of the Reynolds number and volume fraction. These so-called *base curves* pertain when the flow is sufficiently fast for complete mixing to occur and the flow regime becomes homogeneous. We first address these base curves and the issue of homogeneous flow friction. Later, in Section 8.2.3, we comment on the departures from the base curves that occur at lower flow rates when the flow is in the heterogeneous or saltation regimes.

8.2.2 Homogeneous Flow Friction

When the multiphase flow or slurry is thoroughly mixed the pressure drop can be approximated by the friction coefficient for a single-phase flow with the mixture density, ρ [Eq. (1.8)] and the same total volumetric flux, $j = j_S + j_L$, as the multiphase flow. We exemplify this using the slurry pipeline data from the preceding section

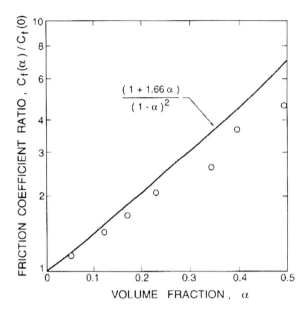

Figure 8.3. The ratio of the *base curve* friction coefficient at finite loading, $C_f(\alpha)$, to the friction coefficient for the continuous phase alone, $C_f(0)$. Equation (8.2) (line) is compared with the data of Lazarus and Neilsen (1978).

assuming that $\alpha = \beta$ (which does tend to be the case in horizontal homogeneous flows) and setting $j = j_L/(1 - \alpha)$. Then the ratio of the base friction coefficient at finite loading, $C_f(\alpha)$, to the friction coefficient for the continuous phase alone, $C_f(0)$, should be given by the following:

$$\frac{C_f(\alpha)}{C_f(0)} = \frac{(1 + \alpha \rho_S/\rho_L)}{(1 - \alpha)^2}. \tag{8.2}$$

A comparison between this expression and the data from the base curves of Lazarus and Neilsen is included in Figure 8.3 and demonstrates a reasonable agreement.

Thus a flow regime that is homogeneous or thoroughly mixed can usually be modeled as a single-phase flow with an effective density, volume flow rate and viscosity. In these circumstances the orientation of the pipe appears to make little difference. Often these correlations also require an effective mixture viscosity. In the above example, an effective kinematic viscosity of the multiphase flow could have been incorporated in Eq. (8.2); however, this has little effect on the comparison in Figure 8.3, especially under the turbulent conditions in which most slurry pipelines operate.

Wallis (1969) includes a discussion of homogeneous flow friction correlations for both laminar and turbulent flow. In laminar flow, most correlations use the mixture density as the effective density and the total volumetric flux, j, as the velocity as we did in the above example. A wide variety of mostly empirical expressions are used for the effective viscosity, μ_e. In low-volume-fraction suspensions of solid particles, Einstein's (1906) classical effective viscosity given by the following:

$$\mu_e = \mu_C(1 + 5\alpha/2) \tag{8.3}$$

8.2 Frictional Loss in Disperse Flow

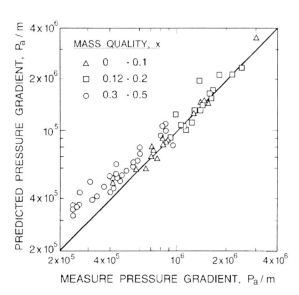

Figure 8.4. Comparison of the measured friction coefficient with that using the homogeneous prediction for steam/water flows of various mass qualities in a 0.3-cm diameter tube. From Owens (1961).

is appropriate, although this expression loses validity for volume fractions greater than a few percentage. In emulsions with droplets of viscosity, μ_D, the extension of Einstein's formula

$$\mu_e = \mu_C \left\{ 1 + \frac{5\alpha}{2} \frac{(\mu_D + 2\mu_C/5)}{(\mu_D + \mu_C)} \right\} \tag{8.4}$$

is the corresponding expression (Happel and Brenner 1965). More empirical expressions for μ_e are typically used at higher volume fractions.

As discussed in Section 1.3.1, turbulence in multiphase flows introduces another set of complicated issues. Nevertheless as was demonstrated by the above example, the effective single-phase approach to pipe friction seems to produce moderately accurate results in homogeneous flows. The comparison in Figure 8.4 shows that the errors in such an approach are about ±25%. The presence of particles, particularly solid particles, can act like surface roughness, enhancing turbulence in many applications. Consequently, turbulent friction factors for homogeneous flow tend to be similar to the values obtained for single-phase flow in rough pipes, values around 0.005 being commonly experienced (Wallis 1969).

8.2.3 Heterogeneous Flow Friction

The most substantial remaining issue is to understand the much larger friction factors that occur when particle segregation predominates. For example, commenting on the data of Figure 8.2, Lazarus and Neilsen show that values larger than the base curves begin when component separation begins to occur and the flow regime changes from the heterogeneous regime to the saltation regime (Section 7.2.3 and Figure 7.5). Another slurry flow example is shown in Figure 8.5. According to Hayden and Stelson (1971)

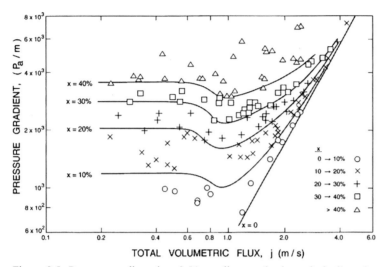

Figure 8.5. Pressure gradients in a 2.54-cm-diameter horizontal pipeline plotted against the total volumetric flux, j, for a slurry of sand with a particle diameter of 0.057 cm. Curves for four specific mass fractions, x (as a percentage) are fitted to the data. Adapted from Hayden and Stelson (1971).

the minima in the fitted curves correspond to the boundary between the heterogeneous and saltation flow regimes. Note that these all occur at essentially the same critical volumetric flux, j_c; this agrees with the criterion of Newitt *et al.* (1955) that was discussed in Section 7.3.1 and is equivalent to a critical volumetric flux, j_c, that is simply proportional to the terminal velocity of individual particles and independent of the loading or mass fraction.

The transition of the flow regime from heterogeneous to saltation results in much of the particle mass being supported directly by particle contacts with the interior surface of the pipe. The frictional forces that this contact produces implies, in turn, a substantial pressure gradient to move the bed. The pressure gradient in the moving bed configuration can be readily estimated as follows. The submerged weight of solids in the packed bed per unit length of the cylindrical pipe of diameter, d, is as follows:

$$\pi d^2 \alpha g(\rho_S - \rho_L), \tag{8.5}$$

where α is the overall effective volume fraction of solids. Therefore, if the effective Coulomb friction coefficient is denoted by η, the longitudinal force required to overcome this friction per unit length of pipe is simply η, times the above expression. The pressure gradient needed to provide this force is therefore

$$-\left(\frac{dp}{ds}\right)_{\text{friction}} = \eta \alpha g(\rho_S - \rho_L). \tag{8.6}$$

With η considered as an adjustable constant, this is the expression for the additional frictional pressure gradient proposed by Newitt *et al.* (1955). The final step is to calculate the volumetric flow rate that occurs with this pressure gradient, part of which

8.2 Frictional Loss in Disperse Flow

proceeds through the packed bed and part of which flows above the bed. The literature contains a number of semiempirical treatments of this problem. One of the first correlations was that of Durand and Condolios (1952) that took the following form:

$$j_c = f(\alpha, D) \left\{ 2gd \frac{\Delta\rho}{\rho_L} \right\}^{\frac{1}{2}}, \tag{8.7}$$

where $f(\alpha, D)$ is some function of the solids fraction, α, and the particle diameter, D. There are both similarities and differences between this expression and that of Newitt et al. (1955). A commonly used criterion that has the same form as Eq. (8.7) but is more specific is that of Zandi and Govatos (1967):

$$j_c = \left\{ \frac{K\alpha dg}{C_D^{\frac{1}{2}}} \frac{\Delta\rho}{\rho_L} \right\}^{\frac{1}{2}}, \tag{8.8}$$

where K is an empirical constant of the order of 10–40. Many other efforts have been made to correlate the friction factor for the heterogeneous and saltation regimes; reviews of these mostly empirical approaches can be found in Zandi (1971) and Lazarus and Neilsen (1978). Fundamental understanding is less readily achieved; perhaps future understanding of the granular flows described in Chapter 13 will provide clearer insights.

8.2.4 Vertical Flow

As indicated by the flow regimes of Section 7.2.2, vertically oriented pipe flow can experience partially separated flows in which large relative velocities develop due to buoyancy and the difference in the densities of the two-phases or components. These large relative velocities complicate the problem of evaluating the pressure gradient. In Section 8.3 we describe the traditional approach used for separated flows in which it is assumed that the phases or components flow in separate but communicating streams. However, even when the multiphase flow has a solid particulate phase or an incompletely separated gas/liquid mixture, partial separation leads to friction factors that exhibit much larger values than would be experienced in a homogeneous flow. One example of that in horizontal flow was described in Section 8.2.1. Here we provide an example from vertical pipe flows. Figure 8.6 contains friction factors (based on the total volumetric flux and the liquid density) plotted against Reynolds number for the flow of air bubbles and water in a 10.2-cm vertical pipe for three ranges of void fraction. Note that these are all much larger than the single phase friction factor. Figure 8.7 presents further details from the same experiments, plotting the ratio of the frictional pressure gradient in the multiphase flow to that in a single phase flow of the same liquid volumetric flux against the volume quality for several ranges of Reynolds number. The data show that for small volume qualities the friction factor can be as much as an order of magnitude larger than the single-phase value. This substantial effect decreases as the

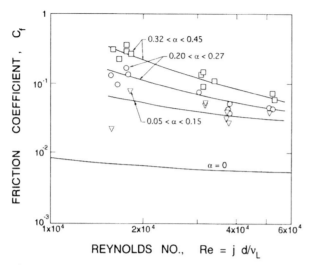

Figure 8.6. Typical friction coefficients (based on total volumetric flux and the liquid density) plotted against Reynolds number (based on the total volumetric flux and the liquid viscosity) for the flow of air bubbles and water in a 10.2-cm vertical pipe flow for three ranges of air volume fraction, α, as shown (Kytömaa 1987).

Reynolds number increases and also decreases at higher volume fractions. To emphasize the importance of this phenomenon in partially separated flows, a line representing the Lockhart–Martinelli correlation for fully separated flow (see Section 8.3.1) is also included in Figure 8.7. As in the case of partially separated horizontal flows discussed in Section 8.2.1, there is, as yet, no convincing explanation of the high values of the

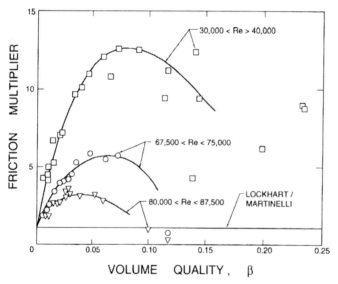

Figure 8.7. Typical friction multiplier data (defined as the ratio of the actual frictional pressure gradient to the frictional pressure gradient that would occur for a single phase flow of the same liquid volume flux) for the flow of air bubbles and water in a 10.2-cm vertical pipe plotted against the volume quality, β, for three ranges of Reynolds number as shown (Kytömaa 1987).

friction at lower Reynolds numbers. But the effect seems to be related to the large unsteady motions caused by the presence of a disperse phase of different density and the effective stresses (similar to Reynolds stresses) that result from the inertia of these unsteady motions.

8.3 Frictional Loss in Separated Flow

Having discussed homogeneous and disperse flows we now turn our attention to the friction in separated flows and, in particular, describe the commonly used Martinelli correlations.

8.3.1 Two-Component Flow

The Lockhart–Martinelli and Martinelli–Nelson correlations attempt to predict the frictional pressure gradient in two-component or two-phase flows in pipes of constant cross-sectional area, A. It is assumed that these multiphase flows consist of two separate cocurrent streams that, for convenience, we refer to as the liquid and the gas, although they could be any two immiscible fluids. The correlations use the results for the frictional pressure gradient in single-phase pipe flows of each of the two fluids. In two-phase flow, the volume fraction is often changing as the mixture progresses along the pipe and such phase change necessarily implies acceleration or deceleration of the fluids. Associated with this acceleration is an acceleration component of the pressure gradient that is addressed in a later section dealing with the Martinelli–Nelson correlation. Obviously, it is convenient to begin with the simpler, two-component case (the Lockhart–Martinelli correlation); this also neglects the effects of changes in the fluid densities with distance, s, along the pipe axis so that the fluid velocities also remain invariant with s. Moreover, in all cases, it is assumed that the hydrostatic pressure gradient has been accounted for so that the only remaining contribution to the pressure gradient, $-dp/ds$, is that due to the wall shear stress, τ_w. A simple balance of forces requires that

$$-\frac{dp}{ds} = \frac{P}{A}\tau_w, \tag{8.9}$$

where P is the perimeter of the cross section of the pipe. For a circular pipe, $P/A = 4/d$, where d is the pipe diameter and, for noncircular cross sections, it is convenient to define a *hydraulic diameter*, $4A/P$. Then, defining the dimensionless friction coefficient, C_f, as

$$C_f = \tau_w / \frac{1}{2}\rho j^2 \tag{8.10}$$

the more general form of Eq. (8.1) becomes the

$$-\frac{dp}{ds} = 2C_f \rho j^2 \frac{P}{4A}. \tag{8.11}$$

In single-phase flow the coefficient, C_f, is a function of the Reynolds number, $\rho d j/\mu$, of the following form:

$$C_f = \mathcal{K} \left\{ \frac{\rho d j}{\mu} \right\}^{-m}, \tag{8.12}$$

where $\mathcal{K}$ is a constant that depends on the roughness of the pipe surface and will be different for laminar and turbulent flow. The index, m, is also different, being 1 in the case of laminar flow and $\frac{1}{4}$ in the case of turbulent flow.

These relations from single phase flow are applied to the two cocurrent streams in the following way. First, we define hydraulic diameters, d_L and d_G, for each of the two streams and define corresponding area ratios, κ_L and κ_G, as follows:

$$\kappa_L = 4A_L/\pi d_L^2; \quad \kappa_G = 4A_G/\pi d_G^2, \tag{8.13}$$

where $A_L = A(1-\alpha)$ and $A_G = A\alpha$ are the actual cross-sectional areas of the two streams. The quantities κ_L and κ_G are shape parameters that depend on the geometry of the flow pattern. In the absence of any specific information on this geometry, one might choose the values pertinent to streams of circular cross section, namely $\kappa_L = \kappa_G = 1$, and the commonly used form of the Lockhart–Martinelli correlation employs these values. However, as an alternative example, we also present data for the case of annular flow in which the liquid coats the pipe wall with a film of uniform thickness and the gas flows in a cylindrical core. When the film is thin, it follows from the annular flow geometry that

$$\kappa_L = 1/(1-\alpha); \quad \kappa_G = 1, \tag{8.14}$$

where it has been assumed that only the exterior perimeter of the annular liquid stream experiences significant shear stress.

In summary, the basic geometric relations yield the following:

$$\alpha = 1 - \kappa_L d_L^2/d^2 = \kappa_G d_G^2/d^2. \tag{8.15}$$

Then, the pressure gradient in each stream is assumed given by the following coefficients taken from single phase pipe flow:

$$C_{fL} = \mathcal{K}_L \left\{ \frac{\rho_L d_L u_L}{\mu_L} \right\}^{-m_L}; \quad C_{fG} = \mathcal{K}_G \left\{ \frac{\rho_G d_G u_G}{\mu_G} \right\}^{-m_G} \tag{8.16}$$

and, because the pressure gradients must be the same in the two streams, this imposes the following relation between the flows:

$$-\frac{dp}{ds} = \frac{2\rho_L u_L^2 \mathcal{K}_L}{d_L} \left\{ \frac{\rho_L d_L u_L}{\mu_L} \right\}^{-m_L} = \frac{2\rho_G u_G^2 \mathcal{K}_G}{d_G} \left\{ \frac{\rho_G d_G u_G}{\mu_G} \right\}^{-m_G}. \tag{8.17}$$

8.3 Frictional Loss in Separated Flow

In the above, m_L and m_G are 1 or $\frac{1}{4}$ depending on whether the stream is laminar or turbulent. It follows that there are four permutations, namely:

1. Both streams are laminar so that $m_L = m_G = 1$, a permutation denoted by the double subscript LL.
2. A laminar liquid stream and a turbulent gas stream so that $m_L = 1, m_G = \frac{1}{4}$ (LT).
3. A turbulent liquid stream and a laminar gas stream so that $m_L = \frac{1}{4}, m_G = 1$ (TL).
4. Both streams are turbulent so that $m_L = m_G = \frac{1}{4}$ (TT).

Equations (8.15) and (8.17) are the basic relations used to construct the Lockhart–Martinelli correlation. However, the solutions to these equations are normally and most conveniently presented in nondimensional form by defining the following dimensionless pressure gradient parameters as follows:

$$\phi_L^2 = \frac{\left(\frac{dp}{ds}\right)_{actual}}{\left(\frac{dp}{ds}\right)_L}; \quad \phi_G^2 = \frac{\left(\frac{dp}{ds}\right)_{actual}}{\left(\frac{dp}{ds}\right)_G}, \quad (8.18)$$

where $(dp/ds)_L$ and $(dp/ds)_G$ are respectively the hypothetical pressure gradients that would occur in the same pipe if only the liquid flow were present and if only the gas flow were present. The ratio of these two hypothetical gradients, Ma^2, given by the following:

$$\text{Ma}^2 = \frac{\phi_G^2}{\phi_L^2} = \frac{\left(\frac{dp}{ds}\right)_L}{\left(\frac{dp}{ds}\right)_G} = \frac{\rho_L}{\rho_G} \frac{G_G^2}{G_L^2} \frac{\mathcal{K}_G}{\mathcal{K}_L} \frac{\left\{\frac{G_G d}{\mu_G}\right\}^{-m_G}}{\left\{\frac{G_L d}{\mu_L}\right\}^{-m_L}}, \quad (8.19)$$

defines the Martinelli parameter, Ma, and allows presentation of the solutions to Eqs. (8.15) and (8.17) in a convenient parametric form. Using the definitions of Eq. (8.18), the nondimensional forms of Eq. 8.15 become the following:

$$\alpha = 1 - \mathcal{K}_L^{-(1+m_L)/(m_L-5)} \phi_L^{4/(m_L-5)} = \mathcal{K}_G^{-(1+m_G)/(m_G-5)} \phi_G^{4/(m_G-5)} \quad (8.20)$$

and the solution of these equations produces the Lockhart–Martinelli prediction of the nondimensional pressure gradient.

To summarize, for given values of

- the fluid properties, $\rho_L, \rho_G, \mu_L,$ and μ_G
- a given type of flow, LL, LT, TL, or TT, along with the single-phase correlation constants, $m_L, m_G, \mathcal{K}_L,$ and $\mathcal{K}_G$
- given values or expressions for the parameters of the flow pattern geometry, κ_L and κ_G
- and a given value of α

Eq. (8.20) can be solved to find the nondimensional solution to the flow, namely the values of ϕ_L^2 and ϕ_G^2. The value of Ma^2 also follows and the rightmost expression in

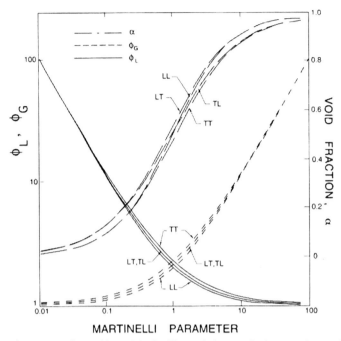

Figure 8.8. The Lockhart–Martinelli correlation results for ϕ_L and ϕ_G and the void fraction, α, as functions of the Martinelli parameter, Ma, for the case, $\kappa_L = \kappa_G = 1$. Results are shown for the four laminar and turbulent stream permutations, LL, LT, TL, and TT.

Eq. (8.19) then yields a relation between the liquid mass flux, G_L, and the gas mass flux, G_G. Thus, if one is also given just *one* mass flux (often this will be the total mass flux, G), the solution will yield the individual mass fluxes, the mass quality, and other flow properties. Alternatively one could begin the calculation with the mass quality rather than the void fraction and find the void fraction as one of the results. Finally the pressure gradient, dp/ds, follows from the values of ϕ_L^2 and ϕ_G^2.

The solutions for the cases $\kappa_L = \kappa_G = 1$ and $\kappa_L = 1/2(1 - \alpha), \kappa_G = 1$ are presented in Figures 8.8 and 8.9 and the comparison of these two figures yields some measure of the sensitivity of the results to the flow geometry parameters, κ_L and κ_G. Similar charts are commonly used in the manner described above to obtain solutions for two-component gas/liquid flows in pipes. A typical comparison of the Lockhart–Martinelli prediction with the experimental data is presented in Figure 8.10. Note that the scatter in the data is significant (about a factor of 3 in ϕ_G) and that the Lockhart–Martinelli prediction often yields an overestimate of the friction or pressure gradient. This is the result of the assumption that the entire perimeter of both phases experiences static wall friction. This is not the case and part of the perimeter of each phase is in contact with the other phase. If the interface is smooth this could result in a decrease in the friction; conversely, a roughened interface could also result in increased interfacial friction.

It is important to recognize that there are many deficiencies in the Lockhart–Martinelli approach. First, it is assumed that the flow pattern consists of two parallel

8.3 Frictional Loss in Separated Flow

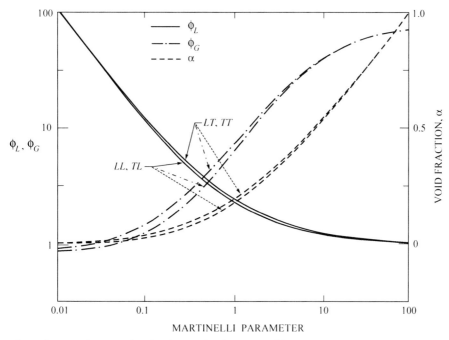

Figure 8.9. As Figure 8.8 but for the annular flow case with $\kappa_L = 1/(1-\alpha)$ and $\kappa_G = 1$.

streams and any departure from this topology could result in substantial errors. In Figure 8.11, the ratios of the velocities in the two streams that are implicit in the correlation [and follow from Eq. (8.19)] are plotted against the Martinelli parameter. Note that large velocity differences appear to be predicted at void fractions close to unity. Because the flow is likely to transition to mist flow in this limit and because the relative velocities in the mist flow are unlikely to become large, it seems inevitable that the correlation would become quite inaccurate at these high void fractions. Similar

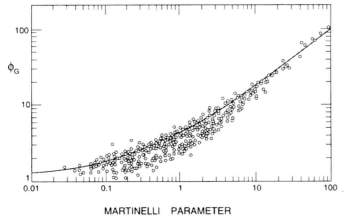

Figure 8.10. Comparison of the Lockhart–Martinelli correlation (the TT case) for ϕ_G (solid line) with experimental data. Adapted from Turner and Wallis (1965).

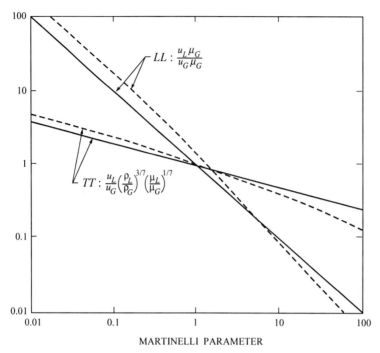

Figure 8.11. Ratios demonstrating the velocity ratio, u_L/u_G, implicit in the Lockhart–Martinelli correlation as functions of the Martinelli parameter, Ma, for the LL and TT cases. Solid lines: $\kappa_L = \kappa_G = 1$; dashed lines: $\kappa_L = 1/(1-\alpha)$, $\kappa_G = 1$.

inaccuracies seem inevitable at low void fraction. Indeed, it appears that the Lockhart–Martinelli correlations work best under conditions that do not imply large velocity differences. Figure 8.11 demonstrates that smaller velocity differences are expected for turbulent flow (TT) and this is mirrored in better correlation with the experimental results in the turbulent flow case (Turner and Wallis 1965).

Second, there is the previously discussed deficiency regarding the suitability of assuming that the perimeters of both phases experience friction that is effectively equivalent to that of a static solid wall. A third source of error arises because the multiphase flows are often unsteady and this yields a multitude of quadratic interaction terms that contribute to the mean flow in the same way that Reynolds stress terms contribute to turbulent single-phase flow.

8.3.2 Flow with Phase Change

The Lockhart–Martinelli correlation was extended by Martinelli and Nelson (1948) to include the effects of phase change. Because the individual mass fluxes are then changing as one moves down the pipe, it becomes convenient to use a different nondimensional pressure gradient

$$\phi_{L0}^2 = \frac{\left(\frac{dp}{ds}\right)_{\text{actual}}}{\left(\frac{dp}{ds}\right)_{L0}}, \qquad (8.21)$$

where $(dp/ds)_{L0}$ is the hypothetical pressure gradient that would occur in the same pipe if a liquid flow with the same total mass flow were present. Such a definition is more practical in this case because the total mass flow is constant. It follows that ϕ_{L0}^2 is simply related to ϕ_L^2 by the following:

$$\phi_{L0}^2 = (1 - \mathcal{X})^{2-m_L} \phi_L^2. \tag{8.22}$$

The Martinelli–Nelson correlation uses the previously described Lockhart–Martinelli results to obtain ϕ_L^2 and, therefore, ϕ_{L0}^2 as functions of the mass quality, $\mathcal{X}$. Then the frictional component of the pressure gradient is given by the following:

$$\left(-\frac{dp}{ds}\right)_{\text{Frictional}} = \phi_{L0}^2 \frac{2G^2 \mathcal{K}_L}{\rho_L d} \left\{\frac{Gd}{\mu_L}\right\}^{-m_L}. \tag{8.23}$$

Note that, although the other quantities in this expression for dp/ds are constant along the pipe, the quantity ϕ_{L0}^2 is necessarily a function of the mass quality, $\mathcal{X}$, and will therefore vary with s. It follows that to integrate Eq. (8.23) to find the pressure drop over a finite pipe length one must know the variation of the mass quality, $\mathcal{X}(s)$. Now, in many boilers, evaporators, or condensers, the mass quality varies linearly with length, s, because

$$\frac{d\mathcal{X}}{ds} = \frac{\mathcal{Q}_\ell}{AG\mathcal{L}}. \tag{8.24}$$

Because the rate of heat supply or removal per unit length of the pipe, $\mathcal{Q}_\ell$, is roughly uniform and the latent heat, $\mathcal{L}$, can be considered roughly constant, it follows that $d\mathcal{X}/ds$ is approximately constant. Then integration of Eq. (8.23) from the location at which $\mathcal{X} = 0$ to the location a distance, ℓ, along the pipe (at which $\mathcal{X} = \mathcal{X}_e$) yields the following:

$$(\Delta p(\mathcal{X}_e))_{\text{Frictional}} = (p)_{\mathcal{X}=0} - (p)_{\mathcal{X}=\mathcal{X}_e} = \frac{2G^2 \ell \mathcal{K}_L}{d\rho_L} \left\{\frac{Gd}{\mu_L}\right\}^{-m_L} \overline{\phi_{L0}^2} \tag{8.25}$$

where

$$\overline{\phi_{L0}^2} = \frac{1}{\mathcal{X}_e} \int_0^{\mathcal{X}_e} \phi_{L0}^2 d\mathcal{X}. \tag{8.26}$$

Given a two-phase flow and assuming that the fluid properties can be estimated with reasonable accuracy by knowing the average pressure level of the flow and finding the saturated liquid and vapor densities and viscosities at that pressure, the results of the previous section can be used to determine ϕ_{L0}^2 as a function of $\mathcal{X}$. Integration of this function yields the required values of $\overline{\phi_{L0}^2}$ as a function of the exit mass quality, $\mathcal{X}_e$, and the prevailing mean pressure level. Typical data for water are exhibited in Figure 8.12 and the corresponding values of the exit void fraction, α_E, are shown in Figure 8.13.

These nondimensional results are used in a more general flow in the following way. If one wishes to determine the pressure drop for a flow with a nonzero inlet quality,

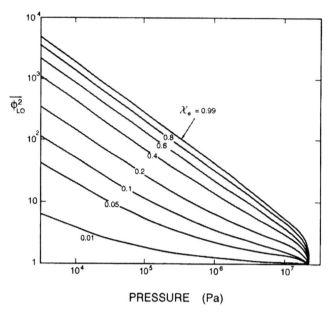

Figure 8.12. The Martinelli–Nelson frictional pressure drop function, $\overline{\phi_{LO}^2}$, for water as a function of the prevailing pressure level and the exit mass quality, $\mathcal{X}_e$. Case shown is for $\kappa_L = \kappa_G = 1.0$ and $m_L = m_G = 0.25$.

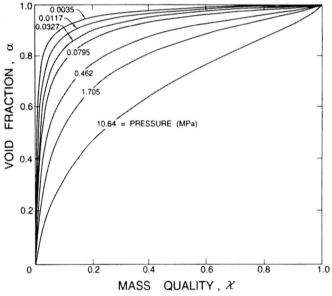

Figure 8.13. The exit void fraction values, α_e, corresponding to the data of Figure 8.12. Case shown is for $\kappa_L = \kappa_G = 1.0$ and $m_L = m_G = 0.25$.

8.3 Frictional Loss in Separated Flow

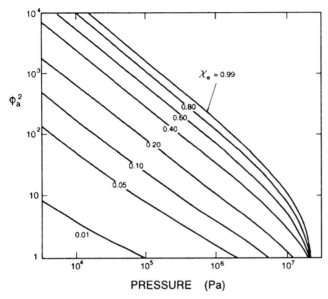

Figure 8.14. The Martinelli–Nelson acceleration pressure drop function, ϕ_a^2, for water as a function of the prevailing pressure level and the exit mass quality, $\mathcal{X}_e$. Case shown is for $\kappa_L = \kappa_G = 1.0$ and $m_L = m_G = 0.25$.

$\mathcal{X}_i$, and an exit quality, $\mathcal{X}_e$, [or, equivalently, a given heat flux because of Eq. (8.24)] then one simply uses Figure 8.12, first, to determine the pressure difference between the hypothetical point upstream of the inlet at which $\mathcal{X} = 0$ and the inlet and, second, to determine the difference between the same hypothetical point and the outlet of the pipe.

But, in addition, to the frictional component of the pressure gradient there is also a contribution caused by the fact that the fluids will be accelerating due to the change in the mixture density caused by the phase change. Using the mixture momentum equation [Eq. (1.50)], it is readily shown that this acceleration contribution to the pressure gradient can be written as follows:

$$\left(-\frac{dp}{ds}\right)_{\text{Acceleration}} = G^2 \frac{d}{ds}\left\{\frac{\mathcal{X}^2}{\rho_G \alpha} + \frac{(1-\mathcal{X})^2}{\rho_L(1-\alpha)}\right\} \tag{8.27}$$

and this can be integrated over the same interval as was used for the frictional contribution to obtain the following:

$$(\Delta p(\mathcal{X}_e))_{\text{Acceleration}} = G^2 \rho_L \phi_a^2(\mathcal{X}_e) \tag{8.28}$$

where

$$\phi_a^2(\mathcal{X}_e) = \left\{\frac{\rho_L \mathcal{X}_e^2}{\rho_G \alpha_e} + \frac{(1-\mathcal{X}_e)^2}{(1-\alpha_e)} - 1\right\}. \tag{8.29}$$

As in the case of $\overline{\phi_{L0}^2}$, $\phi_a^2(\mathcal{X}_e)$ can readily be calculated for a particular fluid given the prevailing pressure. Typical values for water are presented in Figure 8.14. This figure is used in a manner analogous to Figure 8.12 so that, taken together, they allow

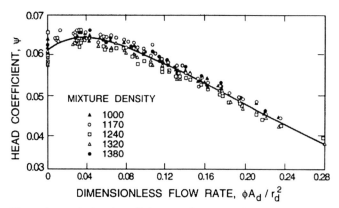

Figure 8.15. The head coefficient, ψ, for a centrifugal dredge pump ingesting silt/clay/water mixtures plotted against a nondimensional flow rate, $\phi A_d / r_d^2$, for various mixture densities (in kilograms per cubic meter). Adapted from Herbich (1975).

prediction of both the frictional and acceleration components of the pressure drop in a two-phase pipe flow with phase change.

8.4 Energy Conversion in Pumps and Turbines

Apart from pipes, most pneumatic or hydraulic systems also involve a whole collection of components such as valves, pumps, turbines, heat exchangers, and so on. The flows in these devices are often complicated and frequently require highly specialized analyses. However, effective single-phase analyses (homogeneous flow analyses) can also yield useful results and we illustrate this here by reference to work on the multiphase flow through rotating impeller pumps (centrifugal, mixed, or axial pumps).

8.4.1 Multiphase Flows in Pumps

Consistent with the usual turbomachinery conventions, the total pressure increase (or decrease) across a pump (or turbine) and the total volumetric flux (based on the discharge area, A_d) are denoted by Δp^T and j, respectively, and these quantities are nondimensionalized to form the head and flow coefficients, ψ and ϕ, for the machine:

$$\psi = \frac{\Delta p^T}{\rho \Omega^2 r_d^2}; \quad \phi = \frac{j}{\Omega r_d}, \qquad (8.30)$$

where Ω and r_d are the rotating speed (in radians per second) and the radius of the impeller discharge respectively and ρ is the mixture density. We note that sometimes in presenting cavitation performance, the impeller inlet area, A_i, is used rather than A_d in defining j, and this leads to a modified flow coefficient based on that inlet area.

The typical centrifugal pump performance with multiphase mixtures is exemplified by Figures 8.15, 8.16, and 8.17. Figure 8.15 from Herbich (1975) presents the performance of a centrifugal dredge pump ingesting silt/clay/water mixtures with mixture

8.4 Energy Conversion in Pumps and Turbines

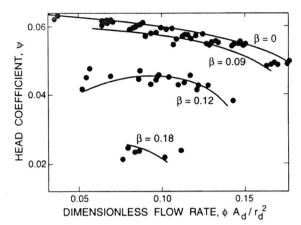

Figure 8.16. The head coefficient, ψ, for a centrifugal dredge pump ingesting air–water mixtures plotted against a nondimensional flow rate, $\phi A_d / r_d^2$, for various volumetric qualities, β. Adapted from Herbich (1975).

densities, ρ, up to 1380 kg/m^3. The corresponding solids fractions therefore range up to about 25% and the figure indicates that, provided ψ is defined using the mixture density, there is little change in the performance even up to such high solids fractions. Herbich also shows that the silt and clay suspensions cause little change in the equivalent homogeneous cavitation performance of the pump.

Data on the same centrifugal pump with air/water mixtures of different volume quality, β, is included in Figure 8.16 (Herbich 1975). Again, there is little difference between the multiphase flow performance and the homogeneous flow prediction at small discharge qualities. However, unlike the solids/liquid case, the air/water performance begins to decline precipitously above some critical volume fraction of gas, in this case a volume fraction consistent with a discharge quality of about 9%. Below this critical value, the homogeneous theory works well; larger volumetric qualities of air produce substantial degradation in performance.

Patel and Runstadler (1978), Murakami and Minemura (1978), and many others present similar data for pumps ingesting air/water and steam/water mixtures. Figure 8.17 presents another example of the air/water flow through a centrifugal pump. In this case the critical inlet volumetric quality is only about $\beta = 3$ or 4%

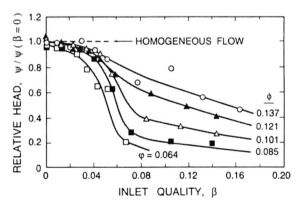

Figure 8.17. The ratio of the pump head with air/water mixtures to the head with water alone, $\psi/\psi(\beta = 0)$, as a function of the inlet volumetric quality, β, for various flow coefficients, ϕ. Data from Patel and Runstadler (1978) for a centrifugal pump.

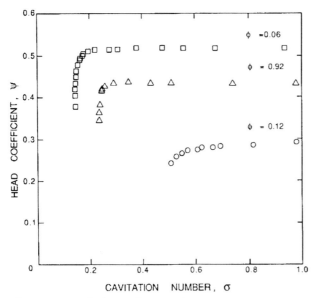

Figure 8.18. Cavitation performance for a typical centrifugal pump (Franz et al. 1990) for three different flow coefficients, $\phi = 0.12$, $\phi = 0.92$, and $\phi = 0.06$.

and the degradation appears to occur at lower volume fractions for lower flow coefficients. Murakami and Minemura (1978) obtained similar data for both axial and centrifugal pumps, although the performance of axial flow pumps appear to fall off at even lower air contents.

A qualitatively similar, precipitous decline in performance occurs in single-phase liquid pumping when cavitation at the inlet to the pump becomes sufficiently extensive. This performance degradation is normally presented dimensionlessly by plotting the head coefficient, ψ, at a given, fixed flow coefficient against a dimensionless inlet pressure, namely the cavitation number, σ (see Section 5.2.1), defined as follows:

$$\sigma = \frac{(p_i - p_V)}{\frac{1}{2}\rho_L \Omega^2 r_i^2}, \tag{8.31}$$

where p_i and r_i are the inlet pressure and impeller tip radius and p_V is the vapor pressure. An example is shown in Figure 8.18, which presents the cavitation performance of a typical centrifugal pump. Note that the performance declines rapidly below a critical cavitation number that usually corresponds to a fairly high vapor volume fraction at the pump inlet.

There appear to be two possible explanations for the decline in performance in gas/liquid flows above a critical volume fraction. The first possible cause, propounded by Murakami and Minemura (1977, 1978), Patel and Runstadler (1978), Furuya (1985), and others is that, when the void fraction exceeds some critical value, the flow in the blade passages of the pump becomes stratified because of the large crossflow pressure gradients. This allows a substantial deviation angle to develop at

8.4 Energy Conversion in Pumps and Turbines

the pump discharge and, as in conventional single-phase turbomachinery analyses (Brennen 1994), an increasing deviation angle implies a decline in performance. The lower critical volume fractions at lower flow coefficients would be consistent with this explanation because the pertinent pressure gradients will increase as the loading on the blades increases. Previously, in Section 7.3.3, we discussed the data on the bubble size in the blade passages compiled by Murakami and Minemura (1977, 1978). Bubble size is critical to the process of stratification because larger bubbles have larger relative velocities and will therefore lead more readily to stratification. But the size of bubbles in the blade passages of a pump is usually determined by the high shear rates to which the inlet flow is subjected and therefore the phenomenon has two key processes, namely shear at inlet that determines bubble size and segregation in the blade passages that governs performance.

The second explanation (and the one most often put forward to explain cavitation performance degradation) is based on the observation that the vapor (or gas) bubbles grow substantially as they enter the pump and subsequently collapse as they are convected into regions of higher pressure within the blade passages of the pump. The displacement of liquid by this volume growth and collapse introduces an additional flow area restriction into the flow, an additional inlet *nozzle* caused by the cavitation. Stripling and Acosta (1962) and others have suggested that the head degradation due to cavitation could be due to a lack of pressure recovery in this effective additional nozzle.

9

Homogeneous Flows

9.1 Introduction

In this chapter we are concerned with the dynamics of multiphase flows in which the relative motion between the phases can be neglected. It is clear that two different streams can readily travel at different velocities, and indeed such relative motion is an implicit part of the study of separated flows. Conversely, it is clear from the results of Section 2.4.2 that any two phases could, in theory, be sufficiently well mixed and therefore the disperse particle size sufficiently small so as to eliminate any significant relative motion. Thus the asymptotic limit of truly homogeneous flow precludes relative motion. Indeed, the term *homogeneous flow* is sometimes used to denote a flow with negligible relative motion. Many bubbly or mist flows come close to this limit and can, to a first approximation, be considered to be homogeneous. In the present chapter some of the properties of homogeneous flows are considered.

9.2 Equations of Homogeneous Flow

In the absence of relative motion the governing mass and momentum conservation equations for inviscid, homogeneous flow reduce to the single-phase form as follows:

$$\frac{\partial \rho}{\partial t} + \frac{\partial}{\partial x_j}(\rho u_j) = 0 \tag{9.1}$$

$$\rho \left[\frac{\partial u_i}{\partial t} + u_j \frac{\partial u_i}{\partial x_j} \right] = -\frac{\partial p}{\partial x_i} + \rho g_i \tag{9.2}$$

where, as before, ρ is the mixture density given by Eq. (1.8). As in single-phase flows the existence of a barotropic relation, $p = f(\rho)$, would complete the system of equations. In some multiphase flows it is possible to establish such a barotropic relation, and this allows one to anticipate (with, perhaps, some minor modification) that the entire spectrum of phenomena observed in single-phase gas dynamics can be expected in such a two-phase flow. In this chapter we do not dwell on this established body of literature. Rather, attention is confined to the identification of a barotropic relation (if any) and focused on some flows in which there are major departures from the conventional gas dynamic behavior.

From a thermodynamic point of view the existence of a barotropic relation, $p = f(\rho)$, and its associated sonic speed,

$$c = \left(\frac{dp}{d\rho}\right)^{\frac{1}{2}}, \tag{9.3}$$

implies that some thermodynamic property is considered to be held constant. In single-phase gas dynamics this quantity is usually the entropy or, occasionally, the temperature. In multiphase flows the alternatives are neither simple nor obvious. In single-phase gas dynamics it is commonly assumed that the gas is in thermodynamic equilibrium at all times. In multiphase flows it is usually the case that the two phases are *not* in thermodynamic equilibrium with each other. These are some of the questions that must be addressed in considering an appropriate homogeneous flow model for a multiphase flow. We begin in the next section by considering the sonic speed of a two-phase or two-component mixture.

9.3 Sonic Speed

9.3.1 Basic Analysis

Consider an infinitesimal volume of a mixture consisting of a disperse phase denoted by the subscript A and a continuous phase denoted by the subscript B. For convenience assume the initial volume to be unity. Denote the initial densities by ρ_A and ρ_B and the initial pressure in the *continuous* phase by p_B. Surface tension, S, can be included by denoting the radius of the disperse phase particles by R. Then the initial pressure in the disperse phase is $p_A = p_B + 2S/R$.

Now consider that the pressure, p_A, is changed to $p_A + \delta p_A$ where the difference δp_A is infinitesimal. Any dynamics associated with the resulting fluid motions will be ignored for the moment. It is assumed that a new equilibrium state is achieved and that, in the process, a mass, δm, is transferred from the continuous to the disperse phase. It follows that the new disperse and continuous phase masses are $\rho_A \alpha_A + \delta m$ and $\rho_B \alpha_B - \delta m$ respectively where, of course, $\alpha_B = 1 - \alpha_A$. Hence the new disperse and continuous phase volumes are respectively

$$(\rho_A \alpha_A + \delta m)/\left[\rho_A + \left.\frac{\partial \rho_A}{\partial p_A}\right|_{QA} \delta p_A\right] \tag{9.4}$$

and

$$(\rho_B \alpha_B - \delta m)/\left[\rho_B + \left.\frac{\partial \rho_B}{\partial p_B}\right|_{QB} \delta p_B\right], \tag{9.5}$$

where the thermodynamic constraints QA and QB are, as yet, unspecified. Adding these together and subtracting unity, one obtains the change in total volume, δV, and hence the sonic velocity, c, as follows:

$$c^{-2} = -\rho \left. \frac{\delta V}{\delta p_B} \right|_{\delta p_B \to 0} \tag{9.6}$$

$$c^{-2} = \rho \left[\frac{\alpha_A}{\rho_A} \left. \frac{\partial \rho_A}{\partial p_A} \right|_{QA} \frac{\delta p_A}{\delta p_B} + \frac{\alpha_B}{\rho_B} \left. \frac{\partial \rho_B}{\partial p_B} \right|_{QB} - \frac{(\rho_B - \rho_A)}{\rho_A \rho_B} \frac{\delta m}{\delta p_B} \right]. \tag{9.7}$$

If it is assumed that no disperse particles are created or destroyed, then the ratio $\delta p_A/\delta p_B$ may be determined by evaluating the new disperse particle size $R + \delta R$ commensurate with the new disperse phase volume and using the relation $\delta p_A = \delta p_B - \frac{2S}{R^2}\delta R$:

$$\frac{\delta p_A}{\delta p_B} = \left[1 - \frac{2S}{3\alpha_A \rho_A R} \frac{\delta m}{\delta p_B} \right] \Big/ \left[1 - \frac{2S}{3\rho_A R} \left. \frac{\partial \rho_A}{\partial p_A} \right|_{QA} \right]. \tag{9.8}$$

Substituting this into Eq. (9.7) and using, for convenience, the notation

$$\frac{1}{c_A^2} = \left. \frac{\partial \rho_A}{\partial p_A} \right|_{QA}; \quad \frac{1}{c_B^2} = \left. \frac{\partial \rho_B}{\partial p_B} \right|_{QB}, \tag{9.9}$$

the result can be written as follows:

$$\frac{1}{\rho c^2} = \frac{\alpha_B}{\rho_B c_B^2} + \frac{\left[\frac{\alpha_A}{\rho_A c_A^2} - \frac{\delta m}{\delta p_B} \left\{ \frac{1}{\rho_A} - \frac{1}{\rho_B} + \frac{2S}{3\rho_A \rho_B c_A^2 R} \right\} \right]}{\left[1 - \frac{2S}{3\rho_A c_A^2 R} \right]}. \tag{9.10}$$

This expression for the sonic speed, c, is incomplete in several respects. First, appropriate thermodynamic constraints QA and QB must be identified. Second, some additional constraint is necessary to establish the relation $\delta m/\delta p_B$. But before entering into a discussion of appropriate practical choices for these constraints (see Section 9.3.3) several simpler versions of Eq. (9.10) should be identified.

First, in the absence of any exchange of mass between the components the result [Eq. (9.10)] reduces to the following:

$$\frac{1}{\rho c^2} = \frac{\alpha_B}{\rho_B c_B^2} + \frac{\frac{\alpha_A}{\rho_A c_A^2}}{\left\{ 1 - \frac{2S}{3\rho_A c_A^2 R} \right\}}. \tag{9.11}$$

In most practical circumstances the surface tension effect can be neglected because $S \ll \rho_A c_A^2 R$; Eq. (9.11) then becomes the following:

$$\frac{1}{c^2} = \{\rho_A \alpha_A + \rho_B \alpha_B\} \left[\frac{\alpha_B}{\rho_B c_B^2} + \frac{\alpha_A}{\rho_A c_A^2} \right]. \tag{9.12}$$

9.3 Sonic Speed

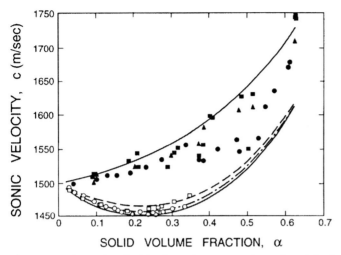

Figure 9.1. The sonic velocities for various suspensions of particles in water: ⊙, frequency of 100 kHz in a suspension of 1-μm Kaolin particles (Hampton 1967) ($2\kappa R = 6.6 \times 10^{-5}$); ◻, frequency of 1 MHz in a suspension of 0.5-μm Kaolin particles (Urick 1948) ($2\kappa R = 3.4 \times 10^{-4}$); solid symbols, frequencies of 100 kHz − 1 MHz in a suspension of 0.5 mm silica particles (Atkinson and Kytömaa 1992) ($2\kappa R = 0.2 − 0.6$). Lines are theoretical predictions for $2\kappa R = 0$, 6.6×10^{-5}, 3.4×10^{-4}, and $2\kappa R = 0.2 − 0.6$ in ascending order (from Atkinson and Kytömaa 1992).

In other words, the acoustic impedance for the mixture, namely $1/\rho c^2$, is given simply by the average of the acoustic impedance of the components weighted according to their volume fractions. Another popular way of expressing Eq. (9.12) is to recognize that ρc^2 is the effective bulk modulus of the mixture and that the inverse of this effective bulk modulus is equal to an average of the inverse bulk moduli of the components ($1/\rho_A c_A^2$ and $1/\rho_B c_B^2$) weighted according to their volume fractions.

Some typical experimental and theoretical data obtained by Hampton (1967), Urick (1948), and Atkinson and Kytömaa (1992) are presented in Figure 9.1. Each set is for a different ratio of the particle size (radius, R) to the wavelength of the sound (given by the inverse of the wavenumber, κ). Clearly the theory described above assumes a continuum and is therefore relevant to the limit $\kappa R \to 0$. The data in the figure shows good agreement with the theory in this low frequency limit. The changes that occur at higher frequency (larger κR) are discussed in the next section.

Perhaps the most dramatic effects occur when one of the components is a gas (subscript G), that is much more compressible than the other component (a liquid or solid, subscript L). In the absence of surface tension ($p = p_G = p_L$), according to Eq. (9.12), it matters not whether the gas is the continuous or the disperse phase. Denoting α_G by α for convenience and assuming the gas is perfect and behaves polytropically according to $\rho_G^k \propto p$, Eq. (9.12) may be written as follows:

$$\frac{1}{c^2} = [\rho_L(1-\alpha) + \rho_G \alpha]\left[\frac{\alpha}{kp} + \frac{(1-\alpha)}{\rho_L c_L^2}\right]. \tag{9.13}$$

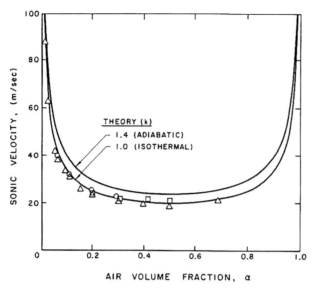

Figure 9.2. The sonic velocity in a bubbly air/water mixture at atmospheric pressure for $k = 1.0$ and 1.4. Experimental data presented are from Karplus (1958) and Gouse and Brown (1964) for frequencies of 1 kHz (○), 0.5 kHz (□), and extrapolated to zero frequency (△).

This is the familiar form for the sonic speed in a two-component gas/liquid or gas/solid flow. In many applications $p/\rho_L c_L^2 \ll 1$ and hence this expression may be further simplified to the following:

$$\frac{1}{c^2} = \frac{\alpha}{kp}[\rho_L(1-\alpha) + \rho_G \alpha]. \tag{9.14}$$

Note, however, that this approximation will not hold for small values of the gas volume fraction α.

Equation (9.13) and its special properties were first identified by Minnaert (1933). It clearly exhibits one of the most remarkable features of the sonic velocity of gas/liquid or gas/solid mixtures. The sonic velocity of the mixture can be very much smaller than that of either of its constituents. This is illustrated in Figure 9.2, where the speed of sound, c, in an air/water bubbly mixture is plotted against the air volume fraction, α. Results are shown for both isothermal ($k = 1$) and adiabatic ($k = 1.4$) bubble behavior using Eq. (9.13) or (9.14), the curves for these two equations being indistinguishable on the scale of the figure. Note that sonic velocities as low as 20 m/s occur.

Also shown in Figure 9.2 is experimental data of Karplus (1958) and Gouse and Brown (1964). Data for frequencies of 1.0 and 0.5 kHz are shown in Figure 9.2, as well as data extrapolated to zero frequency. The last should be compared with the low-frequency analytical results presented here. Note that the data correspond to the isothermal theory, indicating that the heat transfer between the bubbles and the liquid is sufficient to maintain the air in the bubbles at roughly constant temperature.

9.3 Sonic Speed

Further discussion of the acoustic characteristics of dusty gases is presented later in Section 11.4, where the effects of relative motion between the particles and the gas are included. Also, the acoustic characteristics of dilute bubbly mixtures are further discussed in Section 10.3, where the dynamic response of the bubbles are included in the analysis.

9.3.2 Sonic Speeds at Higher Frequencies

Several phenomena can lead to dispersion, that is to say, to an acoustic velocity that is a function of frequency. Among these are the effects of bubble dynamics discussed in the next chapter. Another is the change that occurs at higher frequencies as the wavelength is no longer effectively infinite relative to the size of the particles. Some experimental data on the effect of the ratio of particle size to wavelength (or κR) were presented in Figure 9.1. Note that the minimum in the acoustic velocity at intermediate volume fractions disappears at higher frequencies. Atkinson and Kytömaa (1992) modeled the dynamics at nonzero values of κR using the following set of governing equations: (a) continuity equations [Eq. (1.21)] for both the disperse and continuous phases with no mass exchange ($\mathcal{I}_N = 0$) (b) momentum equations [Eq. (1.45)] for both phases with no gravity terms and no deviatoric stresses $\sigma^D_{C\kappa i} = 0$ and (c) a particle force, F_k [see Eq. (1.55)] that includes the forces on each particle due to the pressure gradient in the continuous phase, the added mass, the Stokes drag and the Basset memory terms [see Section 2.3.4, Eq. (2.67)]. They included a solids fraction dependence in the added mass. The resulting dispersion relation yields sound speeds that depend on frequency, ω, and Reynolds number, $\rho_C \omega R^2 / \mu_C$, but asymptote to constant values at both high and low Reynolds numbers. Typical results are plotted in Figure 9.1 for various κR and exhibit fair agreement with the experimental measurements.

Atkinson and Kytömaa (1992) also compare measured and calculated acoustic attenuation rates given nondimensionally by ζR where the amplitude decays with distance, s, according to $e^{-\zeta s}$. The attenuation results from viscous effects on the relative motion between the particles and the continuous fluid phase. At low frequencies the relative motion and therefore the attenuation is dominated by the contribution from the Stokes drag term in Eq. (2.67); this term is proportional to ω^2. Though the measured data on attenuation is quite scattered, the theory yields values of the dimensionless attenuation, ζR, that are roughly of the correct magnitude as shown by the example in Figure 9.3. Conversely, at high frequencies (large κR) the theoretical attenuation is dominated by the Basset term and is proportional to $(\mu_C \omega)^{\frac{1}{2}}$; it also increases nearly linearly with the solids fraction. However, the measured attenuation rates in this frequency range appear to be about an order of magnitude larger than those calculated.

Weir (2001), following on the work of Gregor and Rumpf (1975), uses a similar perturbation analysis with somewhat different basic equations to generate dispersion relations as a function of frequency and volume fraction. Acknowledging that solutions of this dispersion relation yield a number of propagation velocities, including

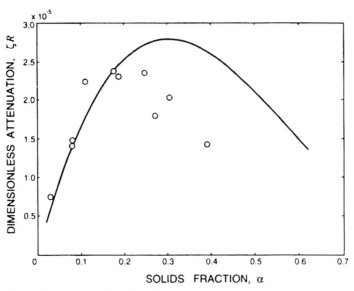

Figure 9.3. An example of the dimensionless attenuation, ζR, at low frequencies as a function of solids fraction, α. The experimental data (○) are for a suspension of Kaolin particles in water with $2\kappa R = 3.4 \times 10^{-4}$ (Urick 1948); the theoretical line is from Atkinson and Kytömaa (1992).

both kinematic and dynamic wave speeds (see Section 15.7.3), Weir chooses to focus on the dynamic or acoustic waves. He demonstrates that, in general, there are two types of dynamic wave. These have the same kinds of high- and low-frequency asymptotes described above. The two low-frequency wave speeds converge to yield a single dynamic wave speed that has a functional dependence on frequency and α that is qualitatively similar to that of Atkinson and Kytömaa (1992). It also agrees well with the measured sound speeds in Musmarra et al. (1995) for suspensions of various types of particles in liquid. Weir also analyzes the wave speeds in fluidized beds and compares them with those in unfluidized or static beds. He also examines the data on wave attenuation; as with the other attenuation data the experimental measurements are quite scattered and do not agree well with the theoretical predictions, particularly at high frequencies.

9.3.3 Sonic Speed with Change of Phase

Turning now to the behavior of a two-phase rather than two-component mixture, it is necessary not only to consider the additional thermodynamic constraint required to establish the mass exchange, δm, but also to reconsider the two thermodynamic constraints, QA and QB, that were implicit in the two-component analysis of Section 9.3.1, in the choice of the polytropic index, k, for the gas and the choice of the sonic speed, c_L, for the liquid. Note that a nonisentropic choice for k (for example, $k = 1$) implies that heat is exchanged between the components, and yet this heat transfer process was not explicitly considered, nor was an overall thermodynamic constraint such as might be placed on the global change in entropy.

9.3 Sonic Speed

We shall see that the two-phase case requires more intimate knowledge of these factors because the results are more sensitive to the thermodynamic constraints. In an ideal, infinitely homogenized mixture of vapor and liquid the phases would everywhere be in such close proximity to each other that heat transfer between the phases would occur instantaneously. The entire mixture of vapor and liquid would then always be in thermodynamic equilibrium. Indeed, one model of the response of the mixture, called the *homogeneous equilibrium model*, assumes this to be the case. In practice, however, there is a need for results for bubbly flows and mist flows in which heat transfer between the phases does not occur so readily. A second common model assumes zero heat transfer between the phases and is known as the *homogeneous frozen model*. In many circumstances the actual response lies somewhere between these extremes. A limited amount of heat transfer occurs between those portions of each phase that are close to the interface. To incorporate this in the analysis, we adopt an approach that includes the homogeneous equilibrium and homogeneous frozen responses as special cases but that requires a minor adjustment to the analysis of Section 9.3.1 to reflect the degree of thermal exchange between the phases. As in Section 9.3.1 the total mass of the phases A and B after application of the incremental pressure, δp, are $\rho_A \alpha_A + \delta m$ and $\rho_B \alpha_B - \delta m$, respectively. We now define the fractions of each phase, ϵ_A and ϵ_B, that, because of their proximity to the interface, exchange heat and therefore approach thermodynamic equilibrium with each other. The other fractions, $(1 - \epsilon_A)$ and $(1 - \epsilon_B)$, are assumed to be effectively insulated so that they behave isentropically. This is, of course, a crude simplification of the actual circumstances, but it permits qualitative assessment of practical flows.

It follows that the volumes of the four fractions following the incremental change in pressure, δp, are as follows:

$$\frac{(1 - \epsilon_A)(\rho_A \alpha_A + \delta m)}{[\rho_A + \delta p(\partial \rho_A/\partial p)_s]}; \quad \frac{\epsilon_A(\rho_A \alpha_A + \delta m)}{[\rho_A + \delta p(\partial \rho_A/\partial p)_e]}$$
$$\frac{(1 - \epsilon_B)(\rho_B \alpha_B - \delta m)}{[\rho_B + \delta p(\partial \rho_B/\partial p)_s]}; \quad \frac{\epsilon_B(\rho_B \alpha_B - \delta m)}{[\rho_B + \delta p(\partial \rho_B/\partial p)_e]}, \quad (9.15)$$

where the subscripts s and e refer to isentropic and phase equilibrium derivatives, respectively. Then the change in total volume leads to the following modified form for Eq. (9.10) in the absence of surface tension:

$$\frac{1}{\rho c^2} = (1 - \epsilon_A)\frac{\alpha_A}{\rho_A}\left(\frac{\partial \rho_A}{\partial p}\right)_s + \epsilon_A \frac{\alpha_A}{\rho_A}\left(\frac{\partial \rho_A}{\partial p}\right)_e + (1 - \epsilon_B)\frac{\alpha_B}{\rho_B}\left(\frac{\partial \rho_B}{\partial p}\right)_s$$
$$+ \epsilon_B \frac{\alpha_B}{\rho_B}\left(\frac{\partial \rho_B}{\partial p}\right)_e - \frac{\delta m}{\delta p}\left(\frac{1}{\rho_A} - \frac{1}{\rho_B}\right). \quad (9.16)$$

The exchange of mass, δm, is now determined by imposing the constraint that the entropy of the whole be unchanged by the perturbation. The entropy prior to δp is as follows:

$$\rho_A \alpha_A s_A + \rho_B \alpha_B s_B. \quad (9.17)$$

where s_A and s_B are the specific entropies of the two phases. Following the application of δp, the entropy is as follows:

$$(1 - \epsilon_A)\{\rho_A\alpha_A + \delta m\}s_A + \epsilon_A\{\rho_A\alpha_A + \delta m\}\{s_A + \delta p(\partial s_A/\partial p)_e\}$$
$$+ (1 - \epsilon_B)\{\rho_B\alpha_B - \delta m\}s_B + \epsilon_B\{\rho_B\alpha_B - \delta m\}\{s_B + \delta p(\partial s_B/\partial p)_e\}. \quad (9.18)$$

Equating Eqs. (9.17) and (9.18) and writing the result in terms of the specific enthalpies h_A and h_B rather than s_A and s_B, one obtains the following:

$$\frac{\delta m}{\delta p} = \frac{1}{(h_A - h_B)}\left[\epsilon_A\alpha_A\left\{1 - \rho_A\left(\frac{\partial h_A}{\partial p}\right)_e\right\} + \epsilon_B\alpha_B\left\{1 - \rho_B\left(\frac{\partial h_B}{\partial p}\right)_e\right\}\right]. \quad (9.19)$$

Note that if the communicating fractions ϵ_A and ϵ_B were both zero, this would imply no exchange of mass. Thus $\epsilon_A = \epsilon_B = 0$ corresponds to the homogeneous frozen model (in which $\delta m = 0$), whereas $\epsilon_A = \epsilon_B = 1$ clearly yields the homogeneous equilibrium model.

Substituting Eq. (9.19) into Eq. (9.16) and rearranging the result, one can write the following:

$$\frac{1}{\rho c^2} = \frac{\alpha_A}{p}[(1 - \epsilon_A)f_A + \epsilon_A g_A] + \frac{\alpha_B}{p}[(1 - \epsilon_B)f_B + \epsilon_B g_B], \quad (9.20)$$

where the quantities f_A, f_B, g_A, and g_B are purely thermodynamic properties of the two phases defined by the following:

$$f_A = \left(\frac{\partial \ln \rho_A}{\partial \ln p}\right)_s; \quad f_B = \left(\frac{\partial \ln \rho_B}{\partial \ln p}\right)_s \quad (9.21)$$

$$g_A = \left(\frac{\partial \ln \rho_A}{\partial \ln p}\right)_e + \left(\frac{1}{\rho_A} - \frac{1}{\rho_B}\right)\left(\rho_A h_A \frac{\partial \ln h_A}{\partial \ln p} - p\right)_e \bigg/ (h_A - h_B)$$

$$g_B = \left(\frac{\partial \ln \rho_B}{\partial \ln p}\right)_e + \left(\frac{1}{\rho_A} - \frac{1}{\rho_B}\right)\left(\rho_B h_B \frac{\partial \ln h_B}{\partial \ln p} - p\right)_e \bigg/ (h_A - h_B).$$

The sensitivity of the results to the, as yet, unspecified quantities ϵ_A and ϵ_B does not emerge until one substitutes vapor and liquid for the phases A and B (A = V, B = L, and $\alpha_A = \alpha$, $\alpha_B = 1 - \alpha$ for simplicity). The functions f_L, f_V, g_L, and g_V then become the following:

$$f_V = \left(\frac{\partial \ln \rho_V}{\partial \ln p}\right)_s; \quad f_L = \left(\frac{\partial \ln \rho_L}{\partial \ln p}\right)_s$$

$$g_V = \left(\frac{\partial \ln \rho_V}{\partial \ln p}\right)_e + \left(1 - \frac{\rho_V}{\rho_L}\right)\left(\frac{h_L}{\mathcal{L}}\frac{\partial \ln h_L}{\partial \ln p} + \frac{\partial \ln \mathcal{L}}{\partial \ln p} - \frac{p}{\mathcal{L}\rho_V}\right)_e$$

$$g_L = \left(\frac{\partial \ln \rho_L}{\partial \ln p}\right)_e + \left(\frac{\rho_L}{\rho_V} - 1\right)\left(\frac{h_L}{\mathcal{L}}\frac{\partial \ln h_L}{\partial \ln p} - \frac{p}{\mathcal{L}\rho_L}\right)_e, \quad (9.22)$$

9.3 Sonic Speed

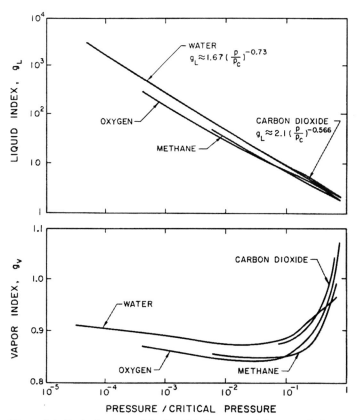

Figure 9.4. Typical values of the liquid index, g_L, and the vapor index, g_V, for various fluids.

where $\mathcal{L} = h_V - h_L$ is the latent heat. It is normally adequate to approximate f_V and f_L by the reciprocal of the ratio of specific heats for the gas and zero respectively. Thus f_V is of the order of unity and f_L is very small. Furthermore, g_L and g_V can readily be calculated for any fluid as functions of pressure or temperature. Some particular values are shown in Figure 9.4. Note that g_V is close to unity for most fluids except in the neighborhood of the critical point. Conversely, g_L can be a large number that varies considerably with pressure. To a first approximation, g_L is given by $g^*(p_C/p)^\eta$, where p_C is the critical pressure and, as indicated in Figure 9.4, g^* and η are respectively 1.67 and 0.73 for water. Thus, in summary, $f_L \approx 0$, f_V and g_V are of the order of unity, and g_L varies significantly with pressure and may be large.

With these magnitudes in mind, we now examine the sensitivity of $1/\rho c^2$ to the interacting fluid fractions ϵ_L and ϵ_V as follows:

$$\frac{1}{\rho c^2} = \frac{\alpha}{p}[(1-\epsilon_V)f_V + \epsilon_V g_V] + \frac{(1-\alpha)}{p}\epsilon_L g_L. \qquad (9.23)$$

Using $g_L = g^*(p_c/p)^\eta$ this is written for future convenience in the following form:

$$\frac{1}{\rho c^2} = \frac{\alpha k_V}{p} + \frac{(1-\alpha)k_L}{p^{1+\eta}}, \qquad (9.24)$$

where $k_V = (1 - \epsilon_V)f_V + \epsilon_V g_V$ and $k_L = \epsilon_L g^*(p_c)^\eta$. Note first that the result is rather insensitive to ϵ_V because f_V and g_V are both of the order of unity. Conversely, $1/\rho c^2$ is sensitive to the interacting liquid fraction ϵ_L, although this sensitivity disappears as α approaches 1; in other words, for mist flow. Thus the choice of ϵ_L is most important at low-vapor-volume fractions (for bubbly flows). In such cases, one possible qualitative estimate is that the interacting liquid fraction, ϵ_L, should be of the same order as the gas volume fraction, α. In Section 9.5.2 we examine the effect of the choice of ϵ_L and ϵ_V on a typical vapor/liquid flow and compare the model with experimental measurements.

9.4 Barotropic Relations

Conceptually, the expressions for the sonic velocity, Eqs. (9.12), (9.13), (9.14), and (9.23), need only be integrated (after substituting $c^2 = dp/d\rho$) to obtain the barotropic relation, $p(\rho)$, for the mixture. In practice this is algebraically complicated except for some of the simpler forms for c^2.

Consider first the case of the two-component mixture in the absence of mass exchange or surface tension as given by Eq. (9.13). It will initially be assumed that the gas volume fraction is not too small so that Eq. (9.14) can be used; we return later to the case of small gas volume fraction. It is also assumed that the liquid or solid density, ρ_L, is constant and that $p \propto \rho_G^\kappa$. Furthermore, it is convenient, as in gas dynamics, to choose reservoir conditions, $p = p_o$, $\alpha = \alpha_o$, $\rho_G = \rho_{Go}$, to establish the integration constants. Then it follows from the integration of Eq. (9.14) that

$$\rho = \rho_o(1-\alpha)/(1-\alpha_o) \tag{9.25}$$

and that

$$\frac{p}{p_o} = \left[\frac{\alpha_o(1-\alpha)}{(1-\alpha_o)\alpha}\right]^k = \left[\frac{\alpha_o \rho}{\rho_o - (1-\alpha_o)\rho}\right]^k, \tag{9.26}$$

where $\rho_o = \rho_L(1-\alpha_o) + \rho_{Go}\alpha_o$. It also follows that, written in terms of α,

$$c^2 = \frac{kp_o}{\rho_o}\frac{(1-\alpha)^{k-1}}{\alpha^{k+1}}\frac{\alpha_o^k}{(1-\alpha_o)^{k-1}}. \tag{9.27}$$

As discussed later, Tangren, Dodge, and Seifert (1949) first made use of a more limited form of the barotropic relation of Eq. (9.26) to evaluate the one-dimensional flow of gas/liquid mixtures in ducts and nozzles.

In the case of very small gas volume fractions, α, it may be necessary to include the liquid compressibility term, $1 - \alpha/\rho_L c_L^2$, in Eq. (9.13). Exact integration then becomes very complicated. However, it is sufficiently accurate at small gas volume fractions to approximate the mixture density ρ by $\rho_L(1-\alpha)$. Integration (assuming $\rho_L c_L^2 = $ constant) then yields the following:

$$\frac{\alpha}{(1-\alpha)} = \left[\frac{\alpha_o}{(1-\alpha_o)} + \frac{k}{k+1}\frac{p_o}{\rho_L c_L^2}\right]\left(\frac{p_o}{p}\right)^{\frac{1}{k}} - \frac{k}{(k+1)}\frac{p_o}{\rho_L c_L^2}\frac{p}{p_o} \quad (9.28)$$

and the sonic velocity can be expressed in terms of p/p_o alone by using Eq. (9.28) and noting that

$$c^2 = \frac{p}{\rho_L}\frac{\left[1 + \frac{\alpha}{(1-\alpha)}\right]^2}{\left[\frac{1}{k}\frac{\alpha}{(1-\alpha)} + \frac{p}{\rho_L c_L^2}\right]}. \quad (9.29)$$

Implicit within Eq. (9.28) is the barotropic relation, $p(\alpha)$, analogous to Eq. (9.26). Note that Eq. (9.28) reduces to Eq. (9.26) when $p_o/\rho_L c_L^2$ is set equal to zero. Indeed, it is clear from Eq. (9.28) that the liquid compressibility has a negligible effect only if $\alpha_o \gg p_o/\rho_L c_L^2$. This parameter, $p_o/\rho_L c_L^2$, is usually quite small. For example, for saturated water at 5×10^7 kg/ms² (500 psi) the value of $p_o/\rho_L c_L^2$ is approximately 0.03. Nevertheless, there are many practical problems in which one is concerned with the discharge of a predominantly liquid medium from high-pressure containers, and under these circumstances it can be important to include the liquid compressibility effects.

Now turning attention to a two-phase rather than two-component homogeneous mixture, the particular form of the sonic velocity given in Eq. (9.24) may be integrated to yield the implicit barotropic relation

$$\frac{\alpha}{1-\alpha} = \left[\frac{\alpha_o}{(1-\alpha_o)} + \frac{k_L p_o^{-\eta}}{(k_V - \eta)}\right]\left(\frac{p_o}{p}\right)^{k_V} - \left[\frac{k_L p_o^{-\eta}}{(k_V - \eta)}\right]\left(\frac{p_o}{p}\right)^{\eta}, \quad (9.30)$$

in which the approximation $\rho \approx \rho_L(1-\alpha)$ has been used. As before, c^2 may be expressed in terms of p/p_o alone by noting that

$$c^2 = \frac{p}{\rho_L}\frac{\left[1 + \frac{\alpha}{(1-\alpha)}\right]^2}{\left[k_V \frac{\alpha}{(1-\alpha)} + k_L p^{-\eta}\right]}. \quad (9.31)$$

Finally, we note that close to $\alpha = 1$ Eqs. (9.30) and (9.31) may fail because the approximation $\rho \approx \rho_L(1-\alpha)$ is not sufficiently accurate.

9.5 Nozzle Flows

9.5.1 One-Dimensional Analysis

The barotropic relations of the last section can be used in conjunction with the steady, one-dimensional continuity and frictionless momentum equations

$$\frac{d}{ds}(\rho A u) = 0 \quad (9.32)$$

and
$$u\frac{du}{ds} = -\frac{1}{\rho}\frac{dp}{ds} \qquad (9.33)$$

to synthesize homogeneous multiphase flow in ducts and nozzles. The predicted phenomena are qualitatively similar to those in one-dimensional gas dynamics. The results for isothermal, two-component flow were first detailed by Tangren, Dodge, and Seifert (1949); more general results for any polytropic index are given in this section.

Using the barotropic relation given by Eq. (9.26) and Eq. (9.25) for the mixture density, ρ, to eliminate p and ρ from the momentum Eq. (9.33), one obtains the following:

$$u\,du = \frac{kp_o}{\rho_o}\frac{\alpha_o^k}{(1-\alpha_o)^{k-1}}\frac{(1-\alpha)^{k-2}}{\alpha^{k+1}}d\alpha, \qquad (9.34)$$

which upon integration and imposition of the reservoir condition, $u_o = 0$, yields the following:

$$u^2 = \frac{2kp_o}{\rho_o}\frac{\alpha_o^k}{(1-\alpha_o)^{k-1}}\left[\frac{1}{k}\left\{\left(\frac{1-\alpha_o}{\alpha_o}\right)^k - \left(\frac{1-\alpha}{\alpha}\right)^k\right\}\right.$$

$$+ \text{either } \frac{1}{(k-1)}\left\{\left(\frac{1-\alpha_o}{\alpha_o}\right)^{k-1} - \left(\frac{1-\alpha}{\alpha}\right)^{k-1}\right\}\right] \quad \text{if } k \neq 1$$

$$\text{or } \ln\left\{\frac{(1-\alpha_o)\alpha}{\alpha_o(1-\alpha)}\right\} \quad \text{if } k = 1. \qquad (9.35)$$

Given the reservoir conditions p_o and α_o as well as the polytropic index k and the liquid density (assumed constant), this relates the velocity, u, at any position in the duct to the gas volume fraction, α, at that location. The pressure, p, density, ρ, and volume fraction, α, are related by Eqs. (9.25) and (9.26). The continuity equation,

$$A = \text{constant}/\rho u = \text{constant}/u(1-\alpha), \qquad (9.36)$$

completes the system of equations by permitting identification of the location where p, ρ, u, and α occur from knowledge of the cross-sectional area, A.

As in gas dynamics the conditions at a throat play a particular role in determining both the overall flow and the mass flow rate. This results from the observation that Eqs. (9.32) and (9.30) may be combined to obtain the following:

$$\frac{1}{A}\frac{dA}{ds} = \frac{1}{\rho}\frac{dp}{ds}\left(\frac{1}{u^2} - \frac{1}{c^2}\right), \qquad (9.37)$$

where $c^2 = dp/d\rho$. Hence at a throat where $dA/ds = 0$, either $dp/ds = 0$, which is true when the flow is entirely subsonic and unchoked, or $u = c$, which is true when the flow is choked. Denoting choked conditions at a throat by the subscript $*$, it follows by equating the right-hand sides of Eqs. (9.27) and (9.35) that the gas volume fraction

9.5 Nozzle Flows

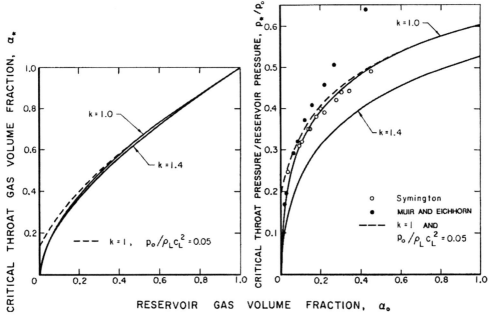

Figure 9.5. Critical or choked flow throat characteristics for the flow of a two-component gas/liquid mixture through a nozzle. On the left is the throat gas volume fraction as a function of the reservoir gas volume fraction, α_o, for gas polytropic indices of $k = 1.0$ and 1.4 and an incompressible liquid (solid lines) and for $k = 1$ and a compressible liquid with $p_o/\rho_L c_L^2 = 0.05$ (dashed line). On the right are the corresponding ratios of critical throat pressure to reservoir pressure. Also shown is the experimental data of Symington (1978) and Muir and Eichhorn (1963).

at the throat, α_*, must be given when $k \neq 1$ by the solution of

$$\frac{(1-\alpha_*)^{k-1}}{2\alpha_*^{k+1}} = \frac{1}{k}\left\{\left(\frac{1-\alpha_o}{\alpha_o}\right)^k - \left(\frac{1-\alpha_*}{\alpha_*}\right)^k\right\}$$
$$+ \frac{1}{(k-1)}\left\{\left(\frac{1-\alpha_o}{\alpha_o}\right)^{k-1} - \left(\frac{1-\alpha_*}{\alpha_*}\right)^{k-1}\right\} \quad (9.38)$$

or, in the case of isothermal gas behavior ($k = 1$), by the solution of the following:

$$\frac{1}{2\alpha_*^2} = \frac{1}{\alpha_o} - \frac{1}{\alpha_*} + \ln\left\{\frac{(1-\alpha_o)\alpha_*}{\alpha_o(1-\alpha_*)}\right\}. \quad (9.39)$$

Thus the throat gas volume fraction, α_*, under choked flow conditions is a function only of the reservoir gas volume fraction, α_o, and the polytropic index. Solutions of Eqs. (9.38) and (9.39) for two typical cases, $k = 1.4$ and $k = 1.0$, are shown in Figure 9.5. The corresponding ratio of the choked throat pressure, p_*, to the reservoir pressure, p_o, follows immediately from Eq. (9.26) given $\alpha = \alpha_*$ and is also shown in Figure 9.5. Finally, the choked mass flow rate, $\dot{m}$, follows as $\rho_* A_* c_*$ where A_* is the

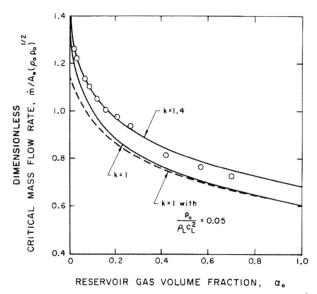

Figure 9.6. Dimensionless critical mass flow rate, $\dot{m}/A_*(p_0\rho_0)^{\frac{1}{2}}$, as a function of α_0 for choked flow of a gas/liquid flow through a nozzle. Solid lines are incompressible liquid results for polytropic indices of 1.4 and 1.0. Dashed line shows effect of liquid compressibility for $p_0/\rho_L c_L^2 = 0.05$. The experimental data (⊙) are from Muir and Eichhorn (1963).

cross-sectional area of the throat and

$$\frac{\dot{m}}{A_*(p_0\rho_0)^{\frac{1}{2}}} = k^{\frac{1}{2}} \frac{\alpha_0^{\frac{k}{2}}}{(1-\alpha_0)^{\frac{k+1}{2}}} \left(\frac{1-\alpha_*}{\alpha_*}\right)^{\frac{k+1}{2}}. \quad (9.40)$$

This dimensionless choked mass flow rate is exhibited in Figure 9.6 for $k = 1.4$ and $k = 1$.

Data from the experiments of Symington (1978) and Muir and Eichhorn (1963) are included in Figures 9.5 and 9.6. Symington's data on the critical pressure ratio (Figure 9.5) is in good agreement with the isothermal ($k = 1$) analysis indicating that, at least in his experiments, the heat transfer between the bubbles and the liquid is large enough to maintain constant gas temperature in the bubbles. Conversely, the experiments of Muir and Eichhorn yielded larger critical pressure ratios and flow rates than the isothermal theory. However, Muir and Eichhorn measured significant slip between the bubbles and the liquid (strictly speaking the abscissa for their data in Figures 9.5 and 9.6 should be the upstream volumetric quality rather than the void fraction), and the discrepancy could be due to the errors introduced into the present analysis by the neglect of possible relative motion (see also van Wijngaarden 1972).

Finally, the pressure, volume fraction, and velocity elsewhere in the duct or nozzle can be related to the throat conditions and the ratio of the area, A, to the throat area, A_*. These relations, which are presented in Figures 9.7 and 9.8 for the case $k = 1$ and various reservoir volume fractions, α_0, are most readily obtained in the following manner. Given α_0 and k, p_*/p_0 and α_* follow from Figure 9.5. Then for p/p_0 or p/p_*,

9.5 Nozzle Flows

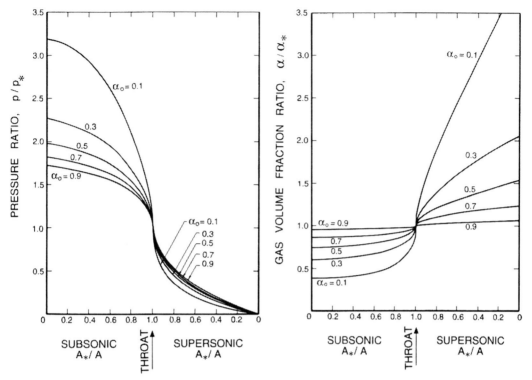

Figure 9.7. (Left) Ratio of the pressure, p, to the throat pressure, p_*, and (right) ratio of the void fraction, α, to the throat void fraction, α_*, for two-component flow in a duct with isothermal gas behavior.

Figure 9.8. Ratio of the velocity, u, to the throat velocity, u_*, for two-component flow in a duct with isothermal gas behavior.

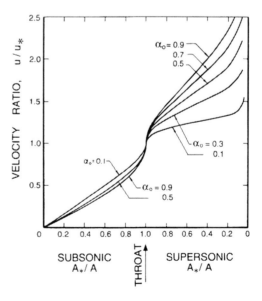

α and u follow from Eqs. (9.26) and (9.35) and the corresponding A/A_* follows by using Eq. (9.36). The resulting charts, Figures 9.7 and 9.8, can then be used in the same way as the corresponding graphs in gas dynamics.

If the gas volume fraction, α_o, is sufficiently small so that it is comparable with $p_o/\rho_L c_L^2$, then the barotropic Eq. (9.28) should be used instead of Eq. (9.26). In cases like this in which it is sufficient to assume that $\rho \approx \rho_L(1-\alpha)$, integration of the momentum Eq. (9.33) is most readily accomplished by writing it in the following form:

$$\frac{\rho_L}{p_o}\frac{u^2}{2} = 1 - \frac{p}{p_o} + \int_{p/p_o}^{1} \left(\frac{\alpha}{1-\alpha}\right) d\left(\frac{p}{p_o}\right). \qquad (9.41)$$

Then substitution of Eq. (9.28) for $\alpha/(1-\alpha)$ leads in the present case to the following:

$$u^2 = \frac{2p_o}{\rho_L}\left[1 - \frac{p}{p_o} + \frac{k}{2(k+1)}\frac{p_o}{\rho_L c_L^2}\left\{\frac{p^2}{p_o^2} - 1\right\}\right.$$

$$+ \text{either} \quad \left.\frac{k}{(k-1)}\left\{\frac{\alpha_o}{1-\alpha_o} + \frac{k}{(k+1)}\frac{p_o}{\rho_L c_L^2}\right\}\left\{1 - \left(\frac{p}{p_o}\right)^{\frac{k-1}{k}}\right\}\right] \quad \text{for } k \neq 1$$

$$\text{or} \quad \left.\left\{\frac{\alpha_o}{1-\alpha_o} + \frac{1}{2}\frac{p_o}{\rho_L c_L^2}\right\}\ln\left(\frac{p_o}{p}\right)\right] \quad \text{for } k = 1. \qquad (9.42)$$

The throat pressure, p_* (or rather p_*/p_o), is then obtained by equating the velocity u for $p = p_*$ from Eq. (9.42) to the sonic velocity c at $p = p_*$ obtained from Eq. (9.29). The resulting relation, though algebraically complicated, is readily solved for the critical pressure ratio, p_*/p_o, and the throat gas volume fraction, α_*, follows from Eq. (9.28). Values of p_*/p_o for $k = 1$ and $k = 1.4$ are shown in Figure 9.5 for the particular value of $p_o/\rho_L c_L^2$ of 0.05. Note that the most significant deviations caused by liquid compressibility occur for gas volume fractions of the order of 0.05 or less. The corresponding dimensionless critical mass flow rates, $\dot{m}/A_*(\rho_o p_o)^{\frac{1}{2}}$, are also readily calculated from the following:

$$\frac{\dot{m}}{A_*(\rho_o p_o)^{\frac{1}{2}}} = \frac{(1-\alpha_*)c_*}{[p_o(1-\alpha_o)/\rho_L]^{\frac{1}{2}}} \qquad (9.43)$$

and sample results are shown in Figure 9.6.

9.5.2 Vapor/Liquid Nozzle Flow

A barotropic relation, Eq. (9.30), was constructed in Section 9.4 for the case of two-phase flow and, in particular, for vapor–liquid flow. This may be used to synthesize nozzle flows in a manner similar to the two-component analysis of the last section. Because the approximation $\rho \approx \rho_L(1-\alpha)$ was used in deriving both Eq. (9.30) and Eq. (9.41), we may eliminate $\alpha/(1-\alpha)$ from these equations to obtain the velocity,

9.5 Nozzle Flows

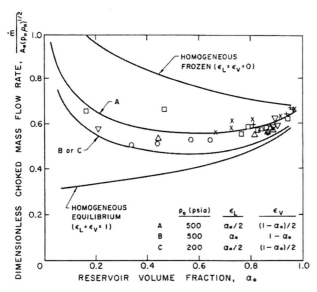

Figure 9.9. The dimensionless choked mass flow rate, $\dot{m}/A_*(p_o\rho_o)^{\frac{1}{2}}$, plotted against the reservoir vapor volume fraction, α_o, for water/steam mixtures. The data shown is from the experiments of Maneely (1962) and Neusen (1962) for $100 \to 200$ psia (+), $200 \to 300$ psia (×), $300 \to 400$ psia (□), $400 \to 500$ psia (Δ), $500 \to 600$ psia (∇), and > 600 psia (*). The theoretical lines use $g^* = 1.67$, $\eta = 0.73$, $g_V = 0.91$, and $f_V = 0.769$ for water.

u, in terms of p/p_o:

$$\frac{\rho_L}{p_o}\frac{u^2}{2} = 1 - \frac{p}{p_o} + \frac{1}{(1-k_V)}\left[\frac{\alpha_o}{(1-\alpha_o)} + \frac{k_L p_o^{-\eta}}{(k_V - \eta)}\right]\left[1 - \left(\frac{p}{p_o}\right)^{1-k_V}\right]$$
$$- \frac{1}{(1-\eta)}\left[\frac{k_L p_o^{-\eta}}{(k_V - \eta)}\right]\left[1 - \left(\frac{p}{p_o}\right)^{1-\eta}\right]. \qquad (9.44)$$

To find the relation for the critical pressure ratio, p_*/p_o, the velocity, u, must equated with the sonic velocity, c, as given by Eq. (9.31) as follows:

$$\frac{c^2}{2} = \frac{p}{\rho_L}\frac{\left[1 + \left\{\frac{\alpha_o}{1-\alpha_o} + k_L\frac{p_o^{-\eta}}{k_V-\eta}\right\}\left(\frac{p_o}{p}\right)^{k_V} - \left\{k_L\frac{p_o^{-\eta}}{(k_V-\eta)}\right\}\left(\frac{p_o}{p}\right)^{\eta}\right]^2}{2\left[k_V\left\{\frac{\alpha_o}{1-\alpha_o} + \frac{k_L p_o^{-\eta}}{(k_V-\eta)}\right\}\left(\frac{p_o}{p}\right)^{k_V} - \eta\left\{\frac{k_L p_o^{-\eta}}{(k_V-\eta)}\right\}\left(\frac{p_o}{p}\right)^{\eta}\right]}. \qquad (9.45)$$

Although algebraically complicated, the equation that results when the right-hand sides of Eqs. (9.44) and (9.45) are equated can readily be solved numerically to obtain the critical pressure ratio, p_*/p_o, for a given fluid and given values of α_o, the reservoir pressure and the interacting fluid fractions ϵ_L and ϵ_V (see Section 9.3.3). Having obtained the critical pressure ratio, the critical vapor volume fraction, α_*, follows from Eq. (9.30) and the throat velocity, c_*, from Eq. (9.45). Then the dimensionless choked mass flow rate follows from the same relation as given in Eq. (9.43).

Sample results for the choked mass flow rate and the critical pressure ratio are shown in Figures 9.9 and 9.10. Results for both homogeneous frozen flow

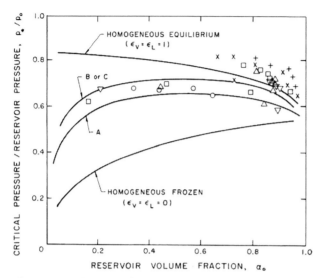

Figure 9.10. The ratio of critical pressure, p_*, to reservoir pressure, p_o, plotted against the reservoir vapor volume fraction, α_o, for water/steam mixtures. The data and the partially frozen model results are for the same conditions as in Figure 9.9.

($\epsilon_L = \epsilon_V = 0$) and for homogeneous equilibrium flow ($\epsilon_L = \epsilon_V = 1$) are presented; note that these results are independent of the fluid or the reservoir pressure, p_o. Also shown in the figures are the theoretical results for various partially frozen cases for water at two different reservoir pressures. The interacting fluid fractions were chosen with the comment at the end of Section 9.3.3 in mind. Because ϵ_L is most important at low-vapor-volume fractions (i.e., for bubbly flows), it is reasonable to estimate that the interacting volume of liquid surrounding each bubble will be of the same order as the bubble volume. Hence $\epsilon_L = \alpha_o$ or $\alpha_o/2$ are appropriate choices. Similarly, ϵ_V is most important at high-vapor-volume fractions (i.e., droplet flows), and it is reasonable to estimate that the interacting volume of vapor surrounding each droplet would be of the same order as the droplet volume; hence $\epsilon_V = (1 - \alpha_o)$ and $(1 - \alpha_o)/2$ are appropriate choices.

Figures 9.9 and 9.10 also include data obtained for water by Maneely (1962) and Neusen (1962) for various reservoir pressures and volume fractions. Note that the measured choked mass flow rates are bracketed by the homogeneous frozen and equilibrium curves and that the appropriately chosen partially frozen analysis is in close agreement with the experiments, despite the neglect (in the present model) of possible slip between the phases. The critical pressure ratio data is also in good agreement with the partially frozen analysis except for some discrepancy at the higher reservoir volume fractions.

It should be noted that the analytical approach described above is much simpler to implement than the numerical solution of the basic equations suggested by Henry and Fauske (1971). The latter does, however, have the advantage that slip between the phases was incorporated into the model.

9.5 Nozzle Flows

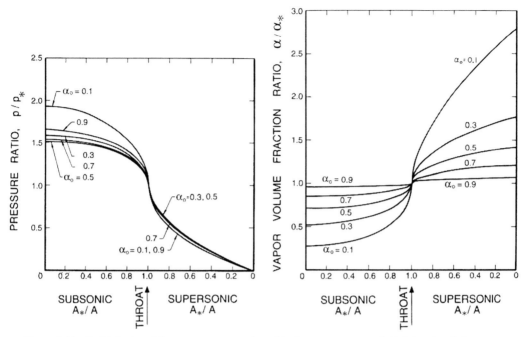

Figure 9.11. (Left) Ratio of the pressure, p, to the critical pressure, p_*, and (right) ratio of the vapor volume fraction, α, to the critical vapor volume fraction, α_*, as functions of the area ratio, A_*/A, for the case of water with $g^* = 1.67$, $\eta = 0.73$, $g_V = 0.91$, and $f_V = 0.769$.

Finally, information on the pressure, volume fraction, and velocity elsewhere in the duct (p/p_*, u/u_*, and α/α_*) as a function of the area ratio A/A_* follows from a procedure similar to that used for the noncondensable case in Section 9.5.1. Typical results for water with a reservoir pressure, p_0, of 500 psia and using the partially frozen analysis with $\epsilon_V = \alpha_0/2$ and $\epsilon_L = (1 - \alpha_0)/2$ are presented in Figures 9.11 and 9.12. In comparing these results with those for the two-component mixture (Figures 9.7 and 9.8) we observe that the pressure ratios are substantially smaller and do not vary monotonically with α_0. The volume fraction changes are smaller, whereas the velocity gradients are larger.

9.5.3 Condensation Shocks

In the preceding sections we investigated nozzle flows in which the two components or phases are present throughout the flow. However, there are also important circumstances in expanding supersonic gas or vapor flows in which the initial expansion is single phase but in which the expansion isentrope subsequently crosses the saturated vapor/liquid line as sketched in Figure 9.13. This can happen either in single-component vapor flows or in gas flows containing some vapor. The result is that liquid droplets form in the flow and this cloud of droplets downstream of nucleation is often visible in the flow. Because of their visibility these condensation fronts came to be

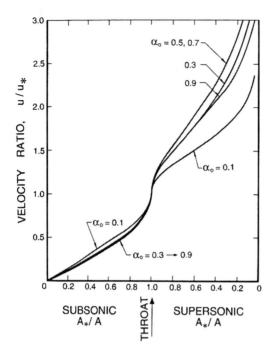

Figure 9.12. Ratio of the velocity, u, to the critical velocity, u_*, as a function of the area ratio for the same case as Figure 9.11.

called condensation *shocks* in the literature. They are not, however, shock waves for no shock wave processes are involved. Indeed, the term is quite misleading because in the flow condensation *fronts* occur during expansion rather than compression.

The detailed structure of condensation fronts and their effect on the overall flow depends on the nucleation dynamics and, as such, is outside the scope of this book. For detailed analyses, the reader is referred to the reviews of Wegener and Mack (1958) and Hill (1966). Unlike the inverse phenomenon of formation of vapor bubbles in a liquid flow (cavitation; see Section 5.2.1), the nucleation of liquid droplets during condensation is governed primarily by homogeneous nucleation rather than

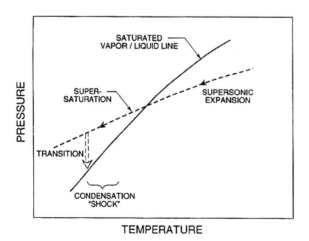

Figure 9.13. The occurence of condensation during expansion in a diffuser.

9.5 Nozzle Flows

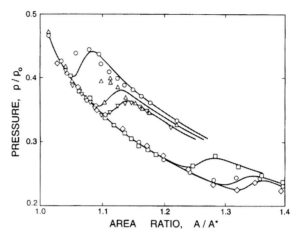

Figure 9.14. Experimental pressure profiles through condensation *fronts* in a diffuser for six different initial conditions. Also shown are the corresponding theoretical results. From Binnie and Green (1942) and Hill (1966).

heterogeneous nucleation on dust particles. In a typical steam expansion $10^{15}/\text{cm}^3$ nuclei are spontaneously formed; this contrasts with the maximum credible concentration of dust particles of about $10^8/\text{cm}^3$ and consequently homogeneous nucleation predominates.

Homogeneous nucleation and the growth of the droplets require time and therefore, as indicated in Figure 9.13, an interval of supersaturation occurs before the two-phase mixture adjusts back toward equilibrium saturated conditions. The rate of nucleation and the rate of growth of these droplets will vary with circumstances and may result in an abrupt or gradual departure from the isentrope and adjustment to saturated conditions. Also, it transpires that the primary effect on the flow is the heating of the

Figure 9.15. Condensation *fronts* in the flow around a transonic F/A-18 Hornet operating in humid conditions. U.S. Navy photograph by Ensign John Gay.

flow due to the release of the latent heat of vaporization inherent in the formation of the droplets (Hill 1966). Typical data on this adjustment process is shown in Figure 9.14 that includes experimental data on the departure from the initial isentrope for a series of six initial conditions. Also shown are the theoretical predictions using homogeneous nucleation theory.

For more recent work computing flows with condensation fronts the reader is referred, by way of example, to Delale *et al.* (1995). It also transpires that flows in diffusers with condensation fronts can generate instabilities that have no equivalent in single-phase flow (Adam and Schnerr 1997).

Condensation fronts occur in both internal and external flows and can often be seen when aircraft operate in humid conditions. Figure 9.15 is a classic photograph of a US Navy F/A-18 Hornet traveling at transonic speeds in which condensation fronts can be observed in the expansion around the cockpit cowling and downstream of the expansion in the flow around the wings. Moreover, the droplets can be seen to be reevaporated when they are compressed as they pass through the recompression shock at the trailing edge of the wings.

10

Flows with Bubble Dynamics

10.1 Introduction

In Chapter 9, the analyses were predicated on the existence of an effective barotropic relation for the homogeneous mixture. Indeed, the construction of the sonic speed in Sections 9.3.1 and 9.3.3 assumes that all the phases are in dynamic equilibrium at all times. For example, in the case of bubbles in liquids, it is assumed that the response of the bubbles to the change in pressure, δp, is an essentially instantaneous change in their volume. In practice this would be the case only if the typical frequencies experienced by the bubbles in the flow are very much smaller than the natural frequencies of the bubbles themselves (see Section 4.4.1). Under these circumstances the bubbles would behave quasistatically and the mixture would be barotropic. However, there are a number of important contexts in which the bubbles are not in equilibrium and in which the nonequilibrium effects have important consequences. One example is the response of a bubbly multiphase mixture to high-frequency excitation. Another is a bubbly cavitating flow where the nonequilibrium bubble dynamics lead to shock waves with substantial noise and damage potential.

In this chapter we therefore examine some flows in which the dynamics of the individual bubbles play an important role. These effects are included by incorporating the Rayleigh–Plesset equation (Rayleigh 1917, Knapp *et al*. 1970, Brennen 1995) into the global conservation equations for the multiphase flow. Consequently the mixture no longer behaves barotropically.

Viewing these flows from a different perspective, we note that analyses of cavitating flows often consist of using a single-phase liquid pressure distribution as input to the Rayleigh–Plesset equation. The result is the history of the size of individual cavitating bubbles as they progress along a streamline in the otherwise purely liquid flow. Such an approach entirely neglects the interactive effects that the cavitating bubbles have on themselves and on the pressure and velocity of the liquid flow. The analysis that follows incorporates these interactions using the equations for nonbarotropic homogeneous flow.

10.2 Basic Equations

In this chapter it is assumed that the ratio of liquid to vapor density is sufficiently large so that the volume of liquid evaporated or condensed is negligible. It is also assumed that bubbles are neither created or destroyed. Then the appropriate continuity equation is as follows:

$$\frac{\partial u_i}{\partial x_i} = \frac{\eta}{(1+\eta v)}\frac{Dv}{Dt}, \qquad (10.1)$$

where η is the population or number of bubbles per unit volume of liquid and $v(x_i, t)$ is the volume of individual bubbles. The above form of the continuity equation assumes that η is uniform; such would be the case if the flow originated from a uniform stream of uniform population and if there were no relative motion between the bubbles and the liquid. Note also that $\alpha = \eta v/(1+\eta v)$ and the mixture density, $\rho \approx \rho_L(1-\alpha) = \rho_L/(1+\eta v)$. This last relation can be used to write the momentum equation [Eq. (9.2)] in terms of v rather than ρ:

$$\rho_L \frac{Du_i}{Dt} = -(1+\eta v)\frac{\partial p}{\partial x_i}. \qquad (10.2)$$

The hydrostatic pressure gradient due to gravity has been omitted for simplicity.

Finally the Rayleigh–Plesset equation [Eq. (4.25)] relates the pressure, p, and the bubble volume, $v = \frac{4}{3}\pi R^3$:

$$R\frac{D^2R}{Dt^2} + \frac{3}{2}\left(\frac{DR}{Dt}\right)^2 = \frac{p_V - p}{\rho_L} + \frac{p_{Go}}{\rho_L}\left(\frac{R_o}{R}\right)^{3k} - \frac{2S}{\rho_L R} - \frac{4\nu_L}{R}\frac{DR}{Dt}, \qquad (10.3)$$

where it is assumed that the mass of gas in the bubble remains constant, p_V is the vapor pressure, p_{Go} is the partial pressure of noncondensable gas at some reference moment in time when $R = R_o$ and k is the polytropic index representing the behavior of the gas.

Equations (10.1), (10.2), and (10.3) can, in theory, be solved to find the unknowns $p(x_i, t)$, $u_i(x_i, t)$, and $v(x_i, t)$ (or $R(x_i, t)$) for any bubbly cavitating flow. In practice the nonlinearities in the Rayleigh–Plesset equation and in the Lagrangian derivative, $D/Dt = \partial/\partial t + u_i\partial/\partial x_i$, present serious difficulties for all flows except those of the simplest geometry. In the following sections several such flows are examined to illustrate the interactive effects of bubbles in cavitating flows and the role played by bubble dynamics in homogeneous flows.

10.3 Acoustics of Bubbly Mixtures

10.3.1 Analysis

One class of phenomena in which bubble dynamics can play an important role is the acoustics of bubble/liquid mixtures. When the acoustic excitation frequency

10.3 Acoustics of Bubbly Mixtures

approaches the natural frequency of the bubbles, the latter no longer respond in the quasistatic manner assumed in Chapter 9, and both the propagation speed and the acoustic attenuation are significantly altered. A review of this subject is given by van Wijngaarden (1972) and we include here only a summary of the key results. This class of problems has the advantage that the magnitude of the perturbations is small so that the equations of the preceding section can be greatly simplified by linearization. Hence the pressure, p, is represented by the following sum:

$$p = \bar{p} + \text{Re}\{\tilde{p}e^{i\omega t}\}, \tag{10.4}$$

where $\bar{p}$ is the mean pressure, ω is the frequency, and $\tilde{p}$ is the small amplitude pressure perturbation. The response of a bubble is similarly represented by a perturbation, φ, to its mean radius, R_o, such that

$$R = R_o[1 + \text{Re}\{\varphi e^{i\omega t}\}] \tag{10.5}$$

and the linearization will neglect all terms of the order of φ^2 or higher.

The literature on the acoustics of dilute bubbly mixtures contains two complementary analytical approaches. Foldy (1945) and Carstensen and Foldy (1947) applied the classical acoustical approach and treated the problem of multiple scattering by randomly distributed point scatterers representing the bubbles. The medium is assumed to be very dilute ($\alpha \ll 1$). The multiple scattering produces both coherent and incoherent contributions. The incoherent part is beyond the scope of this text. The coherent part, which can be represented by Eq. (10.4), was found to satisfy a wave equation and yields a dispersion relation for the wavenumber, κ, of plane waves, that implies a phase velocity, $c_\kappa = \omega/\kappa$, given by (see van Wijngaarden 1972) the following:

$$\frac{1}{c_\kappa^2} = \frac{\kappa^2}{\omega^2} = \frac{1}{c_L^2} + \frac{1}{c_o^2}\left[1 - \frac{i\delta_d\omega}{\omega_n} - \frac{\omega^2}{\omega_n^2}\right]^{-1}. \tag{10.6}$$

Here c_L is the sonic speed in the liquid, c_o is the sonic speed arising from Eq. (9.14) when $\alpha\rho_G \ll (1-\alpha)\rho_L$,

$$c_o^2 = k\bar{p}/\rho_L\alpha(1-\alpha), \tag{10.7}$$

ω_n is the natural frequency of a bubble in an infinite liquid (Section 4.4.1), and δ_d is a dissipation coefficient that will be discussed shortly. It follows from Eq. (10.6) that scattering from the bubbles makes the wave propagation dispersive because c_κ is a function of the frequency, ω.

As described by van Wijngaarden (1972) an alternative approach is to linearize the fluid mechanical Eqs. (10.1), (10.2), and (10.3), neglecting any terms of order φ^2 or higher. In the case of plane wave propagation in the direction x (velocity u) in a frame of reference relative to the mixture (so that the mean velocity is zero), the convective terms in the Lagrangian derivatives, D/Dt, are of order φ^2 and the three governing equations become

$$\frac{\partial u}{\partial x} = \frac{\eta}{(1+\eta v)}\frac{\partial v}{\partial t} \tag{10.8}$$

$$\rho_L \frac{\partial u}{\partial t} = -(1+\eta v)\frac{\partial p}{\partial x} \tag{10.9}$$

$$R\frac{\partial^2 R}{\partial t^2} + \frac{3}{2}\left(\frac{\partial R}{\partial t}\right)^2 = \frac{1}{\rho_L}\left[p_V + p_{Go}\left(\frac{R_o}{R}\right)^{3k} - p\right] - \frac{2S}{\rho_L R} - \frac{4\nu_L}{R}\frac{\partial R}{\partial t}. \tag{10.10}$$

Assuming for simplicity that the liquid is incompressible (ρ_L = constant) and eliminating two of the three unknown functions from these relations, one obtains the following equation for any one of the three perturbation quantities ($Q = \varphi$, $\tilde{p}$, or $\tilde{u}$, the velocity perturbation):

$$3\alpha_o(1-\alpha_o)\frac{\partial^2 Q}{\partial t^2} = \left[\frac{3kp_{Go}}{\rho_L} - \frac{2S}{\rho_L R_o}\right]\frac{\partial^2 Q}{\partial x^2} + R_o^2\frac{\partial^4 Q}{\partial x^2 \partial t^2} + 4\nu_L\frac{\partial^3 Q}{\partial x^2 \partial t}, \tag{10.11}$$

where α_o is the mean void fraction given by $\alpha_o = \eta v_o/(1+\eta v_o)$. This equation governing the acoustic perturbations is given by van Wijngaarden, although we have added the surface tension term. Because the mean state must be in equilibrium, the mean liquid pressure, $\bar{p}$, is related to p_{Go} by the following:

$$\bar{p} = p_V + p_{Go} - \frac{2S}{R_o} \tag{10.12}$$

and hence the term in square brackets in Eq. (10.11) may be written in the following alternate forms:

$$\frac{3kp_{Go}}{\rho_L} - \frac{2S}{\rho_L R_o} = \frac{3k}{\rho_L}(\bar{p} - p_V) + \frac{2S}{\rho_L R_o}(3k-1) = R_o^2\omega_n^2. \tag{10.13}$$

This identifies ω_n, the natural frequency of a single bubble in an infinite liquid (see Section 4.4.1).

Results for the propagation of a plane wave in the positive x direction are obtained by substituting $q = e^{-i\kappa x}$ in Eq. (10.11) to produce the following dispersion relation:

$$c_\kappa^2 = \frac{\omega^2}{\kappa^2} = \frac{\left[\frac{3k}{\rho_L}(\bar{p}-p_V) + \frac{2S}{\rho_L R_o}(3k-1)\right] + 4i\omega\nu_L - \omega^2 R_o^2}{3\alpha_o(1-\alpha_o)}. \tag{10.14}$$

Note that at the low frequencies for which one would expect quasistatic bubble behavior ($\omega \ll \omega_n$) and in the absence of vapor ($p_V = 0$) and surface tension, this reduces to the sonic velocity given by Eq. (9.14) when $\rho_G \alpha \ll \rho_L(1-\alpha)$. Furthermore, Eq. (10.14) may be written as follows:

$$c_\kappa^2 = \frac{\omega^2}{\kappa^2} = \frac{R_o^2\omega_n^2}{3\alpha_o(1-\alpha_o)}\left[1 + i\frac{\delta_d\omega}{\omega_n} - \frac{\omega^2}{\omega_n^2}\right], \tag{10.15}$$

10.3 Acoustics of Bubbly Mixtures

where $\delta_d = 4\nu_L/\omega_n R_o^2$. For the incompressible liquid assumed here this is identical to Eq. (10.6) obtained using the Foldy multiple scattering approach [the difference in sign for the damping term results from using $i(\omega t - \kappa x)$ rather than $i(\kappa x - \omega t)$ and is inconsequential].

In the above derivation, the only damping mechanism that was explicitly included was that due to viscous effects on the radial motion of the bubbles. As Chapman and Plesset (1971) have shown, other damping mechanisms can affect the volume oscillations of the bubble; these include the damping due to temperature gradients caused by evaporation and condensation at the bubble surface and the radiation of acoustic energy due to compressibility of the liquid. However, Chapman and Plesset (1971) and others have demonstrated that, to a first approximation, all of these damping contributions can be included by defining an *effective* damping, δ_d, or, equivalently, an effective liquid viscosity, $\mu_e = \omega_n R_o^2 \delta_d / 4$.

10.3.2 Comparison with Experiments

The real and imaginary parts of κ as defined by Eq. (10.15) lead respectively to a sound speed and an attenuation that are both functions of the frequency of the perturbations. A number of experimental investigations have been carried out (primarily at very small α) to measure the sound speed and attenuation in bubbly gas/liquid mixtures. This data is reviewed by van Wijngaarden (1972) who concentrated on the experiments of Fox, Curley, and Lawson (1955), Macpherson (1957), and Silberman (1957), in which the bubble size distribution was more accurately measured and controlled. In general, the comparison between the experimental and theoretical propagation speeds is good, as illustrated by Figure 10.1. One of the primary experimental difficulties illustrated in both Figures 10.1 and 10.2 is that the results are quite sensitive to the distribution of bubble sizes present in the mixture. This is caused by the fact that the bubble natural frequency is quite sensitive to the mean radius [see Eq. (10.13)]. Hence a distribution in the size of the bubbles yields broadening of the peaks in the data of Figures 10.1 and 10.2.

Though the propagation speed is fairly well predicted by the theory, the same cannot be said of the attenuation, and there remain a number of unanswered questions in this regard. Using Eq. 10.15 the theoretical estimate of the damping coefficient, δ_d, pertinent to the experiments of Fox, Curley, and Lawson (1955) is 0.093. But a much greater value of $\delta_d = 0.5$ had to be used to produce an analytical line close to the experimental data on attenuation; it is important to note that the empirical value, $\delta_d = 0.5$, has been used for the theoretical results in Figure 10.2. Conversely, Macpherson (1957) found good agreement between a measured attenuation corresponding to $\delta_d \approx 0.08$ and the estimated analytical value of 0.079 relevant to his experiments. Similar good agreement was obtained for both the propagation and attenuation by Silberman (1957). Consequently, there appear to be some unresolved issues insofar as the attenuation is concerned. Among the effects that were omitted in the above analysis and that might

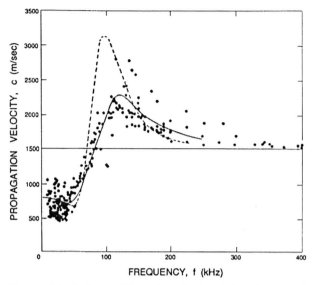

Figure 10.1. Sonic speed for water with air bubbles of mean radius, $R_o = 0.12$ mm, and a void fraction, $\alpha = 0.0002$, plotted against frequency. The experimental data of Fox, Curley, and Larson (1955) is plotted along with the theoretical curve for a mixture with identical $R_o = 0.11$-mm bubbles (dotted line) and with the experimental distribution of sizes (solid line). These lines use $\delta_d = 0.5$.

contribute to the attenuation is the effect of the relative motion of the bubbles. However, Batchelor (1969) has concluded that the viscous effects of translational motion would make a negligible contribution to the total damping.

Finally, it is important to emphasize that virtually all of the reported data on attenuation is confined to very small void fractions of the order of 0.0005 or less. The

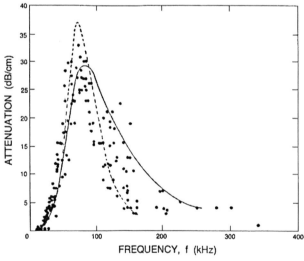

Figure 10.2. Values for the attenuation of sound waves corresponding to the sonic speed data of Figure 10.1. The attenuation in decibels per centimeter is given by 8.69 Im{κ} where κ is expressed per centimeter.

10.4 Shock Waves in Bubbly Flows

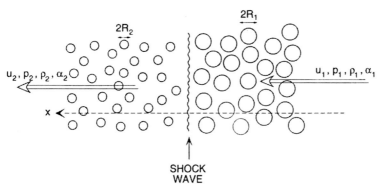

Figure 10.3. Schematic of the flow relative to a bubbly shock wave.

reason for this is clear when one evaluates the imaginary part of κ from Eq. (10.15). At these small void fractions the damping is proportional to α. Consequently, at large void fraction of the order, say, of 0.05, the damping is 100 times greater and therefore more difficult to measure accurately.

10.4 Shock Waves in Bubbly Flows

10.4.1 Shock-Wave Analysis

The propagation and structure of shock waves in bubbly cavitating flows represent a rare circumstance in which fully nonlinear solutions of the governing equations can be obtained. Shock-wave analyses of this kind were investigated by Campbell and Pitcher (1958), Crespo (1969), Noordzij (1973), and Noordzij and van Wijngaarden (1974), among others, and for more detail the reader should consult these works. Because this chapter is confined to flows without significant relative motion, this section does not cover some of the important effects of relative motion on the structural evolution of shocks in bubbly liquids. For this the reader is referred to Noordzij and van Wijngaarden (1974).

Consider a normal shock wave in a coordinate system moving with the shock so that the flow is steady and the shock stationary (Figure 10.3). If x and u represent a coordinate and the fluid velocity normal to the shock, then continuity requires the following:

$$\rho u = \text{constant} = \rho_1 u_1, \qquad (10.16)$$

where ρ_1 and u_1 refer to the mixture density and velocity far upstream of the shock. Hence u_1 is also the velocity of propagation of a shock into a mixture with conditions identical to those upstream of the shock. It is assumed that $\rho_1 \approx \rho_L(1 - \alpha_1) = \rho_L/(1 + \eta v_1)$, where the liquid density is considered constant and α_1, $v_1 = \frac{4}{3}\pi R_1^3$, and η are the void fraction, individual bubble volume, and population of the mixture far upstream respectively.

Substituting for ρ in the equation of motion and integrating, one also obtains the following:

$$p + \frac{\rho_1^2 u_1^2}{\rho} = \text{constant} = p_1 + \rho_1 u_1^2. \tag{10.17}$$

This expression for the pressure, p, may be substituted into the Rayleigh–Plesset equation using the observation that, for this steady flow,

$$\frac{DR}{Dt} = u\frac{dR}{dx} = u_1 \frac{(1+\eta v)}{(1+\eta v_1)} \frac{dR}{dx} \tag{10.18}$$

$$\frac{D^2 R}{Dt^2} = u_1^2 \frac{(1+\eta v)}{(1+\eta v_1)^2}\left[(1+\eta v)\frac{d^2 R}{dx^2} + 4\pi R^2 \eta \left(\frac{dR}{dx}\right)^2\right], \tag{10.19}$$

where $v = \frac{4}{3}\pi R^3$ has been used for clarity. It follows that the structure of the flow is determined by solving the following equation for $R(x)$:

$$u_1^2 \frac{(1+\eta v)^2}{(1+\eta v_1)^2} R\frac{d^2 R}{dx^2} + \frac{3}{2}u_1^2 \frac{(1+3\eta v)(1+\eta v)}{(1+\eta v_1)^2}\left(\frac{dR}{dx}\right)^2 + \frac{2S}{\rho_L R}$$

$$+ \frac{u_1(1+\eta v)}{(1+\eta v_1)}\frac{4 v_L}{R}\left(\frac{dR}{dx}\right) = \frac{(p_B - p_1)}{\rho_L} + \frac{\eta(v - v_1)}{(1+\eta v_1)^2} u_1^2. \tag{10.20}$$

It will be found that dissipation effects in the bubble dynamics strongly influence the structure of the shock. Only one dissipative effect, namely that due to viscous effects (last term on the left-hand side) has been explicitly included in Eq. (10.20). However, as discussed in the last section, other dissipative effects may be incorporated approximately by regarding v_L as a total *effective* viscosity.

The pressure within the bubble is given by the following:

$$p_B = p_V + p_{G1}(v_1/v)^k \tag{10.21}$$

and the equilibrium state far upstream must satisfy the following:

$$p_V - p_1 + p_{G1} = 2S/R_1. \tag{10.22}$$

Furthermore, if there exists an equilibrium state far downstream of the shock (this existence will be explored shortly), then it follows from Eqs. (10.20) and (10.21) that the velocity, u_1, must be related to the ratio, R_2/R_1 (where R_2 is the bubble size downstream of the shock), by the following:

$$u_1^2 = \frac{(1-\alpha_2)}{(1-\alpha_1)(\alpha_1-\alpha_2)}\left[\frac{(p_1-p_V)}{\rho_L}\left\{\left(\frac{R_1}{R_2}\right)^{3k} - 1\right\}\right.$$

$$\left. + \frac{2S}{\rho_L R_1}\left\{\left(\frac{R_1}{R_2}\right)^{3k} - \frac{R_1}{R_2}\right\}\right], \tag{10.23}$$

10.4 Shock Waves in Bubbly Flows

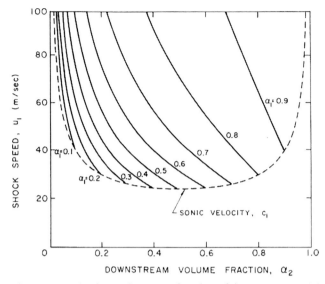

Figure 10.4. Shock speed, u_1, as a function of the upstream and downstream void fractions, α_1 and α_2, for the particular case $(p_1 - p_V)/\rho_L = 100$ m^2/s^2, $2S/\rho_L R_1 = 0.1$ m^2/s^2, and $k = 1.4$. Also shown by the dotted line is the sonic velocity, c_1, under the same upstream conditions.

where α_2 is the void fraction far downstream of the shock and

$$\left(\frac{R_2}{R_1}\right)^3 = \frac{\alpha_2(1-\alpha_1)}{\alpha_1(1-\alpha_2)}. \tag{10.24}$$

Hence the *shock velocity*, u_1, is given by the upstream flow parameters α_1, $(p_1 - p_V)/\rho_L$, and $2S/\rho_L R_1$, the polytropic index, k, and the downstream void fraction, α_2. An example of the dependence of u_1 on α_1 and α_2 is shown in Figure 10.4 for selected values of $(p_1 - p_V)/\rho_L = 100$ m^2/s^2, $2S/\rho_L R_1 = 0.1$ m^2/s^2, and $k = 1.4$. Also displayed by the dotted line in this figure is the sonic velocity of the mixture (at zero frequency), c_1, under the upstream conditions; it is readily shown that c_1 is given by the following:

$$c_1^2 = \frac{1}{\alpha_1(1-\alpha_1)}\left[\frac{k(p_1-p_V)}{\rho_L} + \left(k - \frac{1}{3}\right)\frac{2S}{\rho_L R_1}\right]. \tag{10.25}$$

Alternatively, the presentation conventional in gas dynamics can be adopted. Then the upstream Mach number, u_1/c_1, is plotted as a function of α_1 and α_2. The resulting graphs are functions only of two parameters, the polytropic index, k, and the parameter, $R_1(p_1 - p_V)/S$. An example is included as Figure 10.5 in which $k = 1.4$ and $R_1(p_1 - p_V)/S = 200$. It should be noted that a real shock velocity and a real sonic speed can exist even when the upstream mixture is under tension ($p_1 < p_V$). However, the numerical value of the tension, $p_V - p_1$, for which the values are real is limited to values of the parameter $R_1(p_1 - p_V)/2S > -(1 - 1/3k)$ or -0.762 for $k = 1.4$. Also note that Figure 10.5 does not change much with the parameter, $R_1(p_1 - p_V)/S$.

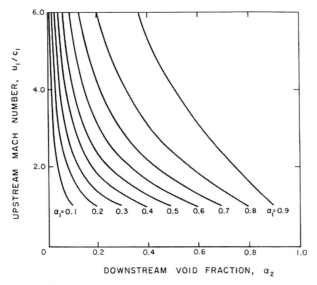

Figure 10.5. The upstream Mach number, u_1/c_1, as a function of the upstream and downstream void fractions, α_1 and α_2, for $k = 1.4$ and $R_1(p_1 - p_V)/S = 200$.

10.4.2 Shock-Wave Structure

Bubble dynamics do not affect the results presented thus far because the speed, u_1, depends only on the equilibrium conditions upstream and downstream. However, the existence and structure of the shock depend on the bubble dynamic terms in Eq. (10.20). That equation is more conveniently written in terms of a radius ratio, $r = R/R_1$, and a dimensionless coordinate, $z = x/R_1$:

$$(1 - \alpha_1 + \alpha_1 r^3)^2 r \frac{d^2 r}{dz^2} + \frac{3}{2}(1 - \alpha_1 + \alpha_1 r^3)(1 - \alpha_1 + 3\alpha_1 r^3)\left(\frac{dr}{dz}\right)^2$$
$$+ (1 - \alpha_1 + \alpha_1 r^3)\frac{4\nu_L}{u_1 R_1}\frac{1}{r}\frac{dr}{dz} + \alpha_1(1 - \alpha_1)(1 - r^3)$$
$$= \frac{1}{u_1^2}\left[\frac{(p_1 - p_V)}{\rho_L}(r^{-3k} - 1) + \frac{2S}{\rho_L R_1}(r^{-3k} - r^{-1})\right]. \quad (10.26)$$

It could also be written in terms of the void fraction, α, because

$$r^3 = \frac{\alpha}{(1 - \alpha)}\frac{(1 - \alpha_1)}{\alpha_1}. \quad (10.27)$$

When examined in conjunction with the expression in Eq. (10.23) for u_1, it is clear that the solution, $r(z)$ or $\alpha(z)$, for the structure of the shock is a function only of α_1, α_2, k, $R_1(p_1 - p_V)/S$, and the effective Reynolds number, $u_1 R_1/\nu_L$, where, as previously mentioned, ν_L should incorporate the various forms of bubble damping.

Equation (10.26) can be readily integrated numerically and typical solutions are presented in Figure 10.6 for $\alpha_1 = 0.3, k = 1.4, R_1(p_1 - p_V)/S \gg 1, u_1 R_1/\nu_L = 100$,

10.4 Shock Waves in Bubbly Flows

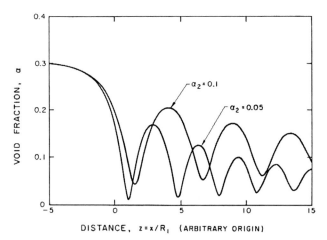

Figure 10.6. The typical structure of a shock wave in a bubbly mixture is illustrated by these examples for $\alpha_1 = 0.3$, $k = 1.4$, $R_1(p_1 - p_V)/S \gg 1$, and $u_1 R_1/\nu_L = 100$.

and two downstream volume fractions, $\alpha_2 = 0.1$ and 0.05. These examples illustrate several important features of the structure of these shocks. First, the initial collapse is followed by many rebounds and subsequent collapses. The decay of these nonlinear oscillations is determined by the damping or $u_1 R_1/\nu_L$. Though $u_1 R_1/\nu_L$ includes an effective kinematic viscosity to incorporate other contributions to the bubble damping, the value of $u_1 R_1/\nu_L$ chosen for this example is probably smaller than would be relevant in many practical applications, in which we might expect the decay to be even smaller. It is also valuable to identify the nature of the solution as the damping is eliminated ($u_1 R_1/\nu_L \to \infty$). In this limit the distance between collapses increases without bound until the structure consists of one collapse followed by a downstream asymptotic approach to a void fraction of α_1 (not α_2). In other words, no solution in which $\alpha \to \alpha_2$ exists in the absence of damping.

Another important feature in the structure of these shocks is the typical interval between the downstream oscillations. This *ringing* will, in practice, result in acoustic radiation at frequencies corresponding to this interval, and it is of importance to identify the relationship between this ring frequency and the natural frequency of the bubbles downstream of the shock. A characteristic ring frequency, ω_r, for the shock oscillations can be defined as follows:

$$\omega_r = 2\pi u_1/\Delta x, \qquad (10.28)$$

where Δx is the distance between the first and second bubble collapses. The natural frequency of the bubbles far downstream of the shock, ω_2, is given by [see Eq. (10.13)] the following:

$$\omega_2^2 = \frac{3k(p_2 - p_V)}{\rho_L R_2^2} + (3k - 1)\frac{2S}{\rho_L R_2^3} \qquad (10.29)$$

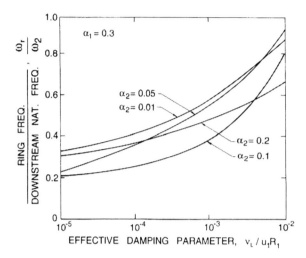

Figure 10.7. The ratio of the ring frequency downstream of a bubbly mixture shock to the natural frequency of the bubbles far downstream as a function of the effective damping parameter, $v_L/u_1 R_1$, for $\alpha_1 = 0.3$ and various downstream void fractions as indicated.

and typical values for the ratio ω_r/ω_2 are presented in Figure 10.7 for $\alpha_1 = 0.3$, $k = 1.4$, $R_1(p_1 - p_V)/S \gg 1$, and various values of α_2. Similar results were obtained for quite a wide range of values of α_1. Therefore note that the frequency ratio is primarily a function of the damping and that ring frequencies up to a factor of 10 less than the natural frequency are to be expected with typical values of the damping in water. This reduction in the typical frequency associated with the collective behavior of bubbles presages the natural frequencies of bubble clouds, which are discussed in the next section.

10.5 Finite Bubble Clouds

10.5.1 Natural Modes of a Spherical Cloud of Bubbles

A second illustrative example of the effect of bubble dynamics on the behavior of a homogeneous bubbly mixture is the study of the dynamics of a finite cloud of bubbles. One of the earliest investigations of the collective dynamics of bubble clouds was the work of van Wijngaarden (1964) on the oscillations of a layer of bubbles near a wall. Later d'Agostino and Brennen (1983) investigated the dynamics of a spherical cloud (see also d'Agostino and Brennen 1989, Omta 1987), and we choose the latter as a example of that class of problems with one space dimension in which analytical solutions may be obtained but only after linearization of the Rayleigh–Plesset equation [Eq. (10.3)].

The geometry of the spherical cloud is shown in Figure 10.8. Within the cloud of radius, $A(t)$, the population of bubbles per unit *liquid* volume, η, is assumed constant and uniform. The linearization assumes small perturbations of the bubbles from an equilibrium radius, R_o:

$$R(r, t) = R_o[1 + \varphi(r, t)], \quad |\varphi| \ll 1. \tag{10.30}$$

10.5 Finite Bubble Clouds

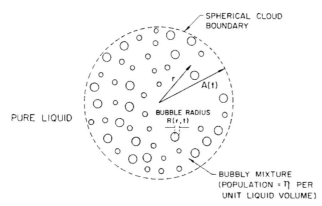

Figure 10.8. Notation for the analysis of a spherical cloud of bubbles.

We seek the response of the cloud to a correspondingly small perturbation in the pressure at infinity, $p_\infty(t)$, which is represented by the following:

$$p_\infty(t) = p(\infty, t) = \bar{p} + \text{Re}\{\tilde{p}e^{i\omega t}\}, \tag{10.31}$$

where $\bar{p}$ is the mean, uniform pressure and $\tilde{p}$ and ω are the perturbation amplitude and frequency, respectively. The solution will relate the pressure, $p(r, t)$, radial velocity, $u(r, t)$, void fraction, $\alpha(r, t)$, and bubble perturbation, $\varphi(r, t)$, to $\tilde{p}$. Because the analysis is linear, the response to excitation involving multiple frequencies can be obtained by Fourier synthesis.

One further restriction is necessary to linearize the governing Eqs. (10.1), (10.2), and (10.3). It is assumed that the mean void fraction in the cloud, α_o, is small so that the term $(1 + \eta v)$ in Eqs. (10.1) and (10.2) is approximately unity. Then these equations become the following:

$$\frac{1}{r^2}\frac{\partial}{\partial r}(r^2 u) = \eta \frac{Dv}{Dt} \tag{10.32}$$

$$\frac{Du}{Dt} = \frac{\partial u}{\partial t} + u\frac{\partial u}{\partial r} = -\frac{1}{\rho}\frac{\partial p}{\partial r}. \tag{10.33}$$

It is readily shown that the velocity u is of the order of φ and hence the convective component of the material derivative is of the order of φ^2; thus the linearization implies replacing D/Dt by $\partial/\partial t$. Then to order φ the Rayleigh–Plesset equation yields the following:

$$p(r, t) = \bar{p} - \rho R_o^2 \left[\frac{\partial^2 \varphi}{\partial t^2} + \omega_n^2 \varphi\right]; \quad r < A(t), \tag{10.34}$$

where ω_n is the natural frequency of an individual bubble if it were alone in an infinite fluid [Eq. (10.13)]. It must be assumed that the bubbles are in stable equilibrium in the mean state so that ω_n is real.

Upon substitution of Eqs. (10.30) and (10.34) into (10.32) and (10.33) and elimination of $u(r, t)$ one obtains the following equation for $\varphi(r, t)$ in the domain $r < A(t)$:

$$\frac{1}{r^2}\frac{\partial}{\partial r}\left[r^2\frac{\partial}{\partial r}\left\{\frac{\partial^2 \varphi}{\partial t^2} + \omega_n^2 \varphi\right\}\right] - 4\pi \eta R_o \frac{\partial^2 \varphi}{\partial t^2} = 0. \tag{10.35}$$

The incompressible liquid flow outside the cloud, $r \geq A(t)$, must have the standard solution of the following form:

$$u(r, t) = \frac{C(t)}{r^2}; \quad r \geq A(t) \tag{10.36}$$

$$p(r, t) = p_\infty(t) + \frac{\rho}{r}\frac{dC(t)}{dt} - \frac{\rho C^2}{2r^4}; \quad r \geq A(t), \tag{10.37}$$

where $C(t)$ is of perturbation order. It follows that, to the first order in $\varphi(r, t)$, the continuity of $u(r, t)$ and $p(r, t)$ at the interface between the cloud and the pure liquid leads to the following boundary condition for $\varphi(r, t)$:

$$\left(1 + A_o\frac{\partial}{\partial r}\right)\left[\frac{\partial^2 \varphi}{\partial t^2} + \omega_n^2 \varphi\right]_{r=A_o} = \frac{\tilde{p} - p_\infty(t)}{\rho R_o^2}. \tag{10.38}$$

The solution of Eq. (10.35) under the above boundary condition is as follows:

$$\varphi(r, t) = -\frac{1}{\rho R_o^2}\text{Re}\left\{\frac{\tilde{p}}{\omega_n^2 - \omega^2}\frac{e^{i\omega t}}{\cos \lambda A_o}\frac{\sin \lambda r}{\lambda r}\right\}; \quad r < A_o, \tag{10.39}$$

where

$$\lambda^2 = 4\pi \eta R_o \frac{\omega^2}{\omega_n^2 - \omega^2}. \tag{10.40}$$

Another possible solution involving $(\cos \lambda r)/\lambda r$ has been eliminated because $\varphi(r, t)$ must clearly be finite as $r \to 0$. Therefore in the domain $r < A_o$:

$$R(r, t) = R_o - \frac{1}{\rho R_o}\text{Re}\left\{\frac{\tilde{p}}{\omega_n^2 - \omega^2}\frac{e^{i\omega t}}{\cos \lambda A_o}\frac{\sin \lambda r}{\lambda r}\right\} \tag{10.41}$$

$$u(r, t) = \frac{1}{\rho}\text{Re}\left\{i\frac{\tilde{p}}{\omega r}\frac{1}{r}\left(\frac{\sin \lambda r}{\lambda r} - \cos \lambda r\right)\frac{e^{i\omega t}}{\cos \lambda A_o}\right\} \tag{10.42}$$

$$p(r, t) = \tilde{p} - \text{Re}\left\{\tilde{p}\frac{\sin \lambda r}{\lambda r}\frac{e^{i\omega t}}{\cos \lambda A_o}\right\}. \tag{10.43}$$

The entire flow has thus been determined in terms of the prescribed quantities A_o, R_o, η, ω, and $\tilde{p}$.

Note first that the cloud has a number of natural frequencies and modes of oscillation. From Eq. (10.39) it follows that if $\tilde{p}$ were zero, oscillations would only occur if

$$\omega = \omega_n \quad \text{or} \quad \lambda A_o = (2m - 1)\frac{\pi}{2}, \quad m = 0, \pm 2 \ldots \tag{10.44}$$

10.5 Finite Bubble Clouds

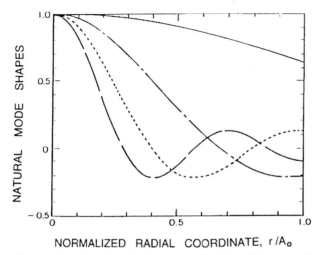

Figure 10.9. Natural mode shapes as a function of the normalized radial position, r/A_o, in the cloud for various orders $m = 1$ (solid line), 2 (dash-dotted line), 3 (dotted line), 4 (broken line). The arbitrary vertical scale represents the amplitude of the normalized undamped oscillations of the bubble radius, the pressure, and the bubble concentration per unit liquid volume. The oscillation of the velocity is proportional to the slope of these curves.

and, therefore, using Eq. (10.40) for λ, the natural frequencies, ω_m, of the cloud are as follows:

1. $\omega_\infty = \omega_n$, the natural frequency of an individual bubble in an infinite liquid, and
2. $\omega_m = \omega_n [1 + 16\eta R_o A_o^2/\pi(2m-1)^2]^{\frac{1}{2}}; m = 1, 2, \ldots$, which is an infinite series of frequencies of which ω_1 is the lowest. The higher frequencies approach ω_n as m tends to infinity.

The lowest natural frequency, ω_1, can be written in terms of the mean void fraction, $\alpha_o = \eta v_o/(1 + \eta v_o)$, as follows:

$$\omega_1 = \omega_n \left[1 + \frac{4}{3\pi^2} \frac{A_o^2}{R_o^2} \frac{\alpha_o}{1 - \alpha_o} \right]^{-\frac{1}{2}}. \quad (10.45)$$

Hence, the natural frequencies of the cloud will extend to frequencies much smaller than the individual bubble frequency, ω_n, if the initial void fraction, α_o, is much larger than the square of the ratio of bubble size to cloud size ($\alpha_o \gg R_o^2/A_o^2$). If the reverse is the case ($\alpha_o \ll R_o^2/A_o^2$), all the natural frequencies of the cloud are contained in a small range just below ω_n.

Typical natural modes of oscillation of the cloud are depicted in Figure 10.9, where normalized amplitudes of the bubble radius and pressure fluctuations are shown as functions of position, r/A_o, within the cloud. The amplitude of the radial velocity oscillation is proportional to the slope of these curves. Because each bubble is supposed to react to a uniform far field pressure, the validity of the model is limited to wavenumbers, m, such that $m \ll A_o/R_o$. Note that the first mode involves almost

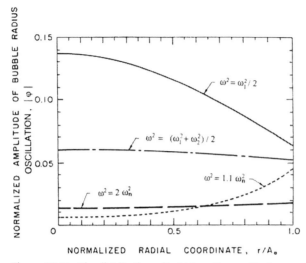

Figure 10.10. The distribution of bubble radius oscillation amplitudes, $|\varphi|$, within a cloud subjected to forced excitation at various frequencies, ω, as indicated (for the case of $\alpha_o(1-\alpha_o)A_o^2/R_o^2 = 0.822$). From d'Agostino and Brennen (1989).

uniform oscillations of the bubbles at all radial positions within the cloud. Higher modes involve amplitudes of oscillation near the center of the cloud that become larger and larger relative to the amplitudes in the rest of the cloud. In effect, an outer shell of bubbles essentially shields the exterior fluid from the oscillations of the bubbles in the central core, with the result that the pressure oscillations in the exterior fluid are of smaller amplitude for the higher modes.

10.5.2 Response of a Spherical Bubble Cloud

The corresponding shielding effects during forced excitation are illustrated in Figure 10.10, which shows the distribution of the amplitude of bubble radius oscillation, $|\varphi|$, within the cloud at various excitation frequencies, ω. Note that, although the entire cloud responds in a fairly uniform manner for $\omega < \omega_n$, only a surface layer of bubbles exhibits significant response when $\omega > \omega_n$. In the latter case the entire core of the cloud is essentially shielded by the outer layer.

The variations in the response at different frequencies are shown in more detail in Figure 10.11, in which the amplitude at the cloud surface, $|\varphi(A_o, t)|$, is presented as a function of ω. The solid line corresponds to the above analysis, which included no bubble damping. Consequently, there are asymptotes to infinity at each of the cloud natural frequencies; for clarity we have omitted the numerous asymptotes that occur just below the bubble natural frequency, ω_n. Also shown in this figure are the corresponding results when a reasonable estimate of the damping is included in the analysis (d'Agostino and Brennen 1989). The attenuation due to the damping is much greater at the higher frequencies so that, when damping is included (Figure 10.11),

10.5 Finite Bubble Clouds

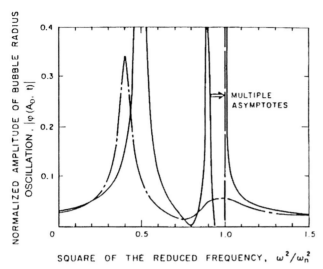

Figure 10.11. The amplitude of the bubble radius oscillation at the cloud surface, $|\varphi(A_o, t)|$, as a function of frequency (for the case of $\alpha_o(1 - \alpha_o)A_o^2/R_o^2 = 0.822$). Solid line is without damping; broken line includes damping. From d'Agostino and Brennen (1989).

the dominant feature of the response is the lowest natural frequency of the cloud. The response at the bubble natural frequency becomes much less significant.

The effect of varying the parameter, $\alpha_o(1 - \alpha_o)A_o^2/R_o^2$, is shown in Figure 10.12. Note that increasing the void fraction causes a reduction in both the amplitude and frequency of the dominant response at the lowest natural frequency of the cloud. d'Agostino and Brennen (1988) have also calculated the acoustical absorption and

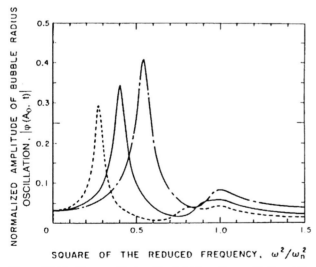

Figure 10.12. The amplitude of the bubble radius oscillation at the cloud surface, $|\varphi(A_o, t)|$, as a function of frequency for damped oscillations at three values of $\alpha_o(1 - \alpha_o)A_o^2/R_o^2$ equal to 0.822 (solid line), 0.411 (dot-dash line), and 1.65 (dashed line). From d'Agostino and Brennen (1989).

scattering cross sections of the cloud that this analysis implies. Not surprisingly, the dominant peaks in the cross sections occur at the lowest cloud natural frequency.

It is important to emphasize that the analysis presented above is purely linear and that there are likely to be very significant nonlinear effects that may have a major effect on the dynamics and acoustics of real bubble clouds. Hanson *et al.* (1981) and Mørch (1980, 1981) visualize that the collapse of a cloud of bubbles involves the formation and inward propagation of a shock wave and that the focusing of this shock at the center of the cloud creates the enhancement of the noise and damage potential associated with cloud collapse. The deformations of the individual bubbles within a collapsing cloud have been examined numerically by Chahine and Duraiswami (1992), who showed that the bubbles on the periphery of the cloud develop inwardly directed reentrant jets.

Numerical investigations of the nonlinear dynamics of cavity clouds have been carried out by Chahine (1982), Omta (1987), and Kumar and Brennen (1991, 1992, 1993). Kumar and Brennen have obtained weakly nonlinear solutions to a number of cloud problems by retaining only the terms that are quadratic in the amplitude. One interesting phenomenon that emerges from this nonlinear analysis involves the interactions between the bubbles of different size that would commonly occur in any real cloud. The phenomenon, called *harmonic cascading* (Kumar and Brennen 1992), occurs when a relatively small number of larger bubbles begins to respond nonlinearly to some excitation. Then the higher harmonics produced will excite the much larger number of smaller bubbles at their natural frequency. The process can then be repeated to even smaller bubbles. In essence, this nonlinear effect causes a cascading of fluctuation energy to smaller bubbles and higher frequencies.

In all of the above we have focused, explicitly or implicitly, on spherical bubble clouds. Solutions of the basic equations for other, more complex, geometries are not readily obtained. However, d'Agostino *et al.* (1988) have examined some of the characteristics of this class of flows past slender bodies (for example, the flow over a wavy surface). Clearly, in the absence of bubble dynamics, one would encounter two types of flow: subsonic and supersonic. Interestingly, the inclusion of bubble dynamics leads to three types of flow. At sufficiently low speeds one obtains the usual elliptic equations of subsonic flow. When the sonic speed is exceeded, the equations become hyperbolic and the flow supersonic. However, with further increase in speed, the time rate of change becomes equivalent to frequencies above the natural frequency of the bubbles. Then the equations become elliptic again and a new flow regime, termed *superresonant*, occurs. d'Agostino *et al.* (1988) explore the consequences of this and other features of these slender body flows.

11

Flows with Gas Dynamics

11.1 Introduction

This chapter addresses the class of compressible flows in which a gaseous continuous phase is seeded with droplets or particles and in which it is necessary to evaluate the relative motion between the disperse and continuous phases for a variety of possible reasons. In many such flows, the motivation is the erosion of the flow boundaries by particles or drops and this is directly related to the relative motion. In other cases, the purpose is to evaluate the change in the performance of the system or device. Still another motivation is the desire to evaluate changes in the instability boundaries caused by the presence of the disperse phase.

Examples include the potential for serious damage to steam turbine blades by impacting water droplets (e.g., Gardner 1963, Smith *et al.* 1967). In the context of aircraft engines, desert sand storms or clouds of volcanic dust can not only cause serious erosion to the gas turbine compressor (Tabakoff and Hussein 1971, Smialek *et al.* 1994, Dunn *et al.* 1996, Tabakoff and Hamed 1986) but can also deleteriously effect the stall margin and cause engine shutdown (Batcho *et al.* 1987). Other examples include the consequences of seeding the fuel of a solid-propelled rocket with metal particles to enhance its performance. This is a particularly complicated example because the particles may also melt and oxidize in the flow (Shorr and Zaehringer 1967).

In recent years considerable advancements have been made in the numerical models and methods available for the solution of dilute particle-laden flows. In this text, we present a survey of the analytical methods and the physical understanding that they generate; for a valuable survey of the numerical methods the reader is referred to Crowe (1982).

11.2 Equations for a Dusty Gas

11.2.1 Basic Equations

First we review the fundamental equations governing the flow of the individual phases or components in a dusty gas flow. The continuity equations [Eqs. (1.21)] may be written as follows:

$$\frac{\partial}{\partial t}(\rho_N \alpha_N) + \frac{\partial(\rho_N \alpha_N u_{Ni})}{\partial x_i} = \mathcal{I}_N, \quad (11.1)$$

where $N = C$ and $N = D$ refer to the continuous and disperse phases respectively. We shall see that it is convenient to define a *loading* parameter, ξ, as follows:

$$\xi = \frac{\rho_D \alpha_D}{\rho_C \alpha_C} \quad (11.2)$$

and that the continuity equations have an important bearing on the variations in the value of ξ within the flow. Note that the mixture density, ρ, is then as follows:

$$\rho = \rho_C \alpha_C + \rho_D \alpha_D = (1 + \xi)\rho_C \alpha_C. \quad (11.3)$$

The momentum and energy equations for the individual phases [Eqs. (1.45) and (1.69)] are respectively

$$\rho_N \alpha_N \left[\frac{\partial u_{Nk}}{\partial t} + u_{Ni} \frac{\partial u_{Nk}}{\partial x_i} \right]$$
$$= \alpha_N \rho_N g_k + \mathcal{F}_{Nk} - \mathcal{I}_N u_{Nk} - \delta_N \left[\frac{\partial p}{\partial x_k} - \frac{\partial \sigma^D_{Cki}}{\partial x_i} \right] \quad (11.4)$$

$$\rho_N \alpha_N c_{vN} \left[\frac{\partial T_N}{\partial t} + u_{Ni} \frac{\partial T_N}{\partial x_i} \right]$$
$$= \delta_N \sigma_{Cij} \frac{\partial u_{Ci}}{\partial x_j} + \mathcal{Q}_N + \mathcal{W}_N + \mathcal{QI}_N + \mathcal{F}_{Ni}(u_{Di} - u_{Ni}) - (e_N^* - u_{Ni} u_{Ni})\mathcal{I}_N$$
$$(11.5)$$

and, when summed over all the phases, these lead to the following combined continuity, momentum, and energy equations [Eqs. (1.24), (1.46), and (1.70)]:

$$\frac{\partial \rho}{\partial t} + \frac{\partial}{\partial x_i}\left(\sum_N \rho_N \alpha_N u_{Ni}\right) = 0 \quad (11.6)$$

$$\frac{\partial}{\partial t}\left(\sum_N \rho_N \alpha_N u_{Nk}\right) + \frac{\partial}{\partial x_i}\left(\sum_N \rho_N \alpha_N u_{Ni} u_{Nk}\right)$$
$$= \rho g_k - \frac{\partial p}{\partial x_k} + \frac{\partial \sigma^D_{Cki}}{\partial x_i} \quad (11.7)$$

$$\sum_N \left[\rho_N \alpha_N c_{vN} \left\{ \frac{\partial T_N}{\partial t} + u_{Ni} \frac{\partial T_N}{\partial x_i} \right\} \right]$$
$$= \sigma_{Cij} \frac{\partial u_{Ci}}{\partial x_j} - \mathcal{F}_{Di}(u_{Di} - u_{Ci}) - \mathcal{I}_D(e_D^* - e_C^*) + \sum_N u_{Ni} u_{Ni} \mathcal{I}_N. \quad (11.8)$$

To these equations of motion, we must add equations of state for both phases. Throughout this chapter it is assumed that the continuous phase is an ideal gas and that the disperse phase is an incompressible solid. Moreover, temperature and velocity gradients in the vicinity of the interface are neglected.

11.2 Equations for a Dusty Gas

11.2.2 Homogeneous Flow with Gas Dynamics

Though the focus in this chapter is on the effect of relative motion, we must begin by examining the simplest case in which both the relative motion between the phases or components and the temperature differences between the phases or components are sufficiently small that they can be neglected. This establishes the base state that, through perturbation methods, can be used to examine flows in which the relative motion and temperature differences are small. As established in Chapter 9, a flow with no relative motion or temperature differences is referred to as *homogeneous*. The effect of mass exchange will also be neglected in the present discussion and, in such a homogeneous flow, the governing equations, (11.6), (11.7), and (11.8), clearly reduce to the following:

$$\frac{\partial \rho}{\partial t} + \frac{\partial}{\partial x_i}(\rho u_i) = 0 \tag{11.9}$$

$$\rho \left[\frac{\partial u_k}{\partial t} + u_i \frac{\partial u_k}{\partial x_i} \right] = \rho g_k - \frac{\partial p}{\partial x_k} + \frac{\partial \sigma_{Cki}^D}{\partial x_i} \tag{11.10}$$

$$\left[\sum_N \rho_N \alpha_N c_{vN} \right] \left\{ \frac{\partial T}{\partial t} + u_i \frac{\partial T}{\partial x_i} \right\} = \sigma_{Cij} \frac{\partial u_i}{\partial x_j}, \tag{11.11}$$

where u_i and T are the velocity and temperature common to all phases.

An important result that follows from the individual continuity Eq. (11.1) in the absence of exchange of mass ($\mathcal{I}_N = 0$) is that

$$\frac{D}{Dt} \left\{ \frac{\rho_D \alpha_D}{\rho_C \alpha_C} \right\} = \frac{D\xi}{Dt} = 0. \tag{11.12}$$

Consequently, if the flow develops from a uniform stream in which the loading ξ is constant and uniform, then ξ is uniform and constant everywhere and becomes a simple constant for the flow. We shall confine the remarks in this section to such flows.

At this point, one particular approximation is very advantageous. Because in many applications the volume occupied by the particles is very small, it is reasonable to set $\alpha_C \approx 1$ in Eq. (11.2) and elsewhere. This approximation has the important consequence that Eqs. (11.9), (11.10), and (11.11) are now those of a single phase flow of an *effective* gas whose thermodynamic and transport properties are as follows. The approximation allows the equation of state of the *effective* gas to be written as follows:

$$p = \rho \mathcal{R} T, \tag{11.13}$$

where $\mathcal{R}$ is the gas constant of the effective gas. Setting $\alpha_C \approx 1$, the thermodynamic properties of the effective gas are given by the following:

$$\rho = \rho_C(1+\xi); \quad \mathcal{R} = \mathcal{R}_C/(1+\xi)$$

$$c_v = \frac{c_{vC} + \xi c_{sD}}{1+\xi}; \quad c_p = \frac{c_{pC} + \xi c_{sD}}{1+\xi}; \quad \gamma = \frac{c_{pC} + \xi c_{sD}}{c_{vC} + \xi c_{sD}} \tag{11.14}$$

and the effective kinematic viscosity is as follows:

$$\nu = \mu_C/\rho_C(1+\xi) = \nu_C/(1+\xi). \tag{11.15}$$

Moreover, it follows from Eq. (11.14), that the relation between the isentropic speed of sound, c, in the effective gas and that in the continuous phase, c_C, is as follows:

$$c = c_C \left[\frac{1 + \xi c_{sD}/c_{pC}}{(1 + \xi c_{sD}/c_{vC})(1+\xi)} \right]^{\frac{1}{2}}. \tag{11.16}$$

It also follows that the Reynolds, Mach, and Prandtl numbers for the effective gas flow, Re, M, and Pr (based on a typical dimension, ℓ, typical velocity, U, and typical temperature, T_0, of the flow) are related to the Reynolds, Mach, and Prandtl numbers for the flow of the continuous phase, Re_C, M_C and Pr_C, by the following:

$$\text{Re} = \frac{U\ell}{\nu} = \text{Re}_C(1+\xi) \tag{11.17}$$

$$M = \frac{U}{c} = M_C \left[\frac{(1+\xi c_{sD}/c_{vC})(1+\xi)}{(1+\xi c_{sD}/c_{pC})} \right]^{\frac{1}{2}} \tag{11.18}$$

$$\text{Pr} = \frac{c_p \mu}{k} = \text{Pr}_C \left[\frac{(1+\xi c_{sD}/c_{pC})}{(1+\xi)} \right]. \tag{11.19}$$

Thus the first step in most investigations of this type of flow is to solve for the effective gas flow using the appropriate tools from single-phase gas dynamics. Here, it is assumed that the reader is familiar with these basic methods. Thus we focus on the phenomena that constitute departures from single-phase flow mechanics and, in particular, on the process and consequences of relative motion or *slip*.

11.2.3 Velocity and Temperature Relaxation

Although the homogeneous model with effective gas properties may constitute a sufficiently accurate representation in some contexts, there are other technological problems in which the velocity and temperature differences between the phases are important either intrinsically or because of their consequences. The rest of the chapter is devoted to these effects. But, to proceed toward this end, it is necessary to stipulate particular forms for the mass, momentum, and energy exchange processes represented by $\mathcal{I}_N$, $\mathcal{F}_{Nk}$, and $\mathcal{QI}_N$ in Eqs. (11.1), (11.4), and (11.5). For simplicity, the remarks in this chapter are confined to flows in which there is no external heat added or work done so that $\mathcal{Q}_N = 0$ and $\mathcal{W}_N = 0$. Moreover, we assume that there is negligible mass exchange so that $\mathcal{I}_N = 0$. It remains, therefore, to stipulate the force interaction, $\mathcal{F}_{Nk}$ and the heat transfer between the components, $\mathcal{QI}_N$. In the present context it is assumed that the relative motion is at low Reynolds numbers so that the simple model of relative motion defined by a relaxation time (see Section 2.4.1) may be used. Then:

$$\mathcal{F}_{Ck} = -\mathcal{F}_{Dk} = \frac{\rho_D \alpha_D}{t_u}(u_{Dk} - u_{Ck}), \tag{11.20}$$

where t_u is the velocity relaxation time given by Eq. (2.73) (neglecting the added mass of the gas) as follows:

$$t_u = m_p / 12\pi R \mu_C. \tag{11.21}$$

It follows that the equation of motion for the disperse phase, Eq. (11.4), becomes the following:

$$\frac{Du_{Dk}}{Dt} = \frac{u_{Ck} - u_{Dk}}{t_u}. \tag{11.22}$$

It is further assumed that the temperature relaxation may be modeled as described in Section 1.2.9 so that

$$\mathcal{QI}_C = -\mathcal{QI}_D = \frac{\rho_D \alpha_D c_{sD} \text{Nu}}{t_T}(T_D - T_C), \tag{11.23}$$

where t_T is the temperature relaxation time given by Eq. (1.76) as follows:

$$t_T = \rho_D c_{sD} R^2 / 3 k_C. \tag{11.24}$$

It follows that the energy equation for the disperse phase is Eq. (1.75) or

$$\frac{DT_D}{Dt} = \frac{\text{Nu}}{2} \frac{(T_C - T_D)}{t_T}. \tag{11.25}$$

In the context of droplet or particle laden gas flows these are commonly assumed forms for the velocity and temperature relaxation processes (Marble 1970). In his review Rudinger (1969) includes some evaluation of the sensitivity of the calculated results to the specifics of these assumptions.

11.3 Normal Shock Wave

Normal shock waves not only constitute a flow of considerable practical interest but also provide an illustrative example of the important role that relative motion may play in particle- or droplet-laden gas flows. In a frame of reference fixed in the shock, the fundamental equations for this steady flow in one Cartesian direction (x with velocity u in that direction) are obtained from Eqs. (11.1) to (11.8) as follows. Neglecting any mass interaction ($\mathcal{I}_N = 0$) and assuming that there is one continuous and one disperse phase, the individual continuity equations (11.1) become the following:

$$\rho_N \alpha_N u_N = \dot{m}_N = \text{constant}, \tag{11.26}$$

where $\dot{m}_C$ and $\dot{m}_D$ are the mass flow rates per unit area. Because the gravitational term and the deviatoric stresses are negligible, the combined phase momentum equation [Eq. (11.7)] may be integrated to obtain the following:

$$\dot{m}_C u_C + \dot{m}_D u_D + p = \text{constant}. \tag{11.27}$$

Also, eliminating the external heat added ($Q = 0$) and the external work done ($\mathcal{W} = 0$) the combined phase energy Eq. (11.8) may be integrated to obtain the following:

$$\dot{m}_C \left(c_{vC} T_C + \frac{1}{2} u_C^2 \right) + \dot{m}_D \left(c_{sD} T_D + \frac{1}{2} u_D^2 \right) + p u_C = \text{constant} \quad (11.28)$$

and can be recast in the following form:

$$\dot{m}_C \left(c_{pC} T_C + \frac{1}{2} u_C^2 \right) + p u_C (1 - \alpha_C) + \dot{m}_D \left(c_{sD} T_D + \frac{1}{2} u_D^2 \right) = \text{constant} \quad (11.29)$$

In lieu of the individual phase momentum and energy equations, we use the velocity and temperature relaxation relations [Eqs. (11.22) and (11.25)] as follows:

$$\frac{D u_D}{Dt} = u_D \frac{d u_D}{dx} = \frac{u_C - u_D}{t_u} \quad (11.30)$$

$$\frac{D T_D}{Dt} = u_D \frac{d T_D}{dx} = \frac{T_C - T_D}{t_T}, \quad (11.31)$$

where, for simplicity, we confine the present analysis to the pure conduction case, $Nu = 2$.

Carrier (1958) was the first to use these equations to explore the structure of a normal shock wave for a gas containing solid particles, a *dusty gas* in which the volume fraction of particles is negligible. Under such circumstances, the initial shock wave in the gas is unaffected by the particles and can have a thickness that is small compared to the particle size. We denote the conditions upstream of this structure by the subscript 1 so that

$$u_{C1} = u_{D1} = u_1; \quad T_{C1} = T_{D1} = T_1 \quad (11.32)$$

The conditions immediately downstream of the initial shock wave in the gas are denoted by the subscript 2. The normal single phase gas dynamic relations allow ready evaluation of u_{C2}, T_{C2}, and p_2 from u_{C1}, T_{C1}, and p_1.

Unlike the gas, the particles pass through this initial shock without significant change in velocity or temperature so that

$$u_{D2} = u_{D1}; \quad T_{D2} = T_{D1}. \quad (11.33)$$

Consequently, at the location 2 there are now substantial velocity and temperature differences, $u_{C2} - u_{D2}$ and $T_{C2} - T_{D2}$, equal to the velocity and temperature differences across the initial shock wave in the gas. These differences take time to decay and do so according to Eqs. (11.30) and (11.31). Thus the structure downstream of the gas dynamic shock consists of a relaxation zone in which the particle velocity decreases and the particle temperature increases, each asymptoting to a final downstream state that is denoted by the subscript 3. In this final state

$$u_{C3} = u_{D3} = u_3; \quad T_{C3} = T_{D3} = T_3. \quad (11.34)$$

As in any similar shock-wave analysis the relations between the initial (1) and final (3) conditions are independent of the structure and can be obtained directly from the basic

11.3 Normal Shock Wave

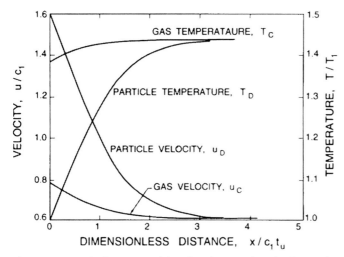

Figure 11.1. Typical structure of the relaxation zone in a shock wave in a dusty gas for $M_1 = 1.6$, $\gamma = 1.4$, $\xi = 0.25$ and $t_u/t_T = 1.0$. In the nondimensionalization, c_1 is the upstream acoustic speed. Adapted from Marble (1970).

conservation equations listed above. Making the small disperse phase volume approximation discussed in Section 11.2.2 and using the definitions shown in Eq. (11.14), the relations that determine both the structure of the relaxation zone and the asymptotic downstream conditions are as follows:

$$\dot{m}_C = \rho_C u_C = \dot{m}_{C1} = \dot{m}_{C2} = \dot{m}_{C3}; \quad \dot{m}_D = \rho_D u_D = \xi \dot{m}_C \tag{11.35}$$

$$\dot{m}_C(u_C + \xi u_D) + p = (1+\xi)\dot{m}_C u_{C1} + p_1 = (1+\xi)\dot{m}_C u_{C3} + p_3 \tag{11.36}$$

$$\left(c_{pC}T_C + \frac{1}{2}u_C^2\right) + \xi\left(c_{sD}T_D + \frac{1}{2}u_D^2\right) = (1+\xi)\left(c_p T_1 + \frac{1}{2}u_1^2\right)$$

$$= (1+\xi)\left(c_p T_3 + \frac{1}{2}u_3^2\right) \tag{11.37}$$

and it is a straightforward matter to integrate Eqs. (11.30), (11.31), (11.35), (11.36), and (11.37) to obtain $u_C(x)$, $u_D(x)$, $T_C(x)$, $T_D(x)$, and $p(x)$ in the relaxation zone.

First, we comment on the typical structure of the shock and the relaxation zone as revealed by this numerical integration. A typical example from the review by Marble (1970) is included as Figure 11.1. This shows the asymptotic behavior of the velocities and temperatures in the case $t_u/t_T = 1.0$. The nature of the relaxation processes is evident in this figure. Just downstream of the shock the particle temperature and velocity are the same as upstream of the shock; but the temperature and velocity of the gas has now changed and, over the subsequent distance, $x/c_1 t_u$, downstream of the shock, the particle temperature rises toward that of the gas and the particle velocity decreases toward that of the gas. The relative motion also causes a pressure rise in the gas, that, in turn, causes a temperature rise and a velocity decrease in the gas.

Clearly, there are significant differences when the velocity and temperature relaxation times are not of the same order. When $t_u \ll t_T$ the velocity equilibration zone will be much thinner than the thermal relaxation zone and when $t_u \gg t_T$ the opposite will be true. Marble (1970) uses a perturbation analysis about the final downstream state to show that the two processes of velocity and temperature relaxation are not closely coupled, at least up to the second order in an expansion in ξ. Consequently, as a first approximation, one can regard the velocity and temperature relaxation zones as uncoupled. Marble also explores the effects of different particle sizes and the collisions that may ensue as a result of relative motion between the different sizes.

This normal shock wave analysis illustrates that the notions of velocity and temperature relaxation can be applied as modifications to the basic gas dynamic structure to synthesize, at least qualitatively, the structure of the multiphase flow.

11.4 Acoustic Damping

Another important consequence of relative motion is the effect it has on the propagation of plane acoustic waves in a dusty gas. Here we examine both the propagation velocity and damping of such waves. To do so we postulate a uniform dusty gas and denote the mean state of this mixture by an overbar so that $\bar{p}, \bar{T}, \bar{\rho}_C, \bar{\xi}$ are respectively the pressure, temperature, gas density, and mass loading of the uniform dusty gas. Moreover, we chose a frame of reference relative to the mean dusty gas so that $\bar{u}_C = \bar{u}_D = 0$. Then we investigate small, linearized perturbations to this mean state denoted by $\tilde{p}, \tilde{T}_C, \tilde{T}_D, \tilde{\rho}_C, \tilde{\alpha}_D, \tilde{u}_C$, and $\tilde{u}_D$. Substituting into the basic continuity, momentum, and energy Eqs. (11.1), (11.4), and (11.5), utilizing the expressions and assumptions of Section 11.2.3, and retaining only terms linear in the perturbations, the equations governing the propagation of plane acoustic waves become the following:

$$\frac{\partial \tilde{u}_C}{\partial x} + \frac{1}{\bar{p}}\frac{\partial \tilde{p}}{\partial t} - \frac{1}{\bar{T}}\frac{\partial \tilde{T}_C}{\partial t} = 0 \tag{11.38}$$

$$\bar{\rho}_D \frac{\partial \tilde{\alpha}_D}{\partial t} + \frac{\partial \tilde{u}_D}{\partial x} = 0 \tag{11.39}$$

$$\frac{\partial \tilde{u}_C}{\partial t} + \frac{\xi \tilde{u}_C}{t_u} - \frac{\xi \tilde{u}_D}{t_u} + \frac{1}{\gamma}\frac{\partial \tilde{p}}{\partial x} = 0 \tag{11.40}$$

$$\frac{\partial \tilde{u}_D}{\partial t} + \frac{\tilde{u}_D}{t_u} - \frac{\tilde{u}_C}{t_u} = 0 \tag{11.41}$$

$$\frac{\partial \tilde{T}_C}{\partial t} + \frac{\xi \tilde{T}_C}{t_T} - \frac{\xi \tilde{T}_D}{t_T} + \frac{(\gamma-1)\bar{p}}{\gamma \bar{T}}\frac{\partial \tilde{p}}{\partial t} = 0 \tag{11.42}$$

$$\frac{\partial \tilde{T}_D}{\partial t} + \frac{c_{pC}\tilde{T}_D}{c_{sD}t_T} - \frac{c_{pC}\tilde{T}_C}{c_{sD}t_T} = 0. \tag{11.43}$$

where $\gamma = \frac{c_{pC}}{c_{vC}}$.

Note that the particle volume fraction perturbation occurs in only one of these, Eq. (11.39); consequently this equation may be set aside and used after the solution has

11.4 Acoustic Damping

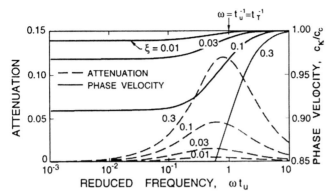

Figure 11.2. Nondimensional attenuation, $\text{Im}\{-\kappa c_C/\omega\}$ (dotted lines), and phase velocity, c_κ/c_C (solid lines), as functions of reduced frequency, ωt_u, for a dusty gas with various loadings, ξ, as shown and $\gamma = 1.4$, $t_T/t_u = 1$ and $c_{pC}/c_{sD} = 0.3$.

been obtained to calculate $\tilde{\alpha}_D$ and therefore the perturbations in the particle loading $\tilde{\xi}$. The basic form of a plane acoustic wave is as follows:

$$Q(x, t) = \bar{Q} + \tilde{Q}(x, t) = \bar{Q} + \text{Re}\{Q(\omega)e^{i\kappa x + i\omega t}\}, \quad (11.44)$$

where $Q(x, t)$ is a generic flow variable, ω is the acoustic frequency and κ is a complex function of ω; clearly the phase velocity of the wave, c_κ, is given by $c_\kappa = \text{Re}\{-\omega/\kappa\}$ and the nondimensional attenuation is given by $\text{Im}\{-\kappa\}$. Then substitution of the expressions 11.44 into the five Eqs. (11.38), (11.40), (11.41), (11.42), and (11.43) yields the following dispersion relation for κ:

$$\left(\frac{\omega}{\kappa c_C}\right)^2 = \frac{(1 + i\omega t_u)\left(\frac{c_{pC}}{c_{sD}} + \xi + i\omega t_T\right)}{(1 + \xi + i\omega t_u)\left(\frac{c_{pC}}{c_{sD}}\gamma\xi + i\omega t_T\right)}, \quad (11.45)$$

where $c_C = (\gamma \mathcal{R}_C \bar{T})^{\frac{1}{2}}$ is the speed of sound in the gas alone. Consequently, the phase velocity is readily obtained by taking the real part of the square root of the right-hand side of Eq. (11.45). It is a function of frequency, ω, as well as the relaxation times, t_u and t_T, the loading, ξ, and the specific heat ratios, γ and c_{pC}/c_{sD}. Typical results are shown in Figures 11.2 and 11.3.

The mechanics of the variation in the phase velocity (acoustic speed) are evident by inspection of Eq. (11.45) and Figures 11.2 and 11.3. At very low frequencies such that $\omega t_u \ll 1$ and $\omega t_T \ll 1$, the velocity and temperature relaxations are essentially instantaneous. Then the phase velocity is simply obtained from the effective properties and is given by Eq. (11.16). These are the phase velocity asymptotes on the left-hand side of Figures 11.2 and 11.3. Conversely, at very high frequencies such that $\omega t_u \gg 1$ and $\omega t_T \gg 1$, there is negligible time for the particles to adjust and they simply do not participate in the propagation of the wave; consequently, the phase velocity is simply the acoustic velocity in the gas alone, c_C. Thus all phase velocity lines asymptote to unity on the right in the figures. Other ranges of frequency may also exist (for example, $\omega t_u \gg 1$ and $\omega t_T \ll 1$ or the reverse) in which other asymptotic expressions for the

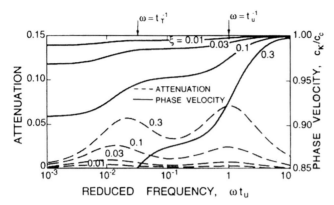

Figure 11.3. Nondimensional attenuation, $\text{Im}\{-\kappa c_C/\omega\}$ (dotted lines), and phase velocity, c_κ/c_C (solid lines), as functions of reduced frequency, ωt_u, for a dusty gas with various loadings, ξ, as shown and $\gamma = 1.4$, $t_T/t_u = 30$ and $c_{pC}/c_{sD} = 0.3$.

acoustic speed can be readily extracted from Eq. (11.45). One such intermediate asymptote can be detected in Figure 11.3. It is also clear that the acoustic speed decreases with increased loading, ξ, although only weakly in some frequency ranges. For small ξ, Eq. (11.45) may be expanded to obtain the linear change in the acoustic speed with loading, ξ, as follows:

$$\frac{c_\kappa}{c_C} = 1 - \frac{\xi}{2}\left[\frac{(\gamma-1)\frac{c_{pC}}{c_{sD}}}{\{(c_{pC}/c_{sD})^2 + (\omega t_T)^2\}} + \frac{1}{\{1+(\omega t_T)^2\}}\right] + \cdots \quad (11.46)$$

This expression shows why, in Figures 11.2 and 11.3, the effect of the loading, ξ, on the phase velocity is small at higher frequencies.

Now we examine the attenuation manifest in the dispersion relation [Eq. (11.45)]. The same expansion for small ξ that led to Eq. (11.46) also leads to the following expression for the attenuation:

$$\text{Im}\{-\kappa\} = \frac{\xi\omega}{2c_C}\left[\frac{(\gamma-1)\omega t_T}{\{(c_{pC}/c_{sD})^2 + (\omega t_T)^2\}} + \frac{\omega t_u}{\{1+(\omega t_T)^2\}}\right] + \cdots \quad (11.47)$$

In Figures 11.2 and 11.3, a dimensionless attenuation, $\text{Im}\{-\kappa c_C/\omega\}$, is plotted against the reduced frequency. This particular nondimensionalization is somewhat misleading because plotted without the ω in the denominator, the attenuation increases monotonically with frequency. However, this presentation is commonly used to demonstrate the enhanced attenuations that occur in the neighborhoods of $\omega = t_u^{-1}$ and $\omega = t_T^{-1}$ and that are manifest in Figures 11.2 and 11.3.

When the gas contains liquid droplets rather than solid particles, the same basic approach is appropriate except for the change that might be caused by the evaporation and condensation of the liquid during the passage of the wave. Marble and Wooten (1970) present a variation of the above analysis that includes the effect of phase change and show that an additional maximum in the attenuation can result as illustrated in

11.5 Other Linear Perturbation Analyses

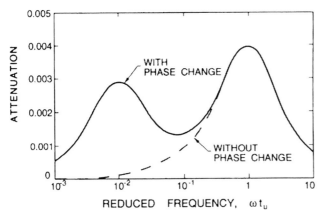

Figure 11.4. Nondimensional attenuation, $\text{Im}\{-\kappa c_C/\omega\}$, as a function of reduced frequency for a droplet-laden gas flow with $\xi = 0.01$, $\gamma = 1.4$, $t_T/t_u = 1$ and $c_{pC}/c_{sD} = 1$. The dashed line is the result without phase change; the solid line is an example of the alteration caused by phase change. Adapted from Marble and Wooten (1970).

Figure 11.4. This additional peak results from another relaxation process embodied in the phase change process. As Marble (1970) points out it is only really separate from the other relaxation times when the loading is small. At higher loadings the effect merges with the velocity and temperature relaxation processes.

11.5 Other Linear Perturbation Analyses

In the preceding section we examined the behavior of small perturbations about a constant and uniform state of the mixture. The perturbation was a plane acoustic wave but the reader will recognize that an essentially similar methodology can be used (and has been) to study other types of flow involving small linear perturbations. An example is steady flow in which the deviation from a uniform stream is small. The equations governing the small deviations in a steady planar flow in, say, the (x, y) plane are then quite analogous to the equations in (x, t) derived in the preceding section.

11.5.1 Stability of Laminar Flow

An important example of this type of solution is the effect that dust might have on the stability of a laminar flow (for instance a boundary layer flow) and, therefore, on the transition to turbulence. Saffman (1962) explored the effect of a small volume fraction of dust on the stability of a parallel flow. As expected and as described in Section 1.3.2, when the response times of the particles are short compared with the typical times associated with the fluid motion, the particles simply alter the effective properties of the fluid, its effective density, viscosity, and compressibility. It follows that under these circumstances the stability is governed by the effective Reynolds

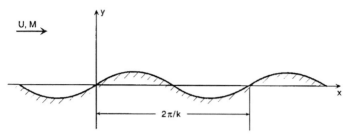

Figure 11.5. Schematic for flow over a wavy wall.

number and effective Mach number. Saffman considered dusty gases at low volume concentrations, α, and low Mach numbers; under those conditions the net effect of the dust is to change the density by $(1 + \alpha \rho_S/\rho_G)$ and the viscosity by $(1 + 2.5\alpha)$. The effective Reynolds number therefore varies like $(1 + \alpha \rho_S/\rho_G)/(1 + 2.5\alpha)$. Because $\rho_S \gg \rho_G$ the effective Reynolds number is increased and, therefore, in the small relaxation time range, the dust is destabilizing. Conversely for large relaxation times, the dust stabilizes the flow.

11.5.2 Flow over a Wavy Wall

A second example of this type of solution that was investigated by Zung (1967) is steady particle-laden flow over a wavy wall of small amplitude (Figure 11.5) so that only the terms that are linear in the amplitude need be retained. The solution takes the form

$$\exp(i\kappa_1 x - i\kappa_2 y), \tag{11.48}$$

where $2\pi/\kappa_1$ is the wavelength of the wall whose mean direction corresponds with the x axis and κ_2 is a complex number whose real part determines the inclination of the characteristics or Mach waves and whose imaginary part determines the attenuation with distance from the wall. The value of κ_2 is obtained in the solution from a dispersion relation that has many similarities to Eq. (11.45). Typical computations of κ_2 are presented in Figure 11.6. The asymptotic values for large t_u that occur on the right in this figure correspond to cases in which the particle motion is constant and unaffected by the waves. Consequently, in subsonic flows ($M = U/c_C < 1$) in which there are no characteristics, the value of $\text{Re}\{\kappa_2/\kappa_1\}$ asymptotes to zero and the waves decay with distance from the wall such that $\text{Im}\{\kappa_2/\kappa_1\}$ tends to $(1 - M^2)^{\frac{1}{2}}$. Conversely, in supersonic flows ($M = U/c_C > 1$) $\text{Re}\{\kappa_2/\kappa_1\}$ asymptotes to the tangent of the Mach wave angle in the gas alone, namely $(M^2 - 1)^{\frac{1}{2}}$, and the decay along these characteristics is zero.

At the other extreme, the asymptotic values as t_u approaches zero correspond to the case of the effective gas whose properties are given in Section 11.2.2. Then the appropriate Mach number, M_0, is that based on the speed of sound in the effective gas [Eq. (11.16)]. In the case of Figure 11.6, $M_0^2 = 2.4M^2$. Consequently, in subsonic

11.6 Small Slip Perturbation

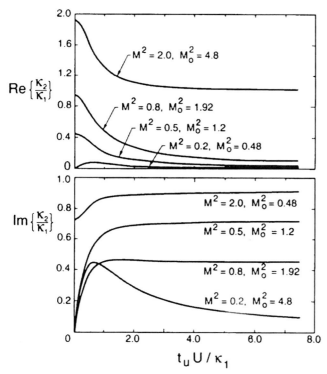

Figure 11.6. Typical results from the wavy wall solution of Zung (1969). Real and imaginary parts of κ_2/κ_1 are plotted against $t_u U/\kappa_1$ for various mean Mach numbers, $M = U/c_C$, for the case of $t_T/t_u = 1$, $c_{pC}/c_{sD} = 1$, $\gamma = 1.4$ and a particle loading, $\xi = 1$.

flows ($M_0 < 1$), the real and imaginary parts of κ_2/κ_1 tend to zero and $(1 - M_0^2)^{\frac{1}{2}}$ respectively as t_u tends to zero. In supersonic flows ($M_0 > 1$), they tend to $(M_0^2 - 1)^{\frac{1}{2}}$ and zero respectively.

11.6 Small Slip Perturbation

The analyses described in the preceding two sections, 11.4 and 11.5, used a linearization about a uniform and constant mean state and assumed that the perturbations in the variables were small compared with their mean values. Another, different linearization known as the small slip approximation can be advantageous in other contexts in which the mean state is more complicated. It proceeds as follows. First, recall that the solutions always asymptote to those for a single effective gas when t_u and t_T tend to zero. Therefore, when these quantities are small and the slip between the particles and the gas is correspondingly small, we can consider constructing solutions in which the flow variables are represented by power series expansions in one of these small quantities, say t_u, and it is assumed that the other (t_T) is of a similar order. Then, generically,

$$Q(x_i, t) = Q^{(0)}(x_i, t) + t_u Q^{(1)}(x_i, t) + t_u^2 Q^{(2)}(x_i, t) + \ldots, \quad (11.49)$$

where Q represents any of the flow quantities, u_{Ci}, u_{Di}, T_C, T_D, p, ρ_C, α_C, α_D, and so on. In addition, it is assumed for the reasons given above that the slip velocity and slip temperature, $(u_{Ci} - u_{Di})$ and $(T_C - T_D)$, are of the order of t_u so that

$$u_{Ci}^{(0)} = u_{Di}^{(0)} = u_i^{(0)}; \quad T_C^{(0)} = T_D^{(0)} = T^{(0)}. \tag{11.50}$$

Substituting these expansions into the basic Eqs. (11.6), (11.7), and (11.8) and gathering together the terms of like order in t_u we obtain the following zeroth order continuity, momentum, and energy relations (omitting gravity):

$$\frac{\partial}{\partial x_i}\left((1+\xi)\rho_C^{(0)} u_i^{(0)}\right) = 0 \tag{11.51}$$

$$(1+\xi)\rho_C^{(0)} u_k^{(0)} \frac{\partial u_i^{(0)}}{\partial x_k} = -\frac{\partial p^{(0)}}{\partial x_i} + \frac{\partial \sigma_{Cik}^{D(0)}}{\partial x_k} \tag{11.52}$$

$$\rho_C^{(0)} u_k^{(0)} (c_{pC} + \xi c_{sD}) \frac{\partial T^{(0)}}{\partial x_k} = u_k^{(0)} \frac{\partial p^{(0)}}{\partial x_k} + \sigma_{Cik}^{D(0)} \frac{\partial u_i^{(0)}}{\partial x_k}. \tag{11.53}$$

Note that Marble (1970) also includes thermal conduction in the energy equation. Clearly the above are just the equations for single-phase flow of the effective gas defined in Section 11.2.2. Conventional single-phase gas dynamic methods can therefore be deployed to obtain their solution.

Next, the relaxation Eqs. (11.22) and (11.25) that are first order in t_u yield the following:

$$u_k^{(0)} \frac{\partial u_i^{(0)}}{\partial x_k} = u_{Ci}^{(1)} - u_{Di}^{(1)} \tag{11.54}$$

$$u_k^{(0)} \frac{\partial T^{(0)}}{\partial x_k} = \left(\frac{t_u}{t_T}\right)\left(\frac{Nu}{2}\right)\left(T_C^{(1)} - T_D^{(1)}\right). \tag{11.55}$$

From these the slip velocity and slip temperature can be calculated once the zeroth order solution is known.

The third step is to evaluate the modification to the effective gas solution caused by the slip velocity and temperature; in other words, to evaluate the first-order terms, $u_{Ci}^{(1)}$, $T_C^{(1)}$, and so on. The relations for these are derived by extracting the $O(t_u)$ terms from the continuity, momentum, and energy equations. For example, the continuity equation yields the following:

$$\frac{\partial}{\partial x_i}\left[\xi\rho_C^{(0)}(u_{Ci}^{(1)} - u_{Di}^{(1)}) + u_i^{(0)}\left(\rho_D \alpha_D^{(1)} - \xi\rho_C^{(1)}\right)\right] = 0. \tag{11.56}$$

This and the corresponding first-order momentum, and energy equations can then be solved to find the $O(t_u)$ slip perturbations to the gas and particle flow variables. For further details the reader is referred to Marble (1970).

A particular useful application of the slip perturbation method is to the one-dimensional steady flow in a convergent/divergent nozzle. The zeroth order, effective

11.6 Small Slip Perturbation

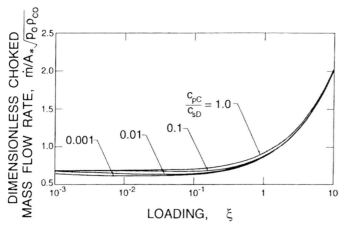

Figure 11.7. The dimensionless choked mass flow rate as a function of loading, ξ, for $\gamma_C = 1.4$ and various specific heat ratios, c_{pC}/c_{sD} as shown.

gas solution leads to pressure, velocity, temperature, and density profiles that are straightforward functions of the Mach number, which is, in turn, derived from the cross-sectional area. This area is used as a surrogate axial coordinate. Here we focus on just one part of this solution, namely the choked mass flow rate, $\dot{m}$, that, according to the single-phase, effective gas analysis will be given by the following:

$$\frac{\dot{m}}{A_*(p_0\rho_{C0})^{\frac{1}{2}}} = (1+\xi)^{\frac{1}{2}}\gamma^{\frac{1}{2}}\left(\frac{2}{1+\gamma}\right)^{(\gamma+1)/2(\gamma-1)}, \qquad (11.57)$$

where p_0 and ρ_{C0} refer to the pressure and gas density in the upstream reservoir, A_* is the throat cross-sectional area, and γ is the effective specific heat ratio as given in Eq. (11.14). The dimensionless choked mass flow rate on the left of Eq. (11.57) is a function only of ξ, γ_C, and the specific heat ratio, c_{pC}/c_{sD}. As shown in Figure 11.7, this is primarily a function of the loading ξ and is only weakly dependent on the specific heat ratio.

12

Sprays

12.1 Introduction

Sprays are an important constituent of many natural and technological processes and range in scale from the very large dimensions of the global air/sea interaction and the dynamics of spillways and plunge pools to the smaller dimensions of fuel injection and ink-jet systems. In this chapter we first examine the processes by which sprays are formed and some of the resulting features of those sprays. Then, because the combustion of liquid fuels in droplet form constitute such an important component of our industrialized society, we focus on the evaporation and combustion of single droplets and follow that with an examination of the features involved in the combustion of sprays.

12.2 Types of Spray Formation

In general, sprays are formed when the interface between a liquid and a gas becomes deformed and droplets of liquid are generated. These then migrate out into the body of the gas. Sometimes the gas plays a negligible role in the kinematics and dynamics of the droplet formation process; this simplifies the analyses of the phenomena. In other circumstances the gas dynamic forces generated can play an important role. This tends to occur when the relative velocity between the gas and the liquid becomes large as is the case, for example, with hurricane-generated ocean spray.

Several prototypical flow geometries are characteristic of the natural and technological circumstances in which spray formation is important. The first prototypical geometry is the flow of a gas over a liquid surface. When the relative velocity is sufficiently large, the interfacial shear stress produces waves on the interface and the breakup of the waves generates a spray that is transported further into the gas phase by the turbulent motions. Ocean spray generated in high wind conditions falls into this category as does annular, vertical two-phase flow. In some fuel injectors a coflowing gas jet is often added to enhance spray formation. Section 12.4.2 provides an overview of this class of spray formation processes.

A second, related configuration is a liquid pool or ocean into which gas is injected so that the bubbles rise up to break through the free surface of the liquid. In the more quiescent version of this configuration, the spray is formed by process of breakthrough (see Section 12.4.1). However, as the superficial gas flux is increased, the induced liquid motions become more violent and spray is formed within the gas bubbles. This spray is then released when the bubbles reach the surface. An example of this is the spray contained within the gas phase of churn-turbulent flow in a vertical pipe.

A third configuration is the formation of a spray due to condensation in a vapor flow. This process is governed by a very different set of physical principles. The nucleation mechanisms involved are beyond the scope of this book.

The fourth configuration is the break up of a liquid jet propelled through a nozzle into a gaseous atmosphere. The unsteady, turbulent motions in the liquid (or the gas) generate ligaments of liquid that project into the gas and the breakup of these ligaments creates the spray. The jet may be laminar or turbulent when it leaves the nozzle and the details of ligament formation, jet breakup, and spray formation are somewhat different in the two cases. Sections 12.4.3 and 12.4.4 summarizes the processes of this flow configuration.

One area in which sprays play a very important role is in the combustion of liquid fuels. We conclude this chapter with brief reviews of the important phenomena associated with the combustion of sprays, beginning with the evaporation of droplets and concluding with droplet and droplet cloud combustion.

12.3 Ocean Spray

Before proceeding with the details of the formation of spray at a liquid/gas interface, a few comments are in order regarding the most widely studied example, namely spray generation on the ocean surface. It is widely accepted that the mixing of the two components, namely air and water, at the ocean surface has important consequences for the global environment (see, for example, Liss and Slinn 1983 or Kraus and Businger 1994). The heat and mass exchange processes that occur as a result of the formation of bubbles in the ocean and of droplets in the atmosphere are critical to many important global balances, including the global balances of many gases and chemicals. For example, the bubbles formed by white caps play an important role in the oceanic absorption of carbon dioxide; on the other side of the interface the spray droplets form salt particles that can be carried high into the atmosphere. They, in turn, are an important contributor to condensation nuclei. Small wonder, then, that ocean surface mixing, the formation of bubbles and droplets has been extensively studied (see, for example, Monahan and Van Patten 1989). But the mechanics of these processes are quite complicated, involving as they do, not only the complexity of wave formation and breaking but also the dynamics of turbulence in the presence of free surfaces. This, in turn, may be affected by free surface contamination or dissolved salts because these effect the surface tension and other free surface properties. Thus, for example, the bubble and droplet size distributions formed in salt water are noticeably different from those formed in fresh water (Monahan and Zietlow, 1969). Here, we do not attempt a comprehensive review of this extensive literature but confine ourselves to some of the basic mechanical processes that are believed to influence these oceanic phenomena.

There appears to be some general concensus regarding the process of spray formation in the ocean (Blanchard 1983, Monahan 1989). This holds that, at relatively low wind speeds, the dominant droplet spray is generated by bubbles rising to breach the surface. The details of the droplet formation process are described in greater detail in the next section. The most prolific source of bubbles are the white caps that can cover up to 10% of the ocean surface (Blanchard 1963). Consequently, an understanding of the droplet formation requires an understanding of bubble formation in breaking waves; this, in itself, is a complex process as illustrated by Wood (1991). What is less clear is the role played by wind shear in ocean spray formation (see Section 12.4.2).

Monahan (1989) provides a valuable survey and rough quantification of ocean spray formation, beginning with the white-cap coverage and proceeding through the bubble size distributions to some estimate of the spray size distribution. Of course, the average droplet size decays with elevation above the surface as the larger droplets settle faster; thus, for example, de Leeuw (1987) found the average droplet diameter at a wind speed of 5.5 m/s dropped from 18 μm at an elevation of 2 m to 15 μm at a 10-m elevation. The size also increases with increasing wind speed due to the greater turbulent velocities in the air.

It is also important to observe that there are substantial differences between spray formation in the ocean and in fresh water. The typical bubbles formed by wave breaking are much smaller in the ocean, although the total bubble volume is similar (Wang and Monahan 1995). Because the bubble size determines the droplet size created when the bubble bursts through the surface, it follows that the spray produced in the ocean has many more, smaller droplets. Moreover, the ocean droplets have a much longer lifetime. Whereas fresh-water droplets evaporate completely in an atmosphere with less than 100% relative humidity, salt-water droplets increase their salinity with evaporation until they reach equilibrium with their surroundings. Parenthetically, it is interesting to note that somewhat similar differences have been observed between cavitation bubbles in salt water and fresh water (Ceccio *et al.* 1997); the bubbles in salt water are smaller and more numerous.

12.4 Spray Formation

12.4.1 Spray Formation by Bubbling

When gas bubbles rise through a pool of liquid and approach the free surface, the various violent motions associated with the break through to the cover gas generate droplets that may persist in the cover gas to constitute a spray. Even in an otherwise

12.4 Spray Formation

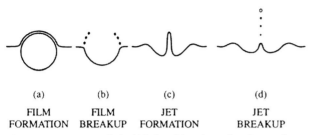

(a) FILM FORMATION   (b) FILM BREAKUP   (c) JET FORMATION   (d) JET BREAKUP

Figure 12.1. Stages of a bubble breaking through a free surface.

quiescent liquid, the details of the bubble breakthrough are surprisingly complicated, as illustrated by the photographs in Figure 12.2. Two of the several important processes are sketched in Figure 12.1. Just prior to breakthrough a film of liquid is formed on the top of the bubble and the disintegration of this film creates one set of droplets. After breakthrough, as surface waves propagate inward (as well as outward) an upward jet is formed in the center of the disruption and the disintegration of this jet also creates droplets. Generally, the largest *jet droplets* are substantially larger than the largest *film droplets*, the latter being about a tenth the diameter of the original bubble.

In both the industrial and oceanic processes, a key question is the range of droplet sizes that will almost immediately fall back into the liquid pool and, conversely, the range of droplet sizes that will be carried high into the atmosphere (cover gas). In the ocean this significant transport above the water surface occurs as a result of turbulent mixing. In the industrial context of a liquid-fluidized bed, the upward transport is often the result of a sufficiently large upward gas flux whose velocity in the cover space exceeds the settling velocity of the droplet (Azbel and Liapis 1983).

12.4.2 Spray Formation by Wind Shear

In annular flows in vertical pipes, the mass of liquid carried as droplets in the gas core is often substantial. Consequently considerable effort has been devoted to studies of the entrainment of droplets from the liquid layer on the pipe wall (Butterworth and Hewitt 1977, Whalley 1987). In many annular flows the droplet concentration in the gas core increases with elevation as illustrated in Figure 12.3.

In steady flow, the mass flux of droplets entrained into the gas core, G_L^E, should be balanced by the mass flux of deposition of droplets onto the wall liquid layer, G_L^D.

Figure 12.2. Photographs by Blanchard (1963) of a bubble breaking through a free surface. Reproduced with permission of the author.

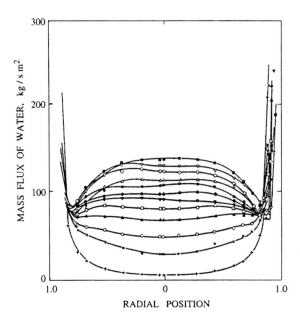

Figure 12.3. Droplet concentration profiles in the gas core of a vertical annular pipe flow (3.2-cm diameter) illustrating the increase with elevation from initiation (lowest line, 15-cm elevation; uppermost line, 531-cm elevation) (from Gill et al. 1963).

Hutchinson and Whalley (1973) observe that droplets are torn from the liquid surface when the wind shear creates and then fractures a surface wave as sketched in Figure 12.4. They suggest that the velocity of ejection of the droplets is related to the friction velocity, $u^* = (\tau_i/\rho_L)^{\frac{1}{2}}$, where τ_i is the interfacial stress and that the entrainment rate, G_L^E, therefore correlates with $(\tau_i \delta/S)^{\frac{1}{2}}$, where δ is the mean liquid layer thickness. They also speculate that the mass deposition rate must be proportional to the core droplet mass concentration, $\rho_L \alpha_L$. As shown in Figure 12.5, the experimental measurements of the concentration do, indeed, appear to correlate with $(\tau_i \delta/S)^{\frac{1}{2}}$ (a typical square root dependence is shown by the solid line in the figure).

McCoy and Hanratty (1977) review the measurements of the deposition mass flux, G_L^D, and the gas core concentration, $\rho_L \alpha_L$, and show that the dimensionless deposition mass transfer coefficient, $G_L^D/\rho_L \alpha_L u^*$, correlates with a dimensionless relaxation time for the droplets defined by $D^2 \rho_L \rho_G u^{*2}/18\mu_G^2$. This correlation is shown in Figure 12.6 and, for a given u^*, can also be considered as a graph with the resulting droplet size,

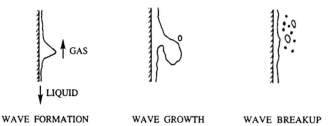

Figure 12.4. Sketch illustrating the ejection of droplets by wind shear in annular flow in a vertical pipe. From Hutchinson and Whalley (1973).

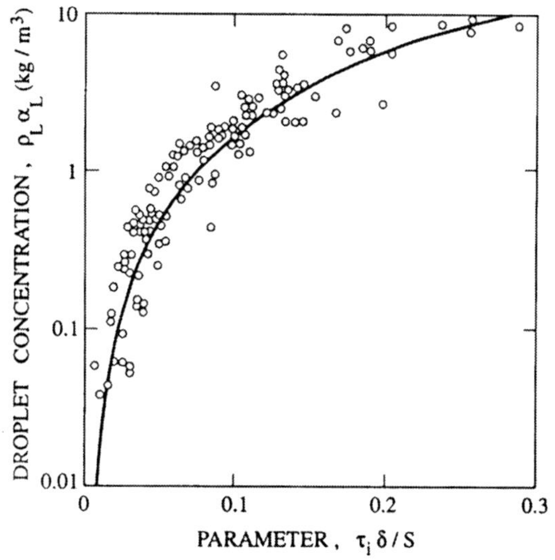

12.4 Spray Formation

Figure 12.5. The mass concentration of liquid droplets in the gas core of an annular flow, $\rho_L \alpha_L$, plotted against $\tau_i \delta / S$. From Hutchinson and Whalley (1973).

D (or rather its square), plotted horizontally; typical droplet sizes are shown in the figure.

12.4.3 Spray Formation by Initially Laminar Jets

In many important technological processes, sprays are formed by the breakup of a liquid jet injected into a gaseous atmosphere. One of the most important of these is fuel injection in power plants, aircraft, and automobile engines and here the character of the spray formed is critical not only for performance but also for pollution control. Consequently much effort has gone into the design of the nozzles (and therefore the jets) that produce sprays with desirable characteristics. *Atomizing* nozzles are those that produce particularly fine sprays. Other examples of technologies in which there is a

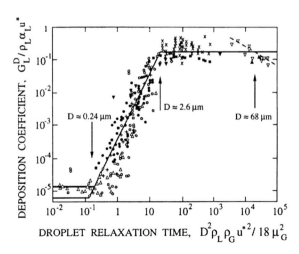

Figure 12.6. The dimensionless deposition mass transfer coefficient, $G_L^D / \rho_L \alpha_L u^*$, for vertical annular flow plotted against a dimensionless relaxation time for the droplets in the core, $D^2 \rho_L \rho_G u^{*2} / 18 \mu_G^2$. A summary of experimental data compiled by McCoy and Hanratty (1977).

Figure 12.7. Photographs of an initially laminar jet emerging from a nozzle. The upper photograph shows the instability wave formation and growth and the lower shows the spray droplet formation at a location 4 diameters further downstream. Figure 12.9 shows the same jet even further downstream. Reproduced from Hoyt and Taylor (1977b) with the permission of the authors.

similar focus on the nature of the spray produced are ink-jet printing and the *scrubbing* of exhaust gases to remove particulate pollutants.

Because of its technological importance, we focus here on the circumstance in which the jet is turbulent when it emerges from the nozzle. However, in passing, we note that the breakup of laminar jets may also be of interest. Two photographs of initially laminar jets taken by Hoyt and Taylor (1977a, b) are reproduced in Figure 12.7. Photographs such as the upper one clearly show that transition to turbulence occurs because the interfacial layer formed when the liquid boundary layer leaves the nozzle becomes unstable. The Tollmein–Schlicting waves (remarkably two-dimensional) exhibit a well-defined wavelength and grow to non-linear amplitudes at which they breakup to form droplets in the gas. Sirignano and Mehring (2000) provide a review of the extensive literature on linear and nonlinear analyses of the stability of liquid jets, not only round jets but also planar and annular jets. The author (Brennen 1970) examined the development of interfacial instability waves in the somewhat different context of cavity flows; this analysis demonstrated that the appropriate length scale is the thickness of the internal boundary layer, δ, on the nozzle walls at the point where the free surface detaches. This is best characterized by the momentum thickness, δ_2, though other measures of the boundary layer thickness have also been used. The stability analysis yields the most unstable wavelength for the Tollmein–Schlicting waves (normalized by δ_2) as a function of the Reynolds number of the interfacial boundary layer (based on the jet velocity and δ_2). At larger Reynolds number, the ratio of

12.4 Spray Formation

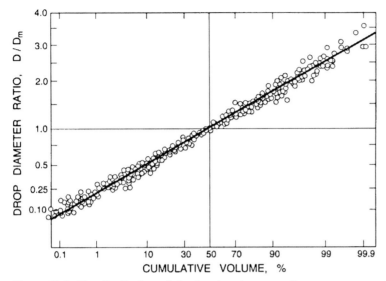

Figure 12.8. The distribution of droplet sizes in sprays from many types of nozzles plotted on a root/normal graph. Adapted from Simmons (1977).

wavelength to δ_2 reaches an asymptotic value of about 25, independent of Reynolds number. Brennen (1970) and Hoyt and Taylor (1977a, b) observe that these predicted wavelengths are in accord with those observed.

A natural extension of this analysis is to argue that the size of the droplets formed by the nonlinear breakup of the instability waves will scale with the wavelength of those waves. Indeed, the pictures of Hoyt and Taylor (1977a, b) exemplified by the lower photograph in Figure 12.7 suggest that this is the case. It follows that at higher Reynolds numbers, the droplet size should scale with the boundary layer thickness, δ_2. Wu, Miranda, and Faeth (1995) have shown that this is indeed the case for the initial drop formation in initially nonturbulent jets.

Further downstream the turbulence spreads throughout the core of the jet and the subsequent jet breakup and droplet formation is then similar to that of jets that are initially turbulent. We now turn to that circumstance.

12.4.4 Spray Formation by Turbulent Jets

Because of the desirability in many technological contexts of nozzles that produce jets that are fully turbulent from the start, there has been extensive testing of many nozzle designed with this objective in mind. Simmons (1977) makes the useful observation that sprays produced by a wide range of nozzle designs have similar droplet size distributions when these are compared in a *root/normal* graph, as shown in Figure 12.8. Here the ordinate corresponds to $(D/D_m)^{\frac{1}{2}}$, where D_m is the mass mean diameter (see Section 1.1.4). The horizontal scale is stretched to correspond to a normal distribution. The straight line to which all the data collapse implies that $(D/D_m)^{\frac{1}{2}}$ follows a normal

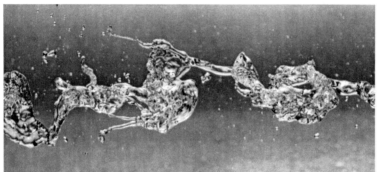

Figure 12.9. A continuation from Figure 12.7 showing two further views of the jet at 72 diameters (above) and 312 diameters (below) downstream from the nozzle. The latter illustrates the final breakup of the jet. Reproduced from Hoyt and Taylor (1977b) with the permission of the authors.

distribution. Because the size distributions from many different nozzles all have the same form, this implies that the sprays from all these nozzles can be characterized by a single diameter, D_m. An alternative measure is the Sauter mean diameter, D_s, because D_s/D_m will have the universal value of 1.2 under these circumstances.

Early studies of liquid jets by Lee and Spencer (1933) and others revealed that the turbulence in a liquid jet was the primary initiator of breakup. Subsequent studies (for example, Phinney 1973, Hoyt and Taylor 1977a, b, Ervine and Falvey 1987, Wu *et al.* 1995, Sarpkaya and Merrill 1998) have examined how this process works. In the early stages of breakup, the turbulent structures in the jet produce ligaments that project into the gaseous phase and then fragment to form droplets as illustrated in Figure 12.7. The studies by Wu *et al.* (1995) and others indicate that the very smallest structures in the turbulence do not have the energy to overcome the restraining forces of surface tension. However, because the smaller turbulent structures distort the free surface more rapidly than the larger structures, the first ligaments and droplets to appear are generated by the smallest scale structures that *are* able to overcome surface tension. This produces small droplets. But these small structures also decay more rapidly with distance from the nozzle. Consequently, further downstream progressively larger structures cause larger ligaments and droplets and therefore add droplets at the higher end of the size distribution. Finally, the largest turbulent structures comparable with the jet diameter or width initiate the final stage of jet decomposition as illustrated in Figure 12.9.

12.4 Spray Formation

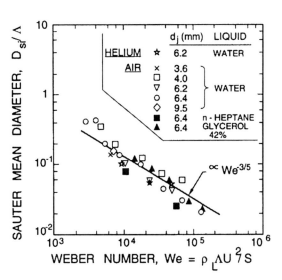

Figure 12.10. The Sauter mean diameter, D_{si}, of the initial droplets formed (divided by the typical dimension of the jet, Λ) in turbulent round jets as a function of the Weber number, We $= \rho_L \Lambda U^2/S$. The points are experimental measurements for various liquids and jet diameters, d_j. Adapted from Wu et al. (1995).

Wu, Miranda, and Faeth (1995) utilized this understanding of the spray formation and jet breakup process to create scaling laws of the phenomenon. With a view to generalizing the results to turbulent jets of other cross sections, the radial integral length scale of the turbulence is denoted by 4Λ, where, in the case of round jets, $\Lambda = d_j/8$, where d_j is the jet diameter. Wu et al. (1995) then argue that the critical condition for the initial formation of a droplet (the so-called primary breakup condition) occurs when the kinetic energy of a turbulent eddy of the critical size is equal to the surface energy required to form a droplet of that size. This leads to the following expression for the Sauter mean diameter of the initial droplets, D_{si}:

$$\frac{D_{si}}{\Lambda} \propto \text{We}^{-\frac{3}{5}} \qquad (12.1)$$

where the Weber number is defined as follows: We $= \rho_L \Lambda U^2$, U being the typical or mean velocity of the jet. Figure 12.10 from Wu et al. (1995) demonstrates that data from a range of experiments with round jets confirm that D_{si}/Λ does appear to be a function only of We and that the correlation is close to the form given in Eq. (12.1).

Wu et al. (1995) further argue that the distance, x_i, from the nozzle to the place where primary droplet formation takes place may be estimated using an eddy convection velocity equal to U and the time required for Rayleigh breakup of a ligament having a diameter equal to the D_{si}. This leads to the following:

$$\frac{x_i}{\Lambda} \propto \text{We}^{-\frac{2}{5}} \qquad (12.2)$$

and, as shown in Figure 12.11, the data for different liquids and jet diameters are in rough accord with this correlation.

Downstream of the point where primary droplet formation occurs, progressively larger eddies produce larger droplets and Wu et al. (1995) use extensions of their

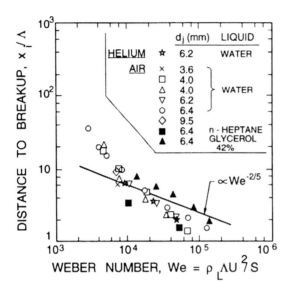

Figure 12.11. The ratio of the distance from the nozzle to the point where turbulent breakup begins (divided by Λ) for turbulent round jets as a function of the Weber number, $We = \rho_L \Lambda U^2 / S$. The points are experimental measurements for various liquids and jet diameters, d_j. Adapted from Wu et al. (1995).

theory to generate the following expression for the Sauter mean diameter, D_s, of the droplets formed at a distance, x, downstream of the nozzle as follows:

$$\frac{D_s}{\Lambda} \propto \left(\frac{x}{\Lambda We^{\frac{1}{2}}} \right)^{\frac{2}{3}}. \qquad (12.3)$$

As shown in Figure 12.12 the experimental measurements show fair agreement with this approximate theory.

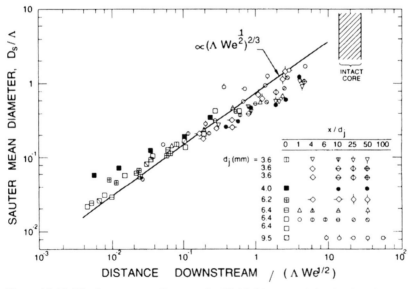

Figure 12.12. The Sauter mean diameter, D_s (divided by Λ), of the droplets formed at a distance, x, from the nozzle for turbulent round jets for various Weber numbers, $We = \rho_L \Lambda U^2 / S$. The points are experimental measurements for various liquids and jet diameters, d_j. Adapted from Wu et al. (1995).

Using this information, the evolution of the droplet size distribution with distance from the nozzle can be constructed as follows. Assuming Simmons size distributions, the droplet size distribution may be characterized by the Sauter mean diameter, D_S. The primary breakup yields droplets characterized by the initial D_{si} of Eq. (12.1). Then, moving downstream along the jet, contributions with progressively larger droplets are added until the jet finally disintegrates completely.

Several footnotes should be added to this picture. First, the evolution described assumes that the gaseous phase plays a negligible role in the dynamics. Wu and Faeth (1993) demonstrate that this will be the case only when $\rho_L/\rho_G > 500$. However, this is frequently the case in practical applications. Second, the above can be extended to other free jet geometries. Dai *et al.* (1998) demonstrate that the simple use of a hydraulic diameter allows the same correlations to be used for plane jets. Conversely, wall jets appear to follow different correlations presumably because the generation of vorticity in wall jets causes a different evolution of the turbulence than occurs in free jets (Dai *et al.* 1997, Sarpkaya and Merrill 1998). Sarpkaya and Merrill's (1998) experiments with wall jets on horizontal smooth and roughened walls exhibit a ligament formation process qualitatively similar to that of free jets. The droplets created by the ligament breakup have a diameter about 0.6 of the wall jet thickness and quite independent of Weber number or plate roughness over the range tested.

Finally, the reader will note that the above characterizations are notably incomplete because they do not address the issue of the total number or mass of droplets produced at each stage in the process. Though this is crucial information in many technological contexts, it has yet to be satisfactorily modeled.

12.5 Single-Droplet Mechanics

12.5.1 Single-Droplet Evaporation

The combustion of liquid fuels in droplet form or of solid fuels in particulate form constitute a very important component of our industrialized society. Spray evaporation is important, in part because it constitutes the first stage in the combustion of atomized liquid fuels in devices such as industrial furnaces, diesel engines, liquid rocket engines, and gas turbines. Consequently, the mechanics of the evaporation and subsequent combustion have been extensively documented and studied (see, for example, Williams 1965, Glassman 1977, Law 1982, Faeth 1983, Kuo 1986) and their air pollution consequences examined in detail (see, for example, Flagan and Seinfeld 1988). It is impossible to present a full review of these subjects within the confines of this book, but it is important and appropriate to briefly review some of the basic multiphase flow phenomena that are central to these processes.

An appropriate place to start is with evaporation of a single droplet in a quiescent environment and we will follow the description given in Flagan and Seinfeld (1988). Heat diffusing inward from the combustion zone, either one surrounding a gas/droplet cloud or one located around an individual droplet, will cause the heating and evaporation of the droplet(s). It transpires that it is adequate for most purposes to model single droplet evaporation as a steady-state process (assuming the droplet radius is only varying slowly). Because the liquid density is much greater than the vapor density, the droplet radius, R, can be assumed constant in the short term and this permits a steady flow analysis in the surrounding gas. Then, because the outward flow of total mass and of vapor mass at every radius, r, is equal to $\dot{m}_V$ and there is no net flux of the other gas, conservation of total mass and conservation of vapor lead through Eqs. (1.21) and (1.29) and Fick's law [Eq. (1.37)] to the following:

$$\frac{\dot{m}_V}{4\pi} = \rho u r^2 = \rho(u)_{r=R} R^2 \tag{12.4}$$

and

$$\frac{\dot{m}_V}{4\pi} = \rho u r^2 x_V - \rho r^2 D \frac{dx_V}{dr}, \tag{12.5}$$

where D is the mass diffusivity. These represent equations to be solved for the mass fraction of the vapor, x_V. Eliminating u and integrating produces

$$\frac{\dot{m}_V}{4\pi} = \rho R D \, \ln\left(1 + \frac{(x_V)_{r=\infty} - (x_V)_{r=R}}{(x_V)_{r=R} - 1}\right). \tag{12.6}$$

Next we examine the heat transfer in this process. The equation governing the radial convection and diffusion of heat is the following:

$$\rho u c_p \frac{dT}{dr} = \frac{1}{r^2} \frac{d}{dr}\left(r^2 k \frac{dT}{dr}\right), \tag{12.7}$$

where c_p and k are representative averages of, respectively, the specific heat at constant pressure and the thermal conductivity of the gas. Substituting for u from Eq. (12.4) this can be integrated to yield the following:

$$\dot{m}_V c_p (T + C) = 4\pi r^2 k \frac{dT}{dr}, \tag{12.8}$$

where C is an integration constant that is evaluated by means of the boundary condition at the droplet surface. The heat required to vaporize a unit mass of fuel whose initial temperature is denoted by T_i is clearly that required to heat it to the saturation temperature, T_e, plus the latent heat, $\mathcal{L}$, or $c_s(T_e - T_i) + \mathcal{L}$. The second contribution is usually dominant so the heat flux at the droplet surface can be set as follows:

$$4\pi R^2 k \left(\frac{dT}{dr}\right)_{r=R} = \dot{m}_V \mathcal{L}. \tag{12.9}$$

Using this boundary condition, C can be evaluated and Eq. (12.8) further integrated to obtain the following:

$$\frac{\dot{m}_V}{4\pi} = \frac{Rk}{c_p} \ln\left\{1 + \frac{c_p(T_{r=\infty} - T_{r=R})}{\mathcal{L}}\right\}. \tag{12.10}$$

12.5 Single-Droplet Mechanics

To solve for $T_{r=R}$ and $(x_V)_{r=R}$ we eliminate $\dot{m}_V$ from Eqs. (12.6) and (12.10) and obtain the following:

$$\frac{\rho D c_p}{k} \ln\left(1 + \frac{(x_V)_{r=\infty} - (x_V)_{r=R}}{(x_V)_{r=R} - 1}\right) = \ln\left(1 + \frac{c_p(T_{r=\infty} - T_{r=R})}{\mathcal{L}}\right). \quad (12.11)$$

Given the transport and thermodynamic properties k, c_p, $\mathcal{L}$, and D (neglecting variations of these with temperature) as well as $T_{r=\infty}$ and ρ, this equation relates the droplet surface mass fraction, $(x_V)_{r=R}$, and temperature $T_{r=R}$. Of course, these two quantities are also connected by the thermodynamic relation

$$(x_V)_{r=R} = \frac{(\rho_V)_{r=R}}{\rho} = \frac{(p_V)_{r=R}}{p} \frac{\mathcal{M}_V}{\mathcal{M}}, \quad (12.12)$$

where $\mathcal{M}_V$ and $\mathcal{M}$ are the molecular weights of the vapor and the mixture. Equation (12.11) can then be solved given the relation 12.12 and the saturated vapor pressure p_V as a function of temperature. Note that because the droplet size does not occur in Eq. (12.11), the surface temperature is independent of the droplet size.

Once the surface temperature and mass fraction are known, the rate of evaporation can be calculated from Eq. (12.7) by substituting $\dot{m}_V = 4\pi \rho_L R^2 dR/dt$ and integrating to obtain the following:

$$R^2 - (R_{t=0})^2 = \left\{\frac{2k}{c_p} \ln\left(1 + \frac{c_p(T_{r=\infty} - T_{r=R})}{\mathcal{L}}\right)\right\} t. \quad (12.13)$$

Thus the time required for complete evaporation, t_{ev}, is as follows:

$$t_{ev} = c_p R_{t=0}^2 \left\{2k \ln\left(1 + \frac{c_p(T_{r=\infty} - T_{r=R})}{\mathcal{L}}\right)\right\}^{-1}. \quad (12.14)$$

This quantity is important in combustion systems. If it approaches the residence time in the combustor this may lead to incomplete combustion, a failure that is usually avoided by using atomizing nozzles that make the initial droplet size, $R_{t=0}$, as small as possible.

Having outlined the form of the solution for an evaporating droplet, albeit in the simplest case, we now proceed to consider the combustion of a single droplet.

12.5.2 Single-Droplet Combustion

For very small droplets of a volatile fuel, droplet evaporation is completed early in the heating process and the subsequent combustion process is unchanged by the fact that the fuel began in droplet form. Conversely, for larger droplets or less volatile fuels, droplet evaporation will be a controlling process during combustion. Consequently, analysis of the combustion of a single droplet begins with the single-droplet evaporation discussed in the preceding section. Then single-droplet combustion consists of the outward diffusion of fuel vapor from the droplet surface and the inward diffusion of oxygen (or other oxidant) from the far field, with the two reacting in a flame front at

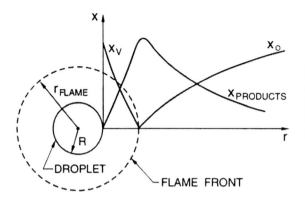

Figure 12.13. Schematic of single droplet combustion indicating the radial distributions of fuel/vapor mass fraction, x_V, oxidant mass fraction, x_O, and combustion products mass fraction.

a certain radius from the droplet. It is usually adequate to assume that this combustion occurs instantaneously in a thin *flame front* at a specific radius, r_{flame}, as indicated in Figure 12.13. As in the last section, a steady-state process will be assumed in which the mass rates of consumption of fuel and oxidant in the flame are denoted by $\dot{m}_{VC}$ and $\dot{m}_{OC}$ respectively. For combustion stoichiometry we therefore have the following:

$$\dot{m}_{VC} = \nu \dot{m}_{OC}, \qquad (12.15)$$

where ν is the mass-based stoichiometric coefficient for complete combustion. Moreover, the rate of heat release due to combustion will be $Q\dot{m}_{VC}$, where Q is the combustion heat release per unit mass of fuel. Assuming the mass diffusivities for the fuel and oxidant and the thermal diffusivity ($k/\rho c_p$) are all the same (a Lewis number of unity) and denoted by D, the thermal and mass conservation equations for this process can then be written as follows:

$$\dot{m}_V \frac{dT}{dr} = \frac{d}{dr}\left(4\pi r^2 \rho D \frac{dT}{dr}\right) + 4\pi r^2 \frac{Q\dot{m}_{VC}}{c_p} \qquad (12.16)$$

$$\dot{m}_V \frac{dx_V}{dr} = \frac{d}{dr}\left(4\pi r^2 \rho D \frac{dx_V}{dr}\right) + 4\pi r^2 \dot{m}_{VC} \qquad (12.17)$$

$$\dot{m}_V \frac{dx_O}{dr} = \frac{d}{dr}\left(4\pi r^2 \rho D \frac{dx_O}{dr}\right) - 4\pi r^2 \dot{m}_{OC}, \qquad (12.18)$$

where x_O is the mass fraction of oxidant.

Using Eq. (12.15) to eliminate the reaction rate terms these become the following:

$$\dot{m}_V \frac{d}{dr}(c_p T + Q x_V) = \frac{d}{dr}\left(4\pi r^2 \rho D \frac{d}{dr}(c_p T + Q x_V)\right) \qquad (12.19)$$

$$\dot{m}_V \frac{d}{dr}(c_p T + \nu Q x_O) = \frac{d}{dr}\left(4\pi r^2 \rho D \frac{d}{dr}(c_p T + \nu Q x_O)\right) \qquad (12.20)$$

$$\dot{m}_V \frac{d}{dr}(x_V - \nu x_O) = \frac{d}{dr}\left(4\pi r^2 \rho D \frac{d}{dr}(x_V - \nu x_O)\right). \qquad (12.21)$$

12.5 Single-Droplet Mechanics

Appropriate boundary conditions on these relations are (1) the droplet surface heat flux condition [Eq. (12.9)], (2) zero droplet surface flux of nonfuel gases from Eqs. (12.4) and (12.5), (3) zero oxidant flux at the droplet surface, (4) zero oxidant mass fraction at the droplet surface, (5) temperature at the droplet surface, $T_{r=R}$, (6) known temperature far from the flame, $T_{r=\infty}$, (7) zero fuel/vapor mass fraction far from the flame, $(x_V)_{r=\infty} = 0$, and (8) a known oxidant mass fraction far from the flame, $(x_O)_\infty$. Using these conditions, Eqs. (12.19), (12.20), and (12.21) may be integrated twice to obtain the following:

$$\frac{\dot{m}_V}{4\pi \rho D r} = \ln\left\{\frac{c_p(T_{r=\infty} - T_{r=R}) + \mathcal{L} - \mathcal{Q}}{c_p(T - T_{r=R}) + \mathcal{L} - \mathcal{Q}(1 - x_V)}\right\} \tag{12.22}$$

$$\frac{\dot{m}_V}{4\pi \rho D r} = \ln\left\{\frac{c_p(T_{r=\infty} - T_{r=R}) + \mathcal{L} + \nu \mathcal{Q}(x_O)_{r=\infty}}{c_p(T - T_{r=R}) + \mathcal{L} + \nu \mathcal{Q} x_O}\right\} \tag{12.23}$$

$$\frac{\dot{m}_V}{4\pi \rho D r} = \ln\left\{\frac{1 + \nu(x_O)_{r=\infty}}{1 - x_V + \nu x_O}\right\} \tag{12.24}$$

and evaluating these expressions at the droplet surface leads to the following:

$$\frac{\dot{m}_V}{4\pi \rho D R} = \ln\left\{\frac{c_p(T_{r=\infty} - T_{r=R}) + \mathcal{L} - \mathcal{Q}}{\mathcal{L} - \mathcal{Q}(1 - (x_V)_{r=R})}\right\} \tag{12.25}$$

$$\frac{\dot{m}_V}{4\pi \rho D R} = \ln\left\{\frac{c_p(T_{r=\infty} - T_{r=R}) + \mathcal{L} + \nu \mathcal{Q}(x_O)_{r=\infty}}{\mathcal{L}}\right\} \tag{12.26}$$

$$\frac{\dot{m}_V}{4\pi \rho D R} = \ln\left\{\frac{1 + \nu(x_O)_{r=\infty}}{1 - (x_V)_{r=R}}\right\} \tag{12.27}$$

and consequently the unknown surface conditions, $T_{r=R}$ and $(x_V)_{r=R}$, may be obtained from the following relations:

$$\frac{1 + \nu(x_O)_{r=\infty}}{1 - (x_V)_{r=R}} = \frac{c_p(T_{r=\infty} - T_{r=R}) + \mathcal{L} + \nu \mathcal{Q}(x_O)_{r=\infty}}{\mathcal{L}}$$
$$= \frac{c_p(T_{r=\infty} - T_{r=R}) + \mathcal{L} - \mathcal{Q}}{\mathcal{L} - \mathcal{Q}(1 - (x_V)_{r=R})}. \tag{12.28}$$

Having solved for these surface conditions, the evaporation rate, $\dot{m}_V$, would follow from any one of Eqs. (12.25) to (12.27). However, a simple, approximate expression for $\dot{m}_V$ follows from Eq. (12.26) because the term $c_p(T_{r=\infty} - T_{r=R})$ is generally small compared with $\mathcal{Q}(x_V)_{r=R}$. Then

$$\dot{m}_V \approx 4\pi R \rho D \ln\left(1 + \frac{\nu \mathcal{Q}(x_O)_{r=\infty}}{\mathcal{L}}\right). \tag{12.29}$$

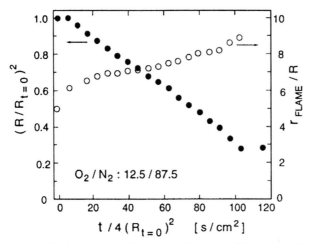

Figure 12.14. Droplet radius, R, and the ratio of the flame radius to the droplet radius, r_{flame}/R, for a burning octane droplet in a 12.5% O_2, 87.5% N_2, 0.15 atm environment. Adapted from Law (1982).

The position of the flame front, $r = r_{\text{flame}}$, follows from Eq. (12.27) by setting $x_V = x_O = 0$ as follows:

$$r_{\text{flame}} = \frac{\dot{m}_V}{4\pi\rho D \ln(1 + \nu(x_O)_{r=\infty})} \approx R \frac{\ln(1 + \nu\mathcal{Q}(x_O)_{r=\infty}/\mathcal{L})}{\ln(1 + \nu(x_O)_{r=\infty})}. \quad (12.30)$$

As one might expect, the radius of the flame front increases rapidly at small oxygen concentrations, $(x_O)_\infty$, because this oxygen is quickly consumed. However, the second expression demonstrates that r_{flame}/R is primarily a function of $\mathcal{Q}/\mathcal{L}$; indeed for small values of $(x_O)_{r=\infty}$ it follows that $r_{\text{flame}}/R \approx \mathcal{Q}/\mathcal{L}$. We discuss the consequences of this in the next section.

Detailed reviews of the corresponding experimental data on single-droplet combustion can be found in numerous texts and review articles, including those listed above. Here we include just two sets of experimental results. Figure 12.14 exemplifies the data on the time history of the droplet radius, R, and the ratio of the flame radius to the droplet radius, r_{flame}/R. Note that after a small initial transient, R^2 decreases quite linearly with time as explicitly predicted by Eq. (12.13) and implicitly contained in the combustion analysis. The slope, $-d(R^2)/dt$, is termed the *burning rate* and examples of the comparison between the theoretical and experimental burning rates are included in Figure 12.15. The flame front location is also shown in Figure 12.14; note that r_{flame}/R is reasonably constant despite the fivefold shrinkage of the droplet.

Further refinements of this simple analysis can also be found in the texts mentioned previously. A few of the assumptions that require further analysis include whether the assumed steady state is pertinent, whether relative motion of the droplets through the gas convectively enhances the heat and mass transfer processes, the role of turbulence

12.6 Spray Combustion

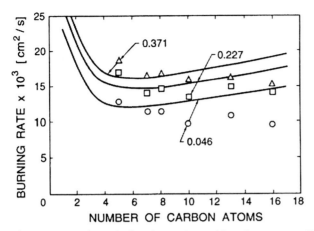

Figure 12.15. Theoretical and experimental burning rates, $-d(R^2)/dt$ (in cm²/s), of various paraffin hydrocarbon droplets ($R = 550$ μm) in a $T_{r=\infty} = 2530°K$ environment with various mass fractions of oxygen, $(x_O)_{r=\infty}$, as shown. Adapted from Faeth and Lazar (1971).

in modifying the heat and mass transfer processes in the gas, whether the chemistry can be modeled by a simple flame front, the complexity introduced by mixtures of liquids of different volatilities, and whether all the diffusivities can be assumed to be similar.

12.6 Spray Combustion

Now consider the combustion of a spray of liquid droplets. When the radius of the flame front around individual droplets is small compared with the distance separating the droplets, each droplet will burn on its own surrounded by a flame front. However, when r_{flame} becomes comparable with the interdroplet separation the flame front will begin to surround a number of droplets and combustion will change to a form of droplet cloud combustion. Figure 12.16 depicts four different spray combustion scenarios as described by Chiu and Croke (1981) (see also Kuo 1986). Because the ratio of the flame front radius to droplet radius is primarily a function of the rate of the combustion heat release per unit mass of fuel to the latent heat of vaporization of the fuel, or $\mathcal{Q}/\mathcal{L}$ as demonstrated in the preceding section, these patterns of droplet cloud combustion occur in different ranges of that parameter. As depicted in Figure 12.16(a), at high values of $\mathcal{Q}/\mathcal{L}$, the flame front surrounds the entire cloud of droplets. Only the droplets in the outer shell of this cloud are heated sufficiently to produce significant evaporation and the outer flow of this vapor fuels the combustion. At somewhat lower values of $\mathcal{Q}/\mathcal{L}$ [Figure 12.16(b)] the entire cloud of droplets is evaporating but the flame front is still outside the droplet cloud. At still lower values of $\mathcal{Q}/\mathcal{L}$ [Figure 12.16(c)], the main flame front is within the droplet cloud and the droplets in the outer shell beyond that main flame front have individual flames surrounding each droplet. Finally at low $\mathcal{Q}/\mathcal{L}$ values [Figure 12.16(d)] every droplet is surrounded by its own flame front. Of

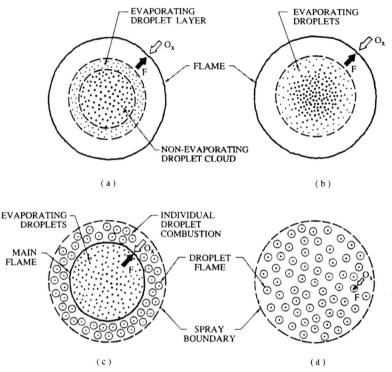

Figure 12.16. Four modes of droplet cloud combustion: (a) Cloud combustion with nonevaporating droplet core, (b) cloud combustion with evaporating droplets, (c) individual droplet combustion shell, (d) single-droplet combustion. Adapted from Chiu and Croke (1981).

course, several of these configurations may be present simultaneously in a particular combustion process. Figure 12.17 depicts one such circumstance occuring in a burning spray emerging from a nozzle.

Note that though we have focused here on the combustion of liquid droplet sprays, the combustion of suspended solid particles is of equal importance. Solid fuels in particulate form are burned both in conventional boilers where they are injected as a dusty gas and in fluidized beds into which granular particles and oxidizing gas are

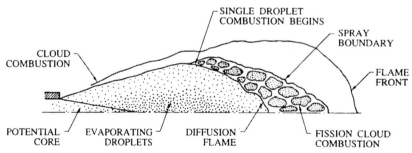

Figure 12.17. An example of several modes of droplet cloud combustion in a burning liquid fuel spray. Adapted from Kuo (1986).

12.6 Spray Combustion

continuously fed. We do not dwell on solid particle combustion because the analysis is very similar to that for liquid droplets. Major differences are the boundary conditions at the particle surface where the devolatilization of the fuel and the oxidation of the char require special attention (see, for example, Gavalas 1982, Flagan and Seinfeld 1988).

13

Granular Flows

13.1 Introduction

Dense fluid-particle flows in which the direct particle/particle interactions are a dominant feature encompass a diverse range of industrial and geophysical contexts (Jaeger *et al.* 1996), including, for example, slurry pipelines (Shook and Roco 1991), fluidized beds (Davidson and Harrison 1971), mining and milling operations, ploughing (Weighardt 1975), abrasive water jet machining, food processing, debris flows (Iverson 1997), avalanches (Hutter 1993), landslides, sediment transport, and earthquake-induced soil liquefaction. In many of these applications, stress is transmitted both by shear stresses in the fluid and by momentum exchange during direct particle/particle interactions. Many of the other chapters in this book analyze flow in which the particle concentration is sufficiently low that the particle-particle momentum exchange is negligible.

In this chapter we address those circumstances, usually at high particle concentrations, in which the direct particle/particle interactions play an important role in determining the flow properties. When those interactions dominate the mechanics, the motions are called *granular flows* and the flow patterns can be quite different from those of conventional fluids. An example is included as Figure 13.1, which shows the downward flow of sand around a circular cylinder. Note the *upstream wake* of stagnant material in front of the cylinder and the empty cavity behind it.

Within the domain of granular flows, there are, as we shall see, several very different types of flow distinguished by the fraction of time for which particles are in contact. For most slow flows, the particles are in contact most of the time. Then large transient structures or assemblages of particles known as *force chains* dominate the rheology and the inertial effects of the random motions of individual particles play little role. Force chains are ephemeral, quasi-linear sequences of particles with large normal forces at their contact points. They momentarily carry much of the stress until they buckle or are superceded by other chains. Force chains were first observed experimentally by Drescher and De Josselin de Jong (1972) and, in computer simulations, by Cundall and Strack (1979).

13.2 Particle Interaction Models

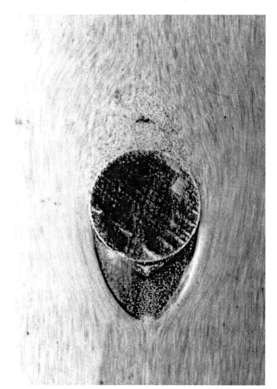

Figure 13.1. Long exposure photograph of the downward flow of sand around a circular cylinder. Reproduced with the permission of R.H. Sabersky.

13.2 Particle Interaction Models

It is self-evident that the rheology of granular flows will be strongly influenced by the dynamics of particle/particle interactions. Consequently, the solid mechanics and dynamics of those interactions must be established prior to a discussion of the rheology of the overall flow. We note that the relation between the rheology and the particle/particle interaction can quite subtle (Campbell 2002, 2003).

Early work on rapid granular material flows often assumed instantaneous, binary collisions between particles, in other words a *hard particle model* (see, for example, Campbell and Brennen 1985a,b). Although this assumption may be valid in some applications, it is now recognized that the high shear rates required to achieve such flow conditions are unusual (Campbell 2002) and that most practical granular flows have more complex particle/particle interactions that, in turn, lead to more complex rheologies. To illustrate this we confine the discussion to the particular form of particle/particle interaction most often used in computer simulations. We refer to the model of the particle/particle dynamics known as the *soft particle model*, depicted in Figure 13.2. First utilized by Cundall and Strack (1979), this admittedly simplistic model consists of a spring, K_n, and dashpot, C, governing the normal motion and a spring, K_s, and Coulomb friction coefficient, μ^*, governing the tangential motion during the contact and deformation of two particles of mass, m_p. The model has been

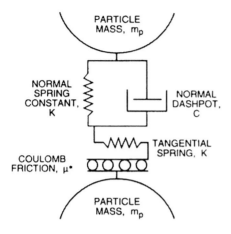

Figure 13.2. Schematic of the soft particle model of particle interaction.

subject to much study and comparison with experiments, for example, by Bathurst and Rothenburg (1988). Although different normal and tangential spring constants are often used, for simplicity we characterize them using a single spring constant (Bathurst and Rothenburg show that K_s/K_n determines the bulk Poisson's ratio) that, neglecting the effects of nonlinear Hertzian-like deformations, are characterized by a simple linear elastic spring constant, K. Note that as described by Bathurst and Rothenburg, the Young's modulus of the bulk material is proportional to K. Note also that K will be a function not only of properties of the solid material but also of the geometry of the contact points. Furthermore, it is clear that the dashpot constant, C, determines the loss of energy during normal collisions and therefore is directly related to the coefficient of restitution for normal collisions. Consequently, appropriate values of C can be determined from known or measured coefficients of restitution, ϵ; the specific relation is as follows:

$$\epsilon = \exp\left(-\pi C/[2m_p K - C^2]^{\frac{1}{2}}\right). \quad (13.1)$$

Note that this particle interaction model leads to a collision time for individual binary collisions, t_c, which is the same for all collisions and is given by

$$t_c = \pi m_p/[2m_p K - C^2]^{\frac{1}{2}}. \quad (13.2)$$

Before leaving the subject of individual particle interactions, several cautionary remarks are appropriate. Models such as that described above and those used in most granular flow simulations are highly simplified and there are many complications whose effects on the granular flow rheology remain to be explored. For example, the spring stiffnesses and the coefficients of restitution are often far from constant and depend on the geometry of the particle/particle contacts and velocity of the impact as well as other factors, such as the surface roughness. The contact stiffnesses may be quite nonlinear, although Hertzian springs (in which the force is proportional to the displacement raised to the 3/2 power) can be readily incorporated into the computer simulations. We also note that velocities greater than a few centimeters per second will normally lead to plastic deformation of the solid at the contact point and to coefficients of restitution that decrease with increasing velocity (Goldsmith 1960, Lun and Savage 1986). Boundary conditions may also involve complications because the coefficient of restitution of particle/wall collisions can depend on the wall thickness in a complicated way (Sondergard *et al.* 1989). Appropriate tangential coefficients are even more difficult to establish. The tangential spring stiffness may be different from the normal stiffness and may depend on whether slippage occurs during contact. This introduces the complications of tangential collisions studied by Maw *et al.* (1976, 1981), Foerster *et al.* (1994) and others. The interstitial fluid can have a major effect on the interaction dynamics; further comment on this is delayed until Section 13.6. The point to emphasize here is that much remains to be done before all the possible effects on the granular flow rheology have been explored.

13.2.1 Computer Simulations

Computer simulations have helped to elucidate the behavior of all types of granular flow. They are useful for two reasons. First there is a dearth of experimental techniques that would allow complete observations of real granular flows and their flow variables such as the local solids fraction; this is particularly the case for interior regions of the flow. Second, it is useful to be able to simplify the particle/particle and particle/wall interactions and therefore learn the features that are most important in determining the flow. The simulations use both *hard particle* models (see, for example, Campbell and Brennen 1985a,b) and *soft particle* models (see, for example, Cundall and Strack (1979), Walton and Braun 1986a,b). The hard particle model is, of course, a limiting case within the soft particle models and, though computationally efficient, is applicable only to rapid granular flows (see Section 13.5). Soft particle models have been particularly useful in helping elucidate granular material flow phenomena, for example the formation and dissipation of force chains (Cundall and Strack 1979) and the complex response of a bed of grains to imposed vertical vibration (Wassgren *et al.* 1996).

13.3 Flow Regimes

13.3.1 Dimensional Analysis

As pointed out by Campbell (2002), given a particle interaction model (such as that described above) characterized by a set of parameters like (K, ϵ, μ^*), it follows from dimensional analysis that the stress, τ, in a typical shearing flow with a shear rate, $\dot{\gamma}$, and a solids volume fraction, α, will be a function of the particle interaction parameters plus $(D, \rho_S, \alpha, \dot{\gamma})$, where the particle density, ρ_S, has been used instead of the particle mass, m_p. Applying dimensional analysis to this function it follows that the dimensionless stress, $\tau D/K$, must be a function of the following dimensionless quantities:

$$\frac{\tau D}{K} = f\left(\alpha, \mu^*, \epsilon, \frac{K}{\rho_S D^3 \dot{\gamma}^2}\right). \tag{13.3}$$

Alternatively one could also use a different form for the nondimensional stress, namely $\tau/\rho_S D^2 \dot{\gamma}^2$, and express this as a function of the same set of dimensionless quantities.

Such a construct demonstrates the importance in granular flows of the parameter, $K/\rho_S D^3 \dot{\gamma}^2$, which is the square of the ratio of the typical time associated with the shearing, $t_{\text{shear}} = 1/\dot{\gamma}$, to a typical collision time, $(m_p/K)^{\frac{1}{2}}$. The shearing time, t_{shear}, will determine the time between collisions for a particular particle though this time will also be heavily influenced by the solids fraction, α. The typical collision time, $(m_p/K)^{\frac{1}{2}}$, will be close to the binary collision time. From these considerations, we can discern two possible flow regimes or asymptotic flow states. The first is identified by instantaneous (and therefore necessarily binary) collisions in which the collision time is very short compared with the shearing time so that $K/\rho_S D^3 \dot{\gamma}^2 \gg 1$. We refer to this as the *inertial regime*. It includes an asymptotic case called *rapid granular flows* in which the collisions are essentially instantaneous and binary. The above dimensional analysis shows that appropriate dimensionless stresses in the inertial regime take the form $\tau/\rho_S D^2 \dot{\gamma}^2$ and should be functions only of

$$\frac{\tau}{\rho_S D^2 \dot{\gamma}^2} = f(\alpha, \mu^*, \epsilon). \tag{13.4}$$

This is the form that Bagnold (1954) surmised in his classic and much-quoted article on granular shear flows.

The second asymptotic flow regime is characterized by contact times that are long compared with the shearing time so that $K/\rho_S D^3 \dot{\gamma}^2 \ll 1$. From computer simulations Campbell (2002) finds that as $K/\rho_S D^3 \dot{\gamma}^2$ is decreased and the flow begins to depart from the inertial regime, the particles are forced to interact with a frequency whose typical time becomes comparable to the binary collision time. Consequently, multiple particle interactions begin to occur and force chains begin to form. Then the dimensional analysis shows that the appropriate dimensionless stresses are $\tau D/K$ and, in this limit, these should be functions only of the following:

$$\frac{\tau D}{K} = f(\alpha, \mu^*, \epsilon). \tag{13.5}$$

Note that this second regime is essentially quasistatic in that the stresses do not depend on any rate quantities. Campbell refers to this as the elastic-quasistatic regime.

13.3.2 Flow Regime Rheologies

Campbell (2002, 2003) has carried out an extensive series of computer simulations of shear flows designed to identify further characteristics of the flow regimes and, in

13.3 Flow Regimes

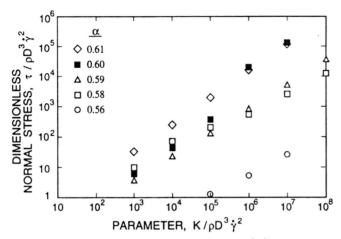

Figure 13.3. Typical nondimensional stress, $\tau/\rho_S D^2 \dot\gamma^2$ (in this case a normal stress), in a uniform shear flow as a function of the parameter, $K/\rho_S D^3 \dot\gamma^2$, for various solids fractions, α, a friction coefficient $\mu^* = 0.5$ and a coefficient of restitution of $\epsilon = 0.7$ (adapted from Campbell 2003).

particular, to identify the boundaries between them. Although his results are complicated because the simulations carried out with the solids fraction fixed seem to exhibit differences from those carried out with the normal stress or overburden fixed, we give here a brief overview of a few key features and results emerging from the fixed normal stress simulations. As one might expect, the flows at low values of $K/\rho_S D^3 \dot\gamma^2$ are dominated by force chains that carry most of the shear stress in the shear flow. These chains form, rotate, and disperse continually during shear (Drescher and De Josselin de Jong 1972, Cundall and Strack 1979). Evaluating the typical particle contact time, Campbell finds that, in this elastic-quasistatic regime the dynamics are not correlated with the binary contact time but are determined by the shear rate. This clearly indicates multiple particle structures (force chains) whose lifetime is determined by their rotation under shear. However, as $K/\rho_S D^3 \dot\gamma^2$ is increased and the flow approaches the rapid granular flow limit, the typical contact time asymptotes to the binary contact time indicating the dominance of simple binary collisions and the disappearance of force chains.

Figure 13.3 is a typical result from Campbell's simulations at fixed normal stress and plots the dimensionless stress $\tau/\rho_S D^2 \dot\gamma^2$ against the parameter $K/\rho_S D^3 \dot\gamma^2$ for various values of the solids fraction, α. Note that at high solids fractions the slopes of the curves approach unity indicating that the ratio, $\tau D/K$, is constant in that part of the parameter space. This is therefore the elastic-quasistatic regime. At lower solids fractions, the dimensionless stress is a more complex function of both solids fraction and the parameter, $K/\rho_S D^3 \dot\gamma^2$, thus indicating the appearance of inertial effects. Another interesting feature is the ratio of the shear to normal stress, τ_s/τ_n, and the manner in which it changes with the change in flow regime. At high $K/\rho_S D^3 \dot\gamma^2$ this ratio asymptotes to a constant value that corresponds to the internal friction angle

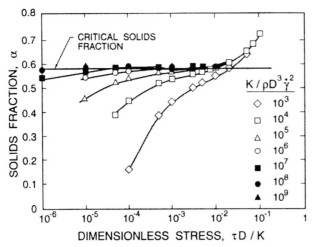

Figure 13.4. The variation of the solids fraction, α, with the dimensionless applied stress, $\tau D/K$, in a uniform shear flow with fixed normal stress for various values of the parameter, $K/\rho_S D^3 \dot\gamma^2$. Computer simulation data from Campbell (2003) for the case of a friction coefficient of $\mu^* = 0.5$ and a coefficient of restitution of $\epsilon = 0.7$.

used in soil mechanics (and is closely related to the interparticle friction coefficient, μ^*). However, as $K/\rho_S D^3 \dot\gamma^2$ is decreased (at constant normal stress) the simulations show τ_s/τ_n increasing with the increases being greater the smaller the normal stress.

Fundamental rheological information such as that given in Figure 13.3 can be used to construct granular flow regime maps. However, it is first necessary to discuss the solids fraction, α, and how that is established in most granular flows. The above analysis assumed, for convenience, that α was known and sought expressions for the stresses, τ, both normal and tangential. In practical granular flows, the normal stress or overburden is usually established by the circumstances of the flow and by the gravitational forces acting on the material. The solids fraction results from the rheology of the flow. Under such circumstances, the data required is the solids fraction, α as a function of the dimensionless overburden, $\tau D/K$ for various values of the parameter, $K/\rho_S D^3 \dot\gamma^2$. An example from Campbell (2003), is shown in Figure 13.4 and illustrates another important feature of granular dynamics. At high values of the overburden and solids fraction, the rate parameter, $K/\rho_S D^3 \dot\gamma^2$ plays little role and the solids fraction simply increases with the overburden. As the solids fraction decreases to facilitate flow, then, for low shear rates or high values of $K/\rho_S D^3 \dot\gamma^2$, the material asymptotes to a *critical* solids fraction of about 0.59 in the case of Figure 13.4. This is the critical state phenomenon familiar to soil mechanicists (see, for example, Schofield and Wroth 1968). However, at higher shear rates, lower values of $K/\rho_S D^3 \dot\gamma^2$, and lower overburdens, the material expands below the critical solids fraction as the material moves into the inertial regime and the collisions and interactions between the particles cause the material to expand. Figure 13.4 therefore displays both the

13.4 Slow Granular Flow

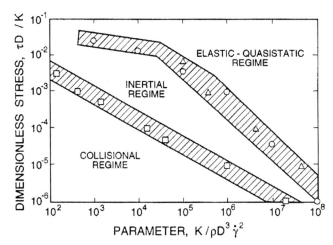

Figure 13.5. Typical flow regime map for uniform shear flow in a plot of the dimensionless overburden or normal stress against the parameter, $K/\rho_S D^3 \dot{\gamma}^2$, as determined from the fixed normal stress computer simulations of Campbell (2003) (for the case of a friction coefficient of $\mu^* = 0.5$ and a coefficient of restitution of $\epsilon = 0.7$).

traditional soil mechanics behavior and the classic kinetic theory behavior that results from the dominance of random, collisional motions. We also see that the traditional critical solids fraction could be considered as the dividing line between the inertial and elastic-quasistatic regimes of flow.

13.3.3 Flow Regime Boundaries

Finally, we include as Figure 13.5 a typical flow regime map as constructed by Campbell (2003) from this computer-modeled rheological information. The regimes are indicated in a map of the overburden or dimensionless stress plotted against the parameter $K/\rho_S D^3 \dot{\gamma}^2$ and the results show the progression at fixed overburden from the elastic-quasistatic regime at low shear rates to the inertial regime. Campbell also indicates that part of the inertial regime in which the flow is purely collisional (rapid granular flow). This occurs at low overburdens but at sufficiently high shear rates that rapid granular flows are uncommon in practice though they have been generated in a number of experimental shear cell devices.

13.4 Slow Granular Flow

13.4.1 Equations of Motion

All of the early efforts to understand granular flow neglected the random kinetic energy of the particles, the granular temperature, and sought to construct equations for the motion as extrapolations of the theories of soil mechanics by including the mean or global inertial effects in the equations of motion. We now recognize that, if these constructs are viable, they apply to the elastic-quasistatic regime of slow granular motion. Notable among these theories were those who sought to construct effective continuum equations of motion for the granular material beginning with the following:

$$\frac{D(\rho_S \alpha)}{Dt} + \rho_S \alpha \frac{\partial u_i}{\partial x_i} = 0 \tag{13.6}$$

$$\rho_S \alpha \frac{Du_k}{dt} = \rho_S \alpha g_k - \frac{\partial \sigma_{ki}}{\partial x_i} \tag{13.7}$$

where Eq. (13.6) is the continuity Eq. (1.25) and Eq. (13.7) is the momentum equation [Eq. (1.46) for a single-phase flow]. It is then assumed that the stress tensor is quasistatic and determined by conventional soil mechanics constructs. A number of models have been suggested but here we focus on the most commonly used approach, namely Mohr–Coulomb models for the stresses.

13.4.2 Mohr–Coulomb Models

As a specific example, the Mohr–Coulomb–Jenike–Shield model (Jenike and Shield 1959) utilizes a Mohr's circle diagram to define a yield criterion and it is assumed that once the material starts to flow, the stresses must continue to obey that yield criterion. For example, in the flow of a cohesionless material, one might utilize a Coulomb friction yield criterion in which it is assumed that the ratio of the principal shear stress to the principal normal stress is simply given by the *internal friction angle*, ϕ, which is considered to be a material property. In a two-dimensional flow, for example, this would imply the following relation between the stress tensor components:

$$\left\{ \left(\frac{\sigma_{xx} - \sigma_{yy}}{2} \right)^2 + \sigma_{xy}^2 \right\}^{\frac{1}{2}} = -\sin\phi \left(\frac{\sigma_{xx} + \sigma_{yy}}{2} \right), \tag{13.8}$$

where the left-hand side would be less than the right in regions where the material is not flowing or deforming.

However, Eqs. (13.6), (13.7), and (13.8) are insufficient and must be supplemented by at least two further relations. In the Mohr–Coulomb–Jenike–Shield model, an assumption of isotropy is also made; this assumes that the directions of principal stress and principal strain rate correspond. For example, in two-dimensional flow, this implies the following:

$$\frac{\sigma_{xx} - \sigma_{yy}}{\sigma_{xy}} = \frac{2\left(\frac{\partial y}{\partial x} - \frac{\partial v}{\partial y}\right)}{\frac{\partial u}{\partial y} + \frac{\partial v}{\partial x}}. \tag{13.9}$$

It should be noted that this part of the model is particularly suspect because experiments have shown substantial departures from isotropy. Finally one must also stipulate some relation for the solids fraction α and typically this has been considered a constant

13.4 Slow Granular Flow

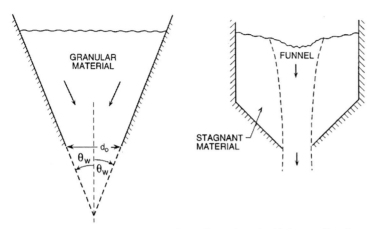

Figure 13.6. Some hopper geometries and notation. (Left) A mass flow hopper. (Right) Funnel flow.

equal to the critical solids fraction or to the maximum shearable solids fraction. This feature is also very questionable because even slow flows such as occur in hoppers display substantial decreases in α in the regions of faster flow.

13.4.3 Hopper Flows

Despite the above criticisms, Mohr–Coulomb models have had some notable successes, particularly in their application to flows in hoppers. Savage (1965, 1967), Morrison and Richmond (1976), Brennen and Pearce (1978), Nguyen et al. (1979), and others utilized Mohr–Coulomb models (and other variants) to find approximate analytical solutions for the flows in hoppers, both conical hoppers and two-dimensional hopper flows. Several types of hopper are shown in Figure 13.6. In narrow *mass flow* hoppers with small opening angles, θ_w, these solutions yield flow rates that agree well with the experimentally measured values for various values of θ_w, various internal friction angles and wall friction angles. An example of the comparison of calculated and experimental flow rates is included in Figure 13.7. These methods also appear to yield roughly the right wall stress distributions. In addition, note that both experimentally and theoretically the flow rate becomes independent of the height of material in the hopper once that height exceeds a few opening diameters; this result was explored by Janssen (1895) in one of the earliest articles dealing with granular flow.

Parenthetically, we note even granular flows as superficially simple as flows in hoppers can be internally quite complex. For example, it is only for narrow hoppers that even low friction granular materials manifest *mass flow*. At larger hopper angles and for more frictional materials, only an internal *funnel* of the granular material actually flows (see Figure 13.8) and the material surrounding that funnel remains at rest. Funnel flows are of considerable practical interest (see, for example, Jenike 1964, Johanson and Colijin 1964) and a substantial literature exists for the heuristic determination of the conditions under which they occur; for a study of the conditions

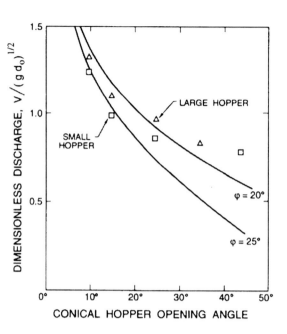

Figure 13.7. Dimensionless discharge, $V/(gd_o)^{\frac{1}{2}}$ (d_o is the opening width and V is the volume-averaged opening velocity), for flows in conical hoppers of various hopper opening angles, θ_w. Experimental data for the flows of glass beads (internal friction angle, $\phi = 25°$, wall friction angle of 15°) in two sizes of hopper are compared with the Mohr–Coulomb–Jenike–Shield calculations of Nguyen et al. (1979) using internal friction angles of 20° and 25°.

that determine these various flow patterns see Nguyen et al. (1980). One interpretation of funnel flow is that the stress state within the funnel is sufficient to allow dilation of the material and therefore flow, whereas the surrounding stagnant material has a stress state in which the solids fraction remains above the critical. It should be possible to generate computer simulations of these complex flows that predict the boundaries between the shearing and nonshearing regions in a granular flow. However, it is clear that some of the experimentally observed flows are even more complex than implied by the above description. With some materials the flow can become quite unsteady; for example, Lee et al. (1974) observed the flow in a two-dimensional hopper to oscillate from side to side with the alternating formation of yield zones within the material.

Figure 13.8. Long exposure photographs of typical granular flows in hoppers showing the streamlines in the flowing material. (Left) Flow of sand without stagnant regions. (Right) A funnel flow of rice with stagnant regions. From Nguyen et al. (1980).

13.5 Rapid Granular Flow

13.5.1 Introduction

Despite the uncommon occurence of truly rapid granular flow, it is valuable to briefly review the substantial literature of analytical results that have been generated in this field. At high shear rates, the inertia of the random motions that result from particle/particle and particle/wall collisions becomes a key feature of the rheology. Those motions can cause a dilation of the material and the granular material begins to behave like a molecular gas. In such a flow, as in kinetic theory, the particle velocities can be decomposed into time-averaged and fluctuating velocity components. The energy associated with the random or fluctuating motions is represented by the granular temperature, T, analogous to the thermodynamic temperature. Various granular temperatures may be defined depending on whether one includes the random energy associated with rotational and vibrational modes as well as the basic translational motions. The basic translational granular temperature used herein is defined as follows:

$$T = \frac{1}{3}\left(<\dot{U}_1^2> + <\dot{U}_2^2> + <\dot{U}_3^2>\right), \tag{13.10}$$

where $\dot{U}_i$ denotes the fluctuating velocity with a zero time average and $<>$ denotes the ensemble average. The kinetic theory of granular material is complicated in several ways. First, instead of tiny point molecules it must contend with a large solids fraction that inhibits the mean free path or flight of the particles. The large particle size also means that momentum is transported both through the flight of the particles (the *streaming* component of the stress tensor) and by the transfer of momentum from the center of one particle to the center of the particle it collides with (the *collisional* component of the stress tensor). Second, the collisions are inelastic and therefore the velocity distributions are not necessarily Maxwellian. Third, the finite particle size means that there may be a significant component of rotational energy, a factor not considered in the above definition. Moreover, the importance of rotation necessarily implies that the communication of rotation from one particle to another may be important and so the tangential friction in particle/particle and particle/wall collisions will need to be considered. All of this means that the development of a practical kinetic theory of granular materials has been long in development.

Early efforts to construct the equations governing rapid granular flow followed the constructs of Bagnold (1954); though his classic experimental observations have recently come under scrutiny (Hunt *et al.* 2002), his qualitative and fundamental understanding of the issues remains valid. Later researchers, building on Bagnold's ideas, used the concept of granular temperature in combination with heuristic but insightful assumptions regarding the random motions of the particles (see, for example, McTigue 1978, Ogawa *et al.* 1980, Haff 1983, Jenkins and Richman 1985, Nakagawa

1988, Babic and Shen 1989) in attempts to construct the rheology of rapid granular flows. Ogawa *et al.* (1978, 1980), Haff (1983) and others suggested that the global shear and normal stresses, τ_s and τ_n, are given by the following:

$$\tau_s = f_s(\alpha)\rho_S\dot{\gamma}T^{\frac{1}{2}} \quad \text{and} \quad \tau_n = f_n(\alpha)\rho_S T \quad (13.11)$$

where f_s and f_n are functions of the solid fraction, α, and some properties of the particles. Clearly the functions, f_s and f_n, would have to tend to zero as $\alpha \to 0$ and become very large as α approaches the maximum shearable solids fraction. The constitutive behavior is then completed by some relation connecting T, α, and, perhaps, other flow properties. Though it was later realized that the solution of a *granular energy* equation would be required to determine T, early dimensional analysis led to speculation that the granular temperature was just a local function of the shear rate, $\dot{\gamma}$, and that $T^{\frac{1}{2}} \propto D\dot{\gamma}$. With some adjustment in f_s and f_n this leads to the following:

$$\tau_s = f_s(\alpha)\rho_S D^2\dot{\gamma}^2 \quad \text{and} \quad \tau_n = f_n(\alpha)\rho_S D^2\dot{\gamma}^2, \quad (13.12)$$

which implies that the effective friction coefficient, τ_s/τ_n should be a function only of α and the particle characteristics.

13.5.2 Example of Rapid Flow Equations

Later, the work of Savage and Jeffrey (1981) and Jenkins and Savage (1983) saw the beginning of a more rigorous application of kinetic theory methods to rapid granular flows and there is now an extensive literature on the subject (see, for example, Gidaspow 1994). The kinetic theories may be best exemplified by quoting the results of Lun *et al.* (1984) who attempted to evaluate both the collisional and streaming contributions to the stress tensor (because momentum is transported both by the collisions of finite-sized particles and by the motions of the particles). In addition to the continuity and momentum equations, Equations 13.6 and 13.7, an *energy equation* must be constructed to represent the creation, transport and dissipation of granular heat; the form adopted is as follows:

$$\frac{3}{2}\rho_S\alpha\frac{DT}{Dt} = -\frac{\partial q_i}{\partial x_i} + \frac{\partial u_j}{\partial x_i}\sigma_{ji} - \Gamma, \quad (13.13)$$

where T is the granular temperature, q_i is the granular heat flux vector, and Γ is the rate of dissipation of granular heat into thermodynamic heat per unit volume. Note that this represents a balance between the granular heat stored in a unit volume (the left-hand side), the conduction of granular heat into the unit volume (first term on the right-hand side), the generation of granular heat (second term on the right-hand side), and the dissipation of granular heat (third term on the right-hand side).

Most of the kinetic theories begin in this way but vary in the expressions obtained for the stress/strain relations, the granular heat flux, and the dissipation term. As an example we quote here the results from the kinetic theory of Lun *et al.* (1984) that

13.5 Rapid Granular Flow

have been subsequently used by a number of authors. Lun *et al.* obtain a stress tensor related to the granular temperature, T [Eq. (13.10)], by the following:

$$\sigma_{ij} = \left(\rho_s g_1 T - \frac{4\pi^{\frac{1}{2}}}{3}\rho_s \alpha^2 (1+\epsilon) g_0 T^{\frac{1}{2}} \frac{\partial u_i}{\partial x_i}\right)\delta_{ij}$$
$$- 2\rho_s D g_2 T^{\frac{1}{2}}\left(\frac{1}{2}(u_{ij}+u_{ji}) - \frac{1}{3}u_{kk}\delta_{ij}\right), \qquad (13.14)$$

an expression for the granular heat flux vector,

$$q_i = -\rho_s D\left(g_3 T^{\frac{1}{2}}\frac{\partial T}{\partial x_i} + g_4 T^{\frac{3}{2}}\frac{\partial \alpha}{\partial x_i}\right) \qquad (13.15)$$

and an expression for the rate of dissipation of granular heat,

$$\Gamma = \rho_s g_5 T^{\frac{3}{2}}/D, \qquad (13.16)$$

where $g_0(\alpha)$, the radial distribution function, is chosen to be as follows:

$$g_0 = (1 - \alpha/\alpha^*)^{-2.5\alpha^*} \qquad (13.17)$$

and α^* is the maximum shearable solids fraction. In Eqs. (13.14), (13.15), and (13.16), the quantities g_1, g_2, g_3, g_4, and g_5, are functions of α and ϵ as follows:

$$g_1(\alpha, \epsilon) = \alpha + 2(1+\epsilon)\alpha^2 g_0$$

$$g_2(\alpha, \epsilon) = \frac{5\pi^{\frac{1}{2}}}{96}\left(\frac{1}{\eta(2-\eta)g_0} + \frac{8(3\eta-1)\alpha}{5(2-\alpha)} + \frac{64\eta\alpha^2 g_0}{25}\left(\frac{(3\eta-2)}{(2-\eta)} + \frac{12}{\pi}\right)\right)$$

$$g_3(\alpha, \epsilon) = \frac{25\pi^{\frac{1}{2}}}{16\eta(41-33\eta)}\left(\frac{1}{g_0} + 2.4\eta\alpha(1-3\eta+4\eta^2)\right.$$
$$\left. + \frac{16\eta^2\alpha^2 g_0}{25}(9\eta(4\eta-3) + 4(41-33\eta)/\pi)\right)$$

$$g_4(\alpha, \epsilon) = \frac{15\pi^{\frac{1}{2}}(2\eta-1)(\eta-1)}{4(41-33\eta)}\left(\frac{1}{\alpha g_0} + 2.4\eta\right)\frac{d}{d\alpha}(\alpha^2 g_0)$$

$$g_5(\alpha, \epsilon) = \frac{48\eta(1-\eta)\alpha^2 g_0}{\pi^{\frac{1}{2}}}, \qquad (13.18)$$

where $\eta = (1+\epsilon)/2$.

For two-dimensional shear flows in the (x, y) plane with a shear $\partial u/\partial y$ and no acceleration in the x direction the Lun *et al.* relations yield stresses given by the following:

$$\sigma_{xx} = \sigma_{yy} = \rho_s g_1 T; \quad \sigma_{xy} = -\rho_s D g_2 T^{\frac{1}{2}}\frac{\partial u}{\partial y} \qquad (13.19)$$

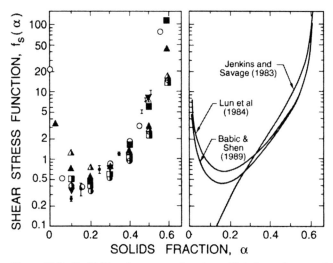

Figure 13.9. (Left) The shear stress function, $f_s(\alpha)$, from the experiments of Savage and Sayed (1984) with glass beads (symbol *I*) and various computer simulations with (open symbols) hard particle model, (solid symbols) soft particle model, and (half solid symbols) Monte Carlo methods. (Right) Several analytical results. Adapted from Campbell (1990).

in accord with Eq. (13.11). They also yield a granular heat flux component in the y direction given by the following:

$$q_y = \rho_S D \left(g_3 T^{\frac{1}{2}} \frac{\partial T}{\partial y} + g_4 T^{\frac{3}{2}} \frac{\partial \alpha}{\partial y} \right). \qquad (13.20)$$

These relations demonstrate the different roles played by the quantities g_1, g_2, g_3, g_4, and g_5: g_1 determines the normal kinetic pressure, g_2 governs the shear stress or *viscosity*, g_3 and g_4 govern the diffusivities controlling the conduction of granular heat from regions of differing temperature and density, and g_5 determines the granular dissipation. Although other kinetic theories may produce different specific expressions for these quantities, all of them seem necessary to model the dynamics of a rapid granular flow.

Figure 13.9 shows typical results for the shear stress function, $f_s(\alpha)$. The left-hand graph includes the data of Savage and Sayed (1984) from shear cell experiments with glass beads as well as a host of computer simulation results using both hard- and soft-particle models and both mechanistic and Monte Carlo methods. The right-hand graph presents some corresponding analytical results. The stress states to the left of the minima in these figures are difficult to observe experimentally, probably because they are unstable in most experimental facilities.

In summary, the governing equations, exemplified by Eqs. (13.6), (13.7), and (13.13), must be solved for the unknowns, α, T, and the three velocity components, u_i, given the expressions for σ_{ij}, q_i, and Γ and the physical constants D, ρ_S, ϵ, α^*, and gravity, g_k.

It was recognized early during research into rapid granular flows that some modification to the purely collisional kinetic theory would be needed to extend the results toward lower shear rates at which frictional stresses become significant. A number of authors explored the consequences of heuristically adding frictional terms to the collisional stress tensor (Savage 1983, Johnson *et al*. 1987, 1990) though it is physically troubling to add contributions from two different flow regimes.

13.5.3 Boundary Conditions

Rheological equations like those given above, also require the stipulation of appropriate boundary conditions and it transpires this is a more difficult issue than in conventional fluid mechanics. Many granular flows change quite drastically with changes in the boundary conditions. For example, the shear cell experiments of Hanes and Inman (1985) yielded stresses about three times those of Savage and Sayed (1984) in a very similar apparatus; the modest differences in the boundary roughnesses employed seem to be responsible for this discrepancy. Moreover, computer simulations in which various particle/wall interaction models have been examined (for example, Campbell and Brennen, 1985a,b) exhibit similar sensitivities. Although the normal velocity at a solid wall must necessarily be zero, the tangential velocities may be nonzero due to wall slip. Perhaps a Coulomb friction condition on the stresses is appropriate. But one must also stipulate a wall boundary condition on the granular temperature and this is particularly complicated for wall slip will imply that work is being done by the wall on the granular material so that the wall is a source of granular heat. At the same time, the particle/wall collisions dissipate energy, so the wall could be either a granular heat source or sink. The reader is referred to the work of Hui *et al*. (1984), Jenkins and Richman (1986), Richman (1988), and Campbell (1993) for further discussion of the boundary conditions.

13.5.4 Computer Simulations

Computer simulations have helped to elucidate the rheology of rapid granular flows and allowed evaluation of some of the approximations inherent in the theoretical kinetic theory models. For example, the shape of the fluctuation velocity distributions begins to deviate from Maxwellian and the velocity fluctuations become more and more nonisotropic as the solids fraction approaches the maximum shearable value. These kinds of details require computer simulations and were explored, for example, in the *hard particle* simulations of shear and chute flows by Campbell and Brennen (1985a,b). More generally, they represent the kinds of organized microstructure that can characterize granular flows close to the maximum shearable solids fraction. Campbell and Brennen (1985a) found that developing microstructure could be readily detected in these shear flow simulations and was manifest in the angular distribution of collision orientations within the shear flow. It is also instructive to observe other phenomenon in the computer simulations such as the conduction of granular temperature that takes place near the bed of a chute flow and helps establish the boundary separating a shearing layer of subcritical solids fraction from the nonshearing, high solids fraction block riding on top of that shearing layer (Campbell and Brennen 1985b).

13.6 Effect of Interstitial Fluid

13.6.1 Introduction

All of the above analysis assumed that the effect of the interstitial fluid was negligible. When the fluid dynamics of the interstitial fluid have a significant effect on the granular flow, analysis of the rheology becomes even more complex and our current understanding is quite incomplete. It was Bagnold (1954) who first attempted to define those circumstances in which the interstitial fluid would begin to effect the rheology of a granular flow. Bagnold introduced a parameter that included the following dimensionless quantity:

$$\text{Ba} = \rho_S D^2 \dot{\gamma} / \mu_L, \tag{13.21}$$

where $\dot{\gamma}$ is the shear rate; we refer to Ba as the Bagnold number. It is simply a measure of the stresses communicated by particle/particle collisions (given according to kinetic theory ideas by $\rho_S V^2$, where V is the typical random velocity of the particles that, in turn, is estimated to be given by $V = D\dot{\gamma}$) to the viscous stress in the fluid, $\mu_L \dot{\gamma}$. Bagnold concluded that when the value of Ba was less than about 40, the viscous fluid stresses dominate and the mixture exhibits a Newtonian rheology in which the shear stress and the strain rate ($\dot{\gamma}$) are linearly related; he called this the viscous regime. Conversely, when Ba is greater than about 400, the direct particle/particle (and particle/wall) interactions dominate and the stresses become proportional to the square of the strain rate. The viscous regime can be considered the dense suspension regime and many other sections of this book are relevant to those circumstances in which the direct particle/particle and particle/wall interactions play a minor role in the mixture rheology. In this chapter we have focused attention on the other limit in which the effect of the interstitial fluid is small and the rheology is determined by the direct interactions of the particles with themselves and with the walls.

13.6.2 Particle Collisions

A necessary prerequisite for the understanding of interstitial fluid effects on granular material flows is the introduction of interstitial fluid effects into particle/particle interaction models such as that described in Section 13.2. But the fluid mechanics of two particles colliding in a viscous fluid is itself a complicated one because of the coupling between the intervening lubrication layer of fluid and the deformation of the solid particles (Brenner 1961, Davis *et al.* 1986, Barnocky and Davis 1988).

13.6 Effect of Interstitial Fluid

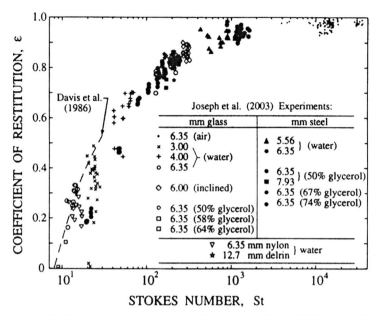

Figure 13.10. Coefficients of restitution for single particles colliding normally with a thick Zerodur wall. The particles are spheres of various diameters and materials suspended in air, water, and water/glycerol mixtures. The experimental data of Joseph *et al.* (2001) is plotted versus the Stokes number, St. Also shown are the theoretical predictions of Davis *et al.* (1986).

Joseph *et al.* (2001) have recently accumulated extensive data on the coefficient of restitution for spheres (diameter, D, and mass, m_p) moving through various liquids and gases to collide with a solid wall. As demonstrated in Figure 13.10, these data show that the coefficient of restitution for collision normal to the wall is primarily a function of the Stokes number, St, defined as $\text{St} = 2m_p V / 3\pi \mu D^2$, where μ is the viscosity of the suspending fluid and V is the velocity of the particle before it begins to be slowed down by interaction with the wall. The data show a strong correlation with St and agreement with the theoretical calculations of Davis *et al.* (1986). It demonstrates that the effect of the interstitial fluid causes a decrease in the coefficient of restitution with decreasing Stokes number and that there is a critical Stokes number of about 8 below which particles do not rebound but come to rest against the wall. It is also evident in Figure 13.10 that some of the data, particularly at low St, show significant scatter. Joseph *et al.* were able to show that the magnitude of the scatter depended on the relation between the size of the typical asperities on the surface of the particles and the estimated minimum thickness of the film of liquid separating the particle and the wall. When the former exceeded the latter, large scatter was understandably observed. Joseph (2003) also accumulated data for oblique collisions that appear to manifest essentially the same dependence of the coefficient of restitution on the Stokes number (based on the normal approach velocity, V) as the normal collisions. He also observed characteristics of the tangential interaction that are similar to those elucidated by Maw *et al.* (1976, 1981) for dry collisions.

Parenthetically, we note that the above descriptions of particle/particle and particle/wall interactions with interstitial fluid effects were restricted to large Stokes numbers and would allow the adaptation of kinetic theory results and simulations to those circumstances in which the interstitial fluid effects are small. However, at lower Stokes and Reynolds numbers, the interstitial fluid effects are no longer small and the particle interactions extend over greater distances. Even, though the particles no longer touch in this regime, their interactions create a more complex multiphase flow, the flow of a concentrated suspension that is challenging to analyze (Sangani *et al.* 1996). Computer simulations have been effectively used to model this rheology (see, for example, Brady 2001) and it is interesting to note that the concept of granular temperature also has value in this regime.

13.6.3 Classes of Interstitial Fluid Effects

We should observe at this point that there clearly several classes of interstitial fluid effects in the dynamics of granular flows. One class of interstitial fluid effect involves a global bulk motion of the interstitial fluid relative to the granular material; these flows are similar to the flow in a porous medium (though one that may be deforming). An example is the flow that is driven through a packed bed in the saltation flow regime of slurry flow in a pipe (see Section 8.2.3). Because of a broad data base of porous media flows, these global flow effects tend to be easier to understand and model though they can still yield unexpected results. An interesting example of unexpected results is the flow in a vertical standpipe (Ginestra *et al.* 1980).

Subtler effects occur when there is no such global relative flow, but there are still interstitial fluid effects on the random particle motions and on the direct particle/particle interactions. One such effect is the transition from inertially dominated to viscously dominated shear flow originally investigated by Bagnold (1954) and characterized by a critical Bagnold number, a phenomena that must still occur despite the criticism of Bagnold's rheological results by Hunt *et al.* (2002). We note a similar transition has been observed to occur in hopper flows, where Zeininger and Brennen (1985) found that the onset of viscous interstitial fluid effects occurred at a consistent critical Bagnold number based on the extensional deformation rate rather than the shear rate.

Consequently, though most of these subtler interstitial fluid effects remain to be fully explored and understood, there are experimental results that provide some guidance, albeit contradictory at times. For example, Savage and McKeown (1983) and Hanes and Inman (1985) both report shear cell experiments with particles in water and find a transition from inertially dominated flow to viscous-dominated flow. Although Hanes and Inman observed behavior similar to Bagnold's experiments, Savage and McKeown found substantial discrepancies.

Several efforts have been made to develop kinetic theory models that incorporate interstitial fluid effects. Tsao and Koch (1995) and Sangani *et al.* (1996) have explored theoretical kinetic theories and simulations in the limit of very small Reynolds

13.6 Effect of Interstitial Fluid

number ($\rho_C \dot{\gamma} D^2/\mu_C \ll 1$) and moderate Stokes number ($m_p \dot{\gamma}/3\pi D\mu_C$; note that if, as expected, V is given roughly by $\dot{\gamma}D$, then this is similar to the Stokes number, St, used in Section 13.6.2). They evaluate an additional contribution to Γ, the dissipation in Eq. (13.13), due to the viscous effects of the interstitial fluid. This supplements the collisional contribution given by a relation similar to Eq. (13.16). The problem is that flows with such Reynolds numbers and Stokes numbers are very rare. Very small Reynolds numbers and finite Stokes numbers require a large ratio of the particle density to the fluid density and therefore apply only to gas/solid suspensions. Gas/solid flows with very low Reynolds numbers are rare. Most dense suspension flows occur at higher Reynolds numbers, where the interstitial fluid flow is complex and often turbulent. Consequently, one must face the issues of the effect of the turbulent fluid motions on the particle motion and granular temperature and, conversely, the effect those particle motions have on the interstitial fluid turbulence. When there is substantial mean motion of the interstitial fluid through the granular material, as in a fluidized bed, that mean motion can cause considerable random motion of the particles coupled with substantial turbulence in the fluid. Zenit *et al.* (1997) have measured the granular temperature generated in such a flow; as expected this temperature is a strong function of the solids fraction, increasing from low levels at low solids fractions to a maximum and then decreasing again to zero at the maximum solids fraction, α_m (see Section 14.3.2). The granular temperature is also a function of the density ratio, ρ_C/ρ_D. Interestingly, Zenit *et al.* find that the granular temperature sensed at the containing wall has two components, one due to direct particle/wall collisions and the other a radiative component generated by particle/particle collisions within the bulk of the bed.

14

Drift Flux Models

14.1 Introduction

In this chapter we consider a class of models of multiphase flows in which the relative motion between the phases is governed by a particular subset of the flow parameters. The members of this subset are called *drift flux models* and were first developed by Zuber (see, for example, Zuber and Findlay 1965) and Wallis (1969) among others. To define the subset consider the one-dimensional flow of a mixture of the two components A and B. From Eqs. (1.4), (1.5), and (1.14), the volumetric fluxes of the two components j_A and j_B are related to the total volumetric flux, j, the drift flux, j_{AB}, and the volume fraction, $\alpha = \alpha_A = 1 - \alpha_B$, by the following:

$$j_A = \alpha j + j_{AB}; \quad j_B = (1-\alpha)j - j_{AB}. \tag{14.1}$$

Frequently, it is necessary to determine the basic kinematics of such a flow, for example, by determining α given j_A and j_B. To do so it is clearly necessary to determine the drift flux, j_{AB}, and, in general, one must consider the dynamics, the forces on the individual phases, to determine the relative motion. In some cases, this requires the introduction and simultaneous solution of momentum and energy equations, a problem that rapidly becomes mathematically complicated. There exists, however, a class of problems in which the dominant relative motion is caused by an external force such as gravity and therefore, to a reasonably good approximation, is a simple function only of the magnitude of that external force (say the acceleration due to gravity, g), of the volume fraction, α, and of the physical properties of the components (densities, ρ_A and ρ_B, and viscosities, μ_A and μ_B). The drift flux models were designed for these circumstances. If the relative velocity, u_{AB}, and, therefore, the drift flux, $j_{AB} = \alpha(1-\alpha)u_{AB}$, are known functions of α and the fluid properties, then it is clear that the solution to the types of kinematic problems described above follow directly from Eq. (14.1). Often this solution is achieved graphically as described in the next section.

Drift flux models are particularly useful in the study of sedimentation, fluidized beds, or other flows in which the relative motion is primarily controlled by buoyancy

14.2 Drift Flux Method

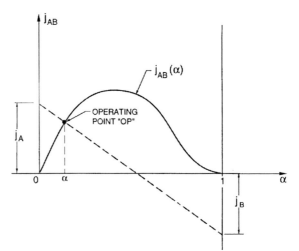

Figure 14.1. Basic graphical schematic or chart of the drift flux model.

forces and the fluid drag. Then, as described in Section 2.4.4, the relative velocity, u_{AB}, is usually a decreasing function of the volume fraction and this function can often be represented by a relation of the following form:

$$u_{AB} = u_{AB0}(1 - \alpha)^{b-1}; \quad j_{AB} = u_{AB0}\alpha(1 - \alpha)^b, \qquad (14.2)$$

where u_{AB0} is the terminal velocity of a single particle of the disperse phase, A, as $\alpha \to 0$ and b is some constant of order 2 or 3 as mentioned in Section 2.4.4. Then, given u_{AB0} and b, the kinematic problem is complete.

Of course, many multiphase flows cannot be approximated by a drift flux model. Most separated flows cannot, because, in such flows, the relative motion is intimately connected with the pressure and velocity gradients in the two phases. But a sufficient number of useful flows can be analyzed using these methods. The drift flux methods also allow demonstration of a number of fundamental phenomena that are common to a wide class of multiphase flows and whose essential components are retained by the equations given above.

14.2 Drift Flux Method

The solution to Eq. (14.1), given the form of the drift flux function, $j_{AB}(\alpha)$, is most conveniently displayed in the graphical form shown in Figure 14.1. Because Eq. (14.1) implies the following:

$$j_{AB} = (1 - \alpha)j_A - \alpha j_B \qquad (14.3)$$

and because the right-hand side of this equation can be plotted as the straight, dashed line in Figure 14.1, it follows that the solution (the values of α and j_{AB}) is given by the intersection of this line and the known $j_{AB}(\alpha)$ curve. We refer to this as the operating point, OP. Note that the straight, dashed line is most readily identified by the intercepts with the vertical axes at $\alpha = 0$ and $\alpha = 1$. The $\alpha = 0$ intercept will be the value of j_A and the $\alpha = 1$ intercept will be the value of $-j_B$.

To explore some of the details of flows modeled in this way, we consider several specific applications in the sections that follow. In the process we identify several phenomena that have broader relevance than the specific examples under consideration.

14.3 Examples of Drift Flux Analyses

14.3.1 Vertical Pipe Flow

Consider first the vertical pipe flow of two generic components, A and B. For ease of visualization, we consider

- that vertically upward is the positive direction so that all fluxes and velocities in the upward direction are positive
- that A is the less dense component and, as a memory aid, we will call A the gas and denote it by A = G. Correspondingly, the denser component B will be termed the liquid and denoted by B = L.
- that, for convenience, $\alpha = \alpha_G = 1 - \alpha_L$.

However, any other choice of components or relative densities are readily accommodated in this example by simple changes in these conventions. We examine the range of phenomena exhibited in such a flow by the somewhat artificial device of fixing the gas flux, j_G, and varying the liquid flux, j_L. Note that in this context Eq. (14.3) becomes the following:

$$j_{GL} = (1 - \alpha)j_G - \alpha j_L. \qquad (14.4)$$

Consider, first, the case of downward or negative gas flux as shown on the left in Figure 14.2. When the liquid flux is also downward the operating point, OP, is usually well defined as illustrated by CASE A in Figure 14.2. However, as one might anticipate, it is impossible to have an upward flux of liquid with a downward flux of gas and this is illustrated by the fact that CASE B has no intersection point and no solution.

The case of upward or positive gas flux, shown on the right in Figure 14.2, is more interesting. For downward liquid flux (CASE C) there is usually just one, unambiguous, operating point, OP. However, for small upward liquid fluxes (CASE D) we see that there are two possible solutions or operating points, OP1 and OP2. Without more information, we have no way of knowing which of these will be manifest in a particular application. In mathematical terms, these two operating points are known as *conjugate states*. Later we show that structures known as *kinematic shocks* or expansion waves may exist and allow transition of the flow from one conjugate state to the other. In many ways, the situation is analogous to gas dynamic flows in pipes where the conjugate states are a subsonic flow and a supersonic flow or to open-channel

14.3 Examples of Drift Flux Analyses

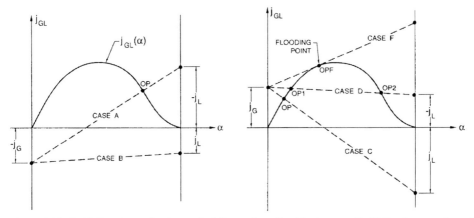

Figure 14.2. Drift flux charts for the vertical flows of gas/liquid mixtures. (Left) For downward gas flux. (Right) For upward gas flux.

flows where the conjugate states are a subcritical flow and a supercritical flow. The structure and propagation of kinematic waves and shocks are discussed later in Chapter 16.

One further phenomenon manifests itself if we continue to increase the downward flux of liquid while maintaining the same upward flux of gas. As shown on the right in Figure 14.2, we reach a limiting condition (CASE F) at which the dashed line becomes tangent to the drift flux curve at the operating point, OPF. We have reached the maximum downward liquid flux that will allow that fixed upward gas flux to move through the liquid. This is known as a *flooded* condition and the point OPF is known as the flooding point. As the reader might anticipate, flooding is quite analogous to choking and might have been better named choking to be consistent with the analogous phenomena that occur in gasdynamics and in open-channel flow.

It is clear that there exists a family of flooding conditions that we denote by j_{Lf} and j_{Gf}. Each member of this family corresponds to a different tangent to the drift flux curve and each has a different volume fraction, α. Indeed, simple geometric considerations allow one to construct the family of flooding conditions in terms of the parameter, α, assuming that the drift flux function, $j_{GL}(\alpha)$, is known as follows:

$$j_{Gf} = j_{GL} - \alpha \frac{dj_{GL}}{d\alpha}; \quad j_{Lf} = -j_{GL} - (1-\alpha)\frac{dj_{GL}}{d\alpha}. \tag{14.5}$$

Often, these conditions are displayed in a flow regime diagram (see Chapter 7) in which the gas flux is plotted against the liquid flux. An example is shown in Figure 14.3. In such a graph it follows from the basic relation [Eq. (14.4)] (and the assumption that j_{GL} is a function only of α) that a contour of constant void fraction, α, will be a straight line similar to the dashed lines in Figure 14.3. The slope of each of these dashed lines is $\alpha/(1-\alpha)$, the intercept with the j_G axis is $j_{GL}/(1-\alpha)$, and the intercept with the j_L axis is $-j_{GL}/\alpha$. It is then easy to see that these dashed lines form an envelope, AB, that defines the flooding conditions in this flow regime diagram. No flow is possible

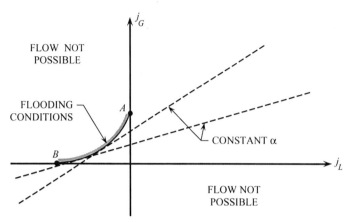

Figure 14.3. Flooding envelope in a flow pattern diagram.

in the fourth quadrant and above and to the left of the flooding envelope. Note that the end points, A and B, may yield useful information. In the case of the drift flux given by Eq. (14.2), the points A and B are given respectively by the following:

$$(j_G)_A = u_{GL0}(1-b)^{1-b}/b^b; \quad (j_L)_B = -u_{GL0}. \tag{14.6}$$

Finally we note that because, in mathematical terms, the flooding curve in Figure 14.3 is simply a mapping of the drift flux curve in Figure 14.2, it is clear that one can construct one from the other and vice versa. Indeed, one of the most convenient experimental methods to determine the drift flux curve is to perform experiments at fixed void fractions and construct the dashed curves in Figure 14.3. These then determine the flooding envelope from which the drift flux curve can be obtained.

14.3.2 Fluidized Bed

As a second example of the use of the drift flux method, we explore a simple model of a fluidized bed. The circumstances are depicted in Figure 14.4. An initially packed bed of solid, granular material (component, A = S) is trapped in a vertical pipe or container. An upward liquid or gas flow (component, B = L) that is less dense than the solid is introduced through the porous base on which the solid material initially rests. We explore the sequence of events as the upward volume flow rate of the gas or liquid is gradually increased from zero. To do so it is first necessary to establish the drift flux chart that would pertain if the particles were freely suspended in the fluid. An example was given earlier in Figure 2.8 and a typical graph of $j_{SL}(\alpha)$ is shown in Figure 14.5 where upward fluxes and velocities are defined as positive so that j_{SL} is negative. In the case of suspensions of solids, the curve must terminate at the maximum packing solids fraction, α_m.

At zero fluid flow rate, the operating point is OPA (see Figure 14.5). At very small fluid flow rates, j_L, we may construct the dashed line labeled CASE B; because this

14.3 Examples of Drift Flux Analyses

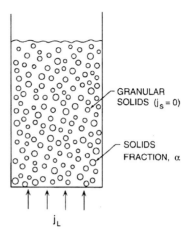

Figure 14.4. Schematic of a fluidized bed.

does not intersect the drift flux curve, the bed remains in its packed state and the operating point remains at $\alpha = \alpha_m$, point OPB in Figure 14.5. Conversely, at higher flow rates such as that represented by CASE D the flow is sufficient to fluidize and expand the bed so that the volume fraction is smaller than α_m. The critical condition, CASE C, at which the bed is just on the verge of fluidization is created when the liquid flux takes the first critical fluidization value, $(j_L)_{C1}$, where

$$(j_L)_{C1} = j_{SL}(\alpha_m)/(1 - \alpha_m). \tag{14.7}$$

As the liquid flux is increased beyond $(j_L)_{C1}$ the bed continues to expand as the volume fraction, α, decreases. However, the process terminates when $\alpha \to 0$, shown as the

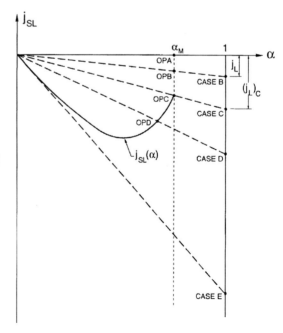

Figure 14.5. Drift flux chart for a fluidized bed.

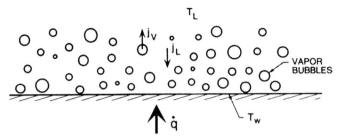

Figure 14.6. Nucleate boiling.

CASE E in Figure 14.5. This occurs at a second critical liquid flux, $(j_L)_{C2}$, given by the following:

$$(j_L)_{C2} = \left(-\frac{dj_{SL}}{d\alpha}\right)_{\alpha=0}. \tag{14.8}$$

At this critical condition the velocity of the particles relative to the fluid cannot maintain the position of the particles and they will be blown away. This is known as the *limit of fluidization*.

Consequently, we see that the drift flux chart provides a convenient device for visualizing the overall properties of a fluidized bed. However, it should be noted that there are many details of the particle motions in a fluidized bed that have not been included in the present description and require much more detailed study. Many of these detailed processes directly affect the form of the drift flux curve and therefore the overall behavior of the bed.

14.3.3 Pool Boiling Crisis

As a third and quite different example of the application of the drift flux method, we examine the two-phase flow associated with pool boiling, the background and notation for which were given in Section 6.2.1. Our purpose here is to demonstrate the basic elements of two possible approaches to the prediction of boiling crisis. Specifically, we follow the approach taken by Zuber, Tribius, and Westwater (1961), who demonstrated that the phenomenon of boiling crisis (the transition from nucleate boiling to film boiling) can be visualized as a flooding phenomenon.

In the first analysis we consider the nucleate boiling process depicted in Figure 14.6 and described in Section 6.2.1. Using that information we can construct a drift flux chart for this flow as shown in Figure 14.7.

It follows that, as illustrated in the figure, the operating point is given by the intersection of the drift flux curve, $j_{VL}(\alpha)$, with the dashed line as follows:

$$j_{VL} = \frac{\dot{q}}{\rho_V \mathcal{L}}\left\{1 - \alpha\left(1 - \frac{\rho_V}{\rho_L}\right)\right\} \approx \frac{\dot{q}}{\rho_V \mathcal{L}}(1-\alpha), \tag{14.9}$$

14.3 Examples of Drift Flux Analyses

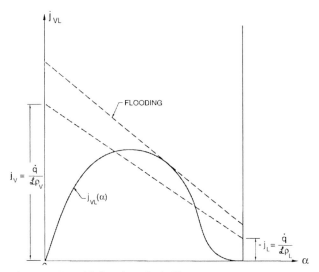

Figure 14.7. Drift flux chart for boiling.

where the second expression is accurate when $\rho_V/\rho_L \ll 1$ as is frequently the case. It also follows that this flow has a maximum heat flux given by the flooding condition sketched in Figure 14.7. If the drift flux took the common form given by Eq. (14.2) and if $\rho_V/\rho_L \ll 1$ it follows that the maximum heat flux, $\dot{q}_{c1}$, is given simply by the following:

$$\frac{\dot{q}_{c1}}{\rho_V \mathcal{L}} = K u_{VL0}, \tag{14.10}$$

where, as before, u_{VL0}, is the terminal velocity of individual bubbles rising alone and K is a constant of order unity. Specifically,

$$K = \frac{1}{b}\left(1 - \frac{1}{b}\right)^{b-1} \tag{14.11}$$

so that, for $b = 2$, $K = 1/4$ and, for $b = 3$, $K = 4/27$.

It remains to determine u_{VL0} for which a prerequisite is knowledge of the typical radius of the bubbles, R. Several estimates of these characteristic quantities are possible. For purposes of an example, we assume that the radius is determined at the moment at which the bubble leaves the wall. If this occurs when the characteristic buoyancy force, $\frac{4}{3}\pi R^3 g(\rho_L - \rho_V)$, is balanced by the typical surface tension force, $2\pi SR$, then an appropriate estimate of the radius of the bubbles is as follows:

$$R = \left\{\frac{3S}{2g(\rho_L - \rho_V)}\right\}^{\frac{1}{2}}. \tag{14.12}$$

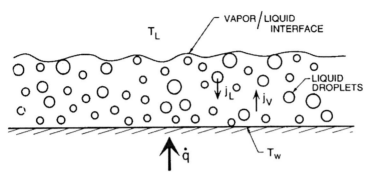

Figure 14.8. Sketch of the conditions close to film boiling.

Moreover, if the terminal velocity, u_{VL0}, is given by a balance between the same buoyancy force and a drag force given by $C_D \pi R^2 \rho_L u_{VL0}^2/2$, then an appropriate estimate of u_{VL0} is as follows:

$$u_{VL0} = \left\{ \frac{8Rg(\rho_L - \rho_V)}{3\rho_L C_D} \right\}^{\frac{1}{2}}. \tag{14.13}$$

Using these relations in Eq. (14.10) for the critical heat flux, $\dot{q}_{c1}$, leads to the following:

$$\dot{q}_{c1} = C_1 \rho_V \mathcal{L} \left\{ \frac{Sg(\rho_L - \rho_V)}{\rho_L^2} \right\}^{\frac{1}{4}}, \tag{14.14}$$

where C_1 is some constant of the order of unity. We delay comment on the relation of this maximum heat flux to the critical heat flux, $\dot{q}_c$, and on the specifics of Eq. (14.14) until the second model calculation is completed.

A second approach to the problem would be to visualize that the flow near the wall is primarily within a vapor layer, but that droplets of water are formed at the vapor/liquid interface and drop through this vapor layer to impinge on the wall and therefore cool it (Figure 14.8). Then, the flow within the vapor film consists of water droplets falling downward through an upward vapor flow rather than the previously envisaged vapor bubbles rising through a downward liquid flow. Up to and including Eq. (14.11), the analytical results for the two models are identical because no reference was made to the flow pattern. However, Eqs. (14.12) and (14.13) must be reevaluated for this second model. Zuber et al. (1961) visualized that the size of the water droplets formed at the vapor/liquid interface would be approximately equal to the most unstable wavelength, λ, associated with this Rayleigh–Taylor unstable surface [see Section 7.5.1, Eq. (7.22)] so that

$$R \approx \lambda \propto \left\{ \frac{S}{g(\rho_L - \rho_V)} \right\}^{\frac{1}{2}}. \tag{14.15}$$

Note that, apart from a constant of the order of unity, this droplet size is functionally identical to the vapor bubble size given by Eq. (14.12). This is reassuring and suggests that both are measures of the *grain size* in this complicated, high void fraction flow.

14.3 Examples of Drift Flux Analyses

The next step is to evaluate the drift flux for this droplet flow or, more explicitly, the appropriate expression for u_{VL0}. Balancing the typical net gravitational force, $\frac{4}{3}\pi R^3 g(\rho_L - \rho_V)$ (identical to that of the previous bubbly flow), with a characteristic drag force given by $C_D \pi R^2 \rho_V u_{VL0}^2 / 2$ (which differs from the previous bubbly flow analysis only in that ρ_V has replaced ρ_L) leads to the following:

$$u_{VL0} = \left\{ \frac{8Rg(\rho_L - \rho_V)}{3\rho_V C_D} \right\}^{\frac{1}{2}}. \tag{14.16}$$

Then, substituting Eqs. (14.15) and (14.16) into Eq. (14.10) leads to a critical heat flux, $\dot{q}_{c2}$, given by the following:

$$\dot{q}_{c2} = C_2 \rho_V \mathcal{L} \left\{ \frac{Sg(\rho_L - \rho_V)}{\rho_V^2} \right\}^{\frac{1}{4}}, \tag{14.17}$$

where C_2 is some constant of the order of unity.

The two model calculations presented above [and leading, respectively, to critical heat fluxes given by Eqs. (14.14) and (14.17)] allow the following interpretation of the pool boiling crisis. The first model shows that the bubbly flow associated with nucleate boiling will reach a critical state at a heat flux given by $\dot{q}_{c1}$ at which the flow will tend to form a vapor film. However, this film is unstable and vapor droplets will continue to be detached and fall through the film to wet and cool the surface. As the heat flux is further increased a second critical heat flux given by $\dot{q}_{c2} = (\rho_L/\rho_V)^{\frac{1}{2}} \dot{q}_{c1}$ occurs beyond which it is no longer possible for the water droplets to reach the surface. Thus, this second value, $\dot{q}_{c2}$, will more closely predict the true boiling crisis limit. Then, the analysis leads to a dimensionless critical heat flux, $(\dot{q}_c)_{nd}$, from Eq. (14.17) given by the following:

$$(\dot{q}_c)_{nd} = \frac{\dot{q}_c}{\rho_V \mathcal{L}} \left\{ \frac{Sg(\rho_L - \rho_V)}{\rho_V^2} \right\}^{-\frac{1}{4}} = C_2. \tag{14.18}$$

Kutateladze (1948) had earlier developed a similar expression using dimensional analysis and experimental data; Zuber et al. (1961) placed it on a firm analytical foundation.

Borishanski (1956), Kutateladze (1952), Zuber et al. (1961), and others have examined the experimental data on critical heat flux to determine the value of $(\dot{q}_c)_{nd}$ (or C_2) that best fits the data. Zuber et al. (1961) estimate that value to be in the range $0.12 \rightarrow 0.15$ though Rohsenow and Hartnett (1973) judge that 0.18 agrees well with most data. Figure 14.9 shows that the values from a wide range of experiments with fluids including water, benzene, ethanol, pentane, heptane, and propane all lie within $0.10 \rightarrow 0.20$. In that figure $(\dot{q}_C)_{nd}$ (or C_2) is presented as a function of the Haberman–Morton number, $Hm = g\mu_L^4(1 - \rho_V/\rho_L)/\rho_L S^3$, because, as was seen in Section 3.2.1, the appropriate type and size of bubble that is likely to form in a given liquid is governed by Hm.

Lienhard and Sun (1970) showed that the correlation could be extended from a simple horizontal plate to more complex geometries such as heated horizontal tubes.

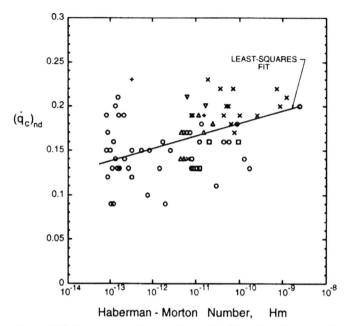

Figure 14.9. Data on the dimensionless critical heat flux, $(\dot{q}_c)_{nd}$ (or C_2), plotted against the Haberman–Morton number, $Hm = g\mu_L^4(1 - \rho_V/\rho_L)/\rho_L S^3$, for water (+), pentane (×), ethanol (□), benzene (△), heptane (▽), and propane (∗) at various pressures and temperatures. Adapted from Borishanski (1956) and Zuber et al. (1961).

However, if the typical dimension of that geometry (say the tube diameter, d) is smaller than λ [Eq. (14.15)], then that dimension should replace λ in the above analysis. Clearly this leads to an alternative correlation in which $(\dot{q}_c)_{nd}$ is a function of d; explicitly Lienhard and Sun recommend the following:

$$(\dot{q}_c)_{nd} = 0.061/K^* \quad \text{where} \quad K^* = d / \left\{ \frac{S}{g(\rho_L - \rho_V)} \right\}^{\frac{1}{2}} \quad (14.19)$$

(the constant, 0.061, was determined from experimental data) and that the result 14.19 should be employed when $K^* < 2.3$. For very small values of K^* (less than 0.24) there is no nucleate boiling regime and film boiling occurs as soon as boiling starts.

For useful reviews of the extensive literature on the critical heat flux in boiling, the reader is referred to Rohsenow and Hartnet (1973), Collier and Thome (1994), Hsu and Graham (1976), and Whalley (1987).

14.4 Corrections for Pipe Flows

Before leaving this discussion of the use of drift flux methods in steady flow, we note that, in many practical applications, the vertical flows under consideration are contained in a pipe. Consequently, instead of being invariant in the horizontal direction as assumed above, the flows may involve significant void fraction and velocity profiles

14.4 Corrections for Pipe Flows

over the pipe cross section. Therefore, the linear relation, Eq. (14.3), used in the simple drift flux method to find the operating point, must be corrected to account for these profile variations. As described in Section 1.4.3, Zuber and Findlay (1965) developed corrections using the profile parameter C_0 [Eq. (1.84)] and suggest that in these circumstances Eq. (14.3) should be replaced by the following:

$$\overline{j_{AB}} = [1 - C_0\bar{\alpha}]\bar{j}_A - C_0\bar{\alpha}\overline{j_B}, \tag{14.20}$$

where the overbar represents an average over the cross section of the pipe.

15

System Instabilities

15.1 Introduction

One of the characteristics of multiphase flows with which the engineer has to contend is that they often manifest instabilities that have no equivalent in single-phase flow (see, for example, Boure *et al.* 1973, Ishii 1982, Gouesbet and Berlemont 1993). Often the result is the occurence of large pressure, flow-rate, or volume-fraction oscillations that, at best, disrupt the expected behavior of the multiphase flow system (and thus decrease the reliability and life of the components, Makay and Szamody 1978) and, at worst, can lead to serious flow stoppage or structural failure (see, for example, NASA 1970, Wade 1974). Moreover, in many systems (such as pump and turbine installations) the trend toward higher rotational speeds and higher power densities increases the severity of the problem because higher flow velocities increase the potential for fluid/structure interaction problems. This chapter focuses on internal flow systems and the multiphase flow instabilities that occur in them.

15.2 System Structure

In the discussion and analysis of system stability, we consider that the system has been divided into its components, each identified by its index, k, as shown in Figure 15.1 where each component is represented by a box. The connecting lines do not depict lengths of pipe that are themselves components. Rather, the lines simply show how the components are connected. More specifically they represent specific locations at which the system has been divided up; these points are called the nodes of the system and are denoted by the index, i.

Typical and common components are pipeline sections, valves, pumps, turbines, accumulators, surge tanks, boilers, and condensers. They can be connected in series and/or in parallel. Systems can be either open loop or closed loop as shown in Figure 15.1. The mass flow rate through a component is denoted by $\dot{m}_k$ and the change in the total head of the flow across the component is denoted by Δp_k^T, defined as the total pressure at inlet minus that at discharge. (When the pressure ratios are large enough

15.2 System Structure

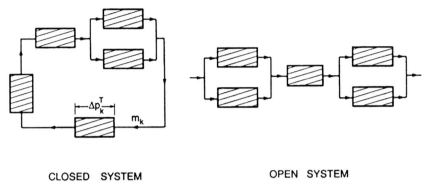

CLOSED SYSTEM   OPEN SYSTEM

Figure 15.1. Flow systems broken into components.

so that the compressibility of one or both of the phases must be accounted for, the analysis can readily be generalized by using total enthalpy rather than total pressure.) Then, each of the components considered in isolation has a performance characteristic in the form of the function $\Delta p_k^T(\dot{m}_k)$ as depicted graphically in Figure 15.2. The shapes of these characteristics are important in identifying and analyzing system instabilities. Some of the shapes are readily anticipated. For example, a typical single-phase flow pipe section (at higher Reynolds numbers) will have a characteristic that is approximately quadratic with $\Delta p_k^T \propto \dot{m}_k^2$. Other components such as pumps, compressors, or fans may have quite nonmonotonic characteristics. The slope of the characteristic, R_k^*, is defined as follows:

$$R_k^* = \frac{1}{\rho g} \frac{d \Delta p_k^T}{d \dot{m}_k}, \qquad (15.1)$$

and is known as the component resistance. However, unlike many electrical components, the resistance of most hydraulic components is almost never constant but varies with the flow, $\dot{m}_k$.

Components can readily be combined to obtain the characteristic of groups of neighboring components or the complete system. A parallel combination of two

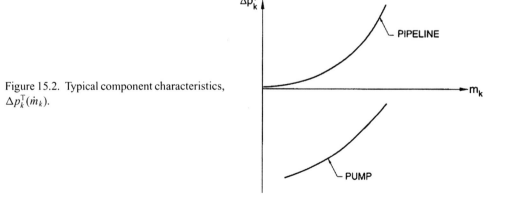

Figure 15.2. Typical component characteristics, $\Delta p_k^T(\dot{m}_k)$.

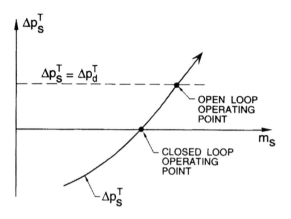

Figure 15.3. Typical system characteristic, $\Delta p_s^T(\dot{m}_s)$, and operating point.

components simply requires one to add the flow rates at the same Δp^T, whereas a series combination simply requires that one add the Δp^T values of the two components at the same flow rate. In this way one can synthesize the total pressure drop, $\Delta p_s^T(\dot{m}_s)$, for the whole system as a function of the flow rate, $\dot{m}_s$. Such a system characteristic is depicted in Figure 15.3. For a closed system, the equilibrium operating point is then given by the intersection of the characteristic with the horizontal axis because one must have $\Delta p_s^T = 0$. An open system driven by a total pressure difference of Δp_d^T (inlet total pressure minus discharge) would have an operating point where the characteristic intersects the horizontal line at $\Delta p_s^T = \Delta p_d^T$. Because these are trivially different we can confine the discussion to the closed loop case without any loss of generality.

In many discussions, this system equilibrium is depicted in a slightly different but completely equivalent way by dividing the system into two series elements, one of which is the *pumping* component, k = pump, and the other is the *pipeline* component, k = line. Then the operating point is given by the intersection of the *pipeline* characteristic, Δp_{line}^T, and the *pump* characteristic, $-\Delta p_{pump}^T$, as shown graphically in Figure 15.4. Note that because the total pressure increases across a pump, the values of $-\Delta p_{pump}^T$ are normally positive. In most single-phase systems, this depiction has the advantage that one can usually construct a series of quadratic pipeline characteristics depending on the valve settings. These pipeline characteristics are usually simple quadratics. Conversely, the pump or compressor characteristic can be quite complex.

15.3 Quasistatic Stability

Using the definitions of the last section, a quasistatic analysis of the stability of the equilibrium operating point is usually conducted in the following way. We consider perturbing the system to a new mass flow rate $d\dot{m}$ greater than that at the operating point as shown in Figure 15.4. Then, somewhat heuristically, one argues from

15.3 Quasistatic Stability

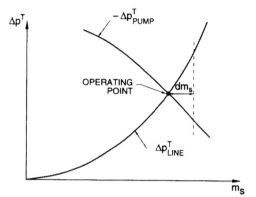

Figure 15.4. Alternate presentation of Figure 15.3.

Figure 15.4 that the total pressure rise across the pumping component is now less than the total pressure drop across the pipeline and therefore the flow rate will decline back to its value at the operating point. Consequently, the particular relationship of the characteristics in Figure 15.4 implies a stable operating point. If, however, the slopes of the two components are reversed (for example, Pump B of Figure 15.5(a) or the operating point C of Figure 15.5(b)) then the operating point is unstable because the increase in the flow has resulted in a pump total pressure that now exceeds the total pressure drop in the pipeline. These arguments lead to the conclusion that the operating point is stable when the slope of the system characteristic at the operating point (Figure 15.3) is positive or

$$\frac{d\Delta p_s^T}{d\dot{m}_s} > 0 \quad \text{or} \quad R_s^* > 0. \tag{15.2}$$

The same criterion can be derived in a somewhat more rigorous way by using an energy argument. Note that the net flux of flow energy from each component is $\dot{m}_k \Delta p_k^T$. In a straight pipe this energy is converted to heat through the action of viscosity. In a pump $\dot{m}_k(-\Delta p_k^T)$ is the work done on the flow by the pump impeller. Thus the net energy flux from the whole system $\dot{m}_s \Delta p_s^T$ and, at the operating point, this is zero (for simplicity we discuss a closed-loop system) because $\Delta p_s^T = 0$. Now, suppose, that

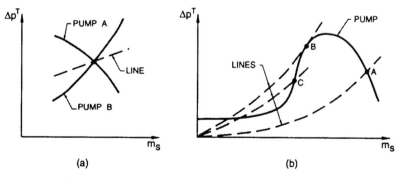

Figure 15.5. Quasistatically stable and unstable flow systems.

the flow rate is perturbed by an amount $d\dot{m}_s$. Then, the new net energy flux from the system is ΔE, where

$$\Delta E = (\dot{m}_s + d\dot{m}_s) \left\{ \Delta p_s^T + d\dot{m}_s \frac{d\Delta p_s^T}{d\dot{m}_s} \right\} \approx \dot{m}_s d\dot{m}_s \frac{d\Delta p_s^T}{d\dot{m}_s}. \tag{15.3}$$

Then we argue that if $d\dot{m}_s$ is positive and the perturbed system therefore dissipates more energy, then it must be stable. Under those circumstances one would have to add to the system a device that injected more energy into the system so as to sustain operation at the perturbed state. Hence Eq. (15.2) for quasistatic stability is reproduced.

15.4 Quasistatic Instability Examples

15.4.1 Turbomachine Surge

Perhaps the most widely studied instabilities of this kind are the surge instabilities that occur in pumps, fans, and compressors when the turbomachine has a characteristic of the type shown in Figure 15.5(b). When the machine is operated at points such as A the operation is stable. However, when the turbomachine is throttled (the resistance of the rest of the system is increased), the operating point will move to smaller flow rates and, eventually, reach the point B at which the system is neutrally stable. Further decrease in the flow rate will result in operating conditions such as the point C that are quasistatically unstable. In compressors and pumps, unstable operation results in large, limit-cycle oscillations that not only lead to noise, vibration, and lack of controllability but may also threaten the structural integrity of the machine. The phenomenon is known as compressor, fan, or pump surge and for further details the reader is referred to Emmons *et al.* (1955), Greitzer (1976, 1981), and Brennen (1994).

15.4.2 Ledinegg Instability

Two-phase flows can exhibit a range of similar instabilities. Usually, however, the instability is the result of a nonmonotonic pipeline characteristic rather than a complex pump characteristic. Perhaps the best known example is the Ledinegg instability (Ledinegg 1983), which is depicted in Figure 15.6. This occurs in boiler tubes through which the flow is forced either by an imposed pressure difference or by a normally stable pump as sketched in Figure 15.6. If the heat supplied to the boiler tube is roughly independent of the flow rate, then, at high flow rates, the flow will remain mostly liquid since, as discussed in Section 8.3.2, $d\mathcal{X}/ds$ is inversely proportional to the flow rate [see Eq. (8.24)]. Therefore $\mathcal{X}$ remains small. Conversely, at low flow rates, the flow may become mostly vapor because $d\mathcal{X}/ds$ is large. To construct the $\Delta p_k^T(\dot{m}_k)$ characteristic for such a flow it is instructive to begin with the two hypothetical characteristics for all-vapor flow and for all-liquid flow. The rough form of these are shown in Figure 15.6; because the frictional losses at high Reynolds numbers are

15.4 Quasistatic Instability Examples

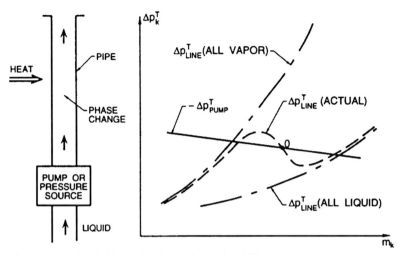

Figure 15.6. Sketch illustrating the Ledinegg instability.

proportional to $\rho u^2 = \dot{m}_k^2/\rho$, the all-vapor characteristic lies above the all-liquid line because of the different density. However, as the flow rate, $\dot{m}_k$, increases, the actual characteristic must make a transition from the all-vapor line to the all-liquid line and may therefore have the nonmonotonic form sketched in Figure 15.6. This may lead to unstable operating points such as the point O. This is the Ledinegg instability and is familiar to most as the phenomenon that occurs in a coffee percolator.

15.4.3 Geyser Instability

The geyser instability that is so familiar to visitors to Yellowstone National Park and other areas of geothermal activity, has some similarities to the Ledinegg instability, but also has important differences. It has been studied in some detail in smaller scale laboratory experiments (see, for example, Nakanishi *et al.* 1978) where the parametric variations are more readily explored.

The geyser instability requires the basic components sketched in Figure 15.7, namely a buried reservoir that is close to a large heat source, a vertical conduit and a near-surface supply of water that can drain into the conduit and reservoir. The geyser limit cycle proceeds as follows. During the early dormant phase of the cycle, the reservoir and conduit are filled with water that is being heated by the geothermal source. Once the water begins to boil the vapor bubbles rise up through the conduit. The hydrostatic pressure in the conduit and reservoir then drop rapidly due to the reduced mixture density in the conduit. This pressure reduction leads to explosive boiling and the eruption so widely publicized by "Old Faithful." The eruption ends when almost all the water in the conduit and reservoir has been ejected. The reduced flow then allows subcooled water to drain into and refill the reservoir and conduit. Due to the resistance to heat transfer in the rock surrounding the reservoir, there is a

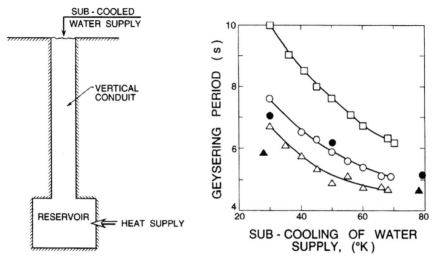

Figure 15.7. (Left) The basic components for a geyser instability. (Right) Laboratory measurements of geysering period as a function of heat supply (200 W: □, 330 W: ○, 400 W: ◊) from experiments (open symbols) and numerical simulations (solid symbols). Adapted from Tae-il et al. (1993).

significant time delay before the next load of water is heated to boiling temperatures. The long cycle times are mostly the result of low thermal conductivity of the rock (or other solid material) surrounding the reservoir and the consequent low rate of transfer of heat available to heat the subcooled water to its boiling temperature.

The dependence of the geysering period on the strength of the heat source and on the temperature of the subcooled water in the water supply is exemplified in Figure 15.7, which presents results from the small-scale laboratory experiments of Tae-il et al. (1993). That figure includes both the experimental data and the results of a numerical simulation. Note that, as expected, the geysering period decreases with increase in the strength of the heat source and with the increase in the temperature of the water supply.

15.5 Concentration Waves

There is one phenomenon that is sometimes listed in discussions of multiphase flow instabilities even though it is not, strictly speaking, an instability. We refer to the phenomenon of concentration wave oscillations and it is valuable to include mention of the phenomenon here before proceeding to more complex matters.

Often in multiphase flow processes, one encounters a circumstance in which one part of the circuit contains a mixture with a concentration that is somewhat different from that in the rest of the system. Such an inhomogeneity may be created during startup or during an excursion from the normal operating point. It is depicted in Figure 15.8, in which the closed loop has been arbitrarily divided into a pipeline component and a pump component. As indicated, a portion of the flow has a mass quality

15.5 Concentration Waves

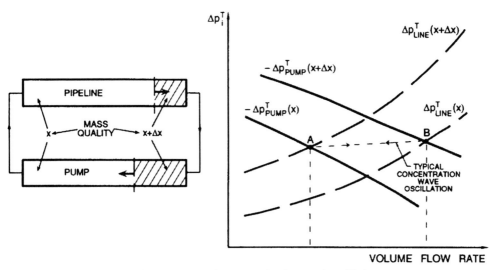

Figure 15.8. Sketch illustrating a concentration wave (density wave) oscillation.

that is larger by $\Delta \mathcal{X}$ than the mass quality in the rest of the system. Such a perturbation could be termed a concentration wave, although it is also called a density wave or a continuity wave; more generally, it is known as a kinematic wave (see Chapter 16). Clearly, the perturbation will move round the circuit at a speed that is close to the mean mixture velocity though small departures can occur in vertical sections in which there is significant relative motion between the phases. The mixing processes that would tend to homogenize the fluid in the circuit are often quite slow so that the perturbation may persist for an extended period.

It is also clear that the pressures and flow rates may vary depending on the location of the perturbation within the system. These fluctuations in the flow variables are termed *concentration wave oscillations* and they arise from the inhomogeneity of the fluid rather than from any instability in the flow. The characteristic frequency of the oscillations is simply related to the time taken for the flow to complete one circuit of the loop (or some multiple if the number of perturbed fluid pockets is greater than unity). This frequency is usually small and its calculation often allows identification of the phenomenon.

One way in which concentration oscillations can be incorporated in the graphical presentation we have used in this chapter is to identify the component characteristics for both the mass quality, $\mathcal{X}$, and the perturbed quality, $\mathcal{X} + \Delta \mathcal{X}$ and to plot them using the volume flow rate rather than the mass flow rate as the abscissa. We do this because, if we neglect the compressibility of the individual phases, then the volume flow rate is constant around the circuit at any moment in time, whereas the mass flow rate differs according to the mass quality. Such a presentation is shown in Figure 15.8. Then, if the perturbed body of fluid were wholly in the pipeline section, the operating point would be close to the point A. Conversely, if the perturbed body of fluid were wholly in the pump, the operating point would be close to the point B. Thus we can see that the operating point will vary along a trajectory such as that shown by the dotted line and that this will result in oscillations in the pressure and flow rate.

In closing, we should note that concentration waves also play an important role in other more complex unsteady flow phenomena and instabilities.

15.6 Dynamic Multiphase Flow Instabilities

15.6.1 Dynamic Instabilities

The descriptions of the preceding sections were predicated on the frequency of the oscillations being sufficiently small for all the components to track up and down their steady-state characteristics. Thus the analysis is only applicable to those instabilities whose frequencies are low enough to lie within some quasistatic range. At higher frequency, the effective resistance could become a complex function of frequency and could depart significantly from the quasistatic resistance. It follows that there may be operating points at which the total *dynamic* resistance over some range of frequencies is negative. Then the system would be dynamically unstable even though it may be quasistatically stable. Such a description of dynamic instability is instructive but overly simplistic and a more systematic approach to this issue will be detailed in Section 15.7. It is nevertheless appropriate at this point to describe two examples of dynamic instabilities so that reference to these examples can be made during the description of the transfer function methodology.

15.6.2 Cavitation Surge in Cavitating Pumps

In many installations involving a pump that cavitates, violent oscillations in the pressure and flow rate in the entire system can occur when the cavitation number is decreased to a value at which the volume of vapor bubbles within the pump becomes sufficient to cause major disruption of the flow and therefore a decrease in the total pressure rise across the pump (see Section 8.4.1). Although most of the detailed investigations have focused on axial pumps and inducers (Sack and Nottage 1965, Miller and Gross 1967, Kamijo *et al.* 1977, Braisted and Brennen 1980) the phenomenon has also been observed in centrifugal pumps (Yamamoto 1991). In the past this surge phenomenon was called *auto-oscillation* though the modern term *cavitation surge* is more appropriate. The phenomenon is described in detail in Brennen (1994). It can lead to very large flow rate and pressure fluctuations. For example, in boiler feed systems, discharge pressure oscillations with amplitudes as high as 14 bar have been reported informally. It is a genuinely dynamic instability in the sense described in Section 15.6.1, for it occurs when the slope of the pump total pressure rise/flow rate characteristic is still strongly negative and the system is therefore quasistatically stable.

15.6 Dynamic Multiphase Flow Instabilities

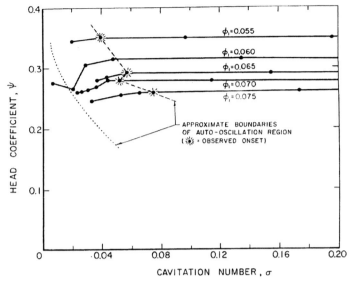

Figure 15.9. Cavitation performance of a SSME low-pressure LOX pump model showing the approximate boundaries of the cavitation surge region for a pump speed of 6000 rpm (from Braisted and Brennen 1980). The flow coefficient, ϕ_1, is based on the impeller inlet area.

As previously stated, cavitation surge occurs when the region of cavitation head loss is approached as the cavitation number is decreased. Figure 15.9 provides an example of the limits of cavitation surge taken from the work of Braisted and Brennen (1980). However, because the onset is sensitive to the detailed dynamic characteristics of the system, it would not even be wise to quote any approximate guideline for onset. Our current understanding is that the methodologies of Section 15.7 are essential for any prediction of cavitation surge.

Unlike compressor surge, the frequency of cavitation surge, ω_i, scales with the shaft speed of the pump, Ω (Braisted and Brennen 1980). The ratio, ω_i/Ω, varies with the cavitation number, σ [see Eq. (8.31)], the flow coefficient, ϕ [see Eq. (8.30)], and the type of pump as illustrated in Figure 15.10. The most systematic variation is with the cavitation number and it appears that the following empirical expression:

$$\omega_i/\Omega = (2\sigma)^{\frac{1}{2}} \qquad (15.4)$$

provides a crude estimate of the cavitation surge frequency. Yamamoto (1991) demonstrated that the frequency also depends on the length of the suction pipe, thus reinforcing the understanding of cavitation surge as a system instability.

15.6.3 Chugging and Condensation Oscillations

As a second example of a dynamic instability involving a two-phase flow we describe the oscillations that occur when steam is forced down a vent into a pool of water. The

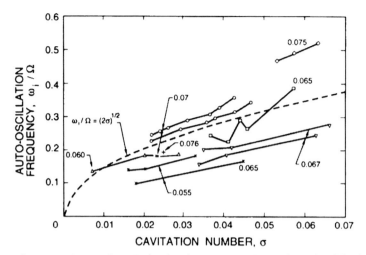

Figure 15.10. Data from Braisted and Brennen (1980) on the ratio of the frequency of cavitation surge, ω_i, to the frequency of shaft rotation, Ω, for several axial flow pumps: for SSME low-pressure LOX pump models: 7.62-cm diameter: × (9000 rpm) and + (12000 rpm), 10.2-cm diameter: ⊙ (4000 rpm) and ⊡ (6000 rpm); for 9° helical inducers: 7.58-cm diameter: ∗ (9000 rpm): 10.4-cm diameter: ∇ (with suction line flow straightener) and △ (without suction line flow straightener). The flow coefficients, ϕ_1, are based on the impeller inlet area.

situation is sketched in Figure 15.11. These instabilities, forms of which are known as *chugging* and *condensation oscillations*, have been most extensively studied in the context of the design of pressure suppression systems for nuclear reactors (see, for example, Wade 1974, Koch and Karwat 1976, Class and Kadlec 1976, Andeen and

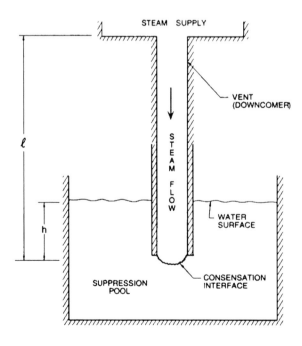

Figure 15.11. Components of a pressure suppression system.

15.6 Dynamic Multiphase Flow Instabilities

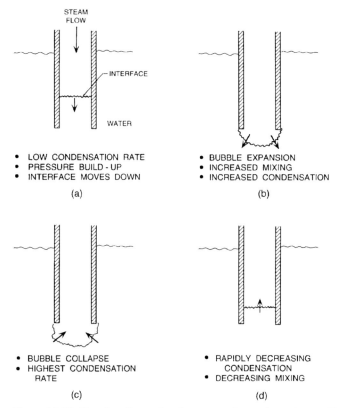

Figure 15.12. Sketches illustrating the stages of a condensation oscillation.

Marks 1978). The intent of the device is to condense steam that has escaped as a result of the rupture of a primary coolant loop and, thereby, to prevent the buildup of pressure in the containment that would have occurred as a result of uncondensed steam.

The basic components of the system are as shown in Figure 15.11 and consist of a vent or pipeline of length, ℓ, the end of which is submerged to a depth, h, in the pool of water. The basic instability is illustrated in Figure 15.12. At relatively low steam flow rates the rate of condensation at the steam/water interface is sufficiently high that the interface remains within the vent. However, at higher flow rates the pressure in the steam increases and the interface is forced down and out of the end of the vent. When this happens both the interface area and the turbulent mixing in the vicinity of the interface increase dramatically. This greatly increases the condensation rate, which, in turn, causes a marked reduction in the steam pressure. Thus the interface collapses back into the vent, often with the same kind of violence that results from cavitation bubble collapse. Then the cycle of growth and collapse, of oscillation of the interface from a location inside the vent to one outside the end of the vent, is repeated. The phenomenon is termed condensation instability and, depending on the dominant frequency, the violent oscillations are known as chugging or condensation oscillations (Andeen and Marks 1978).

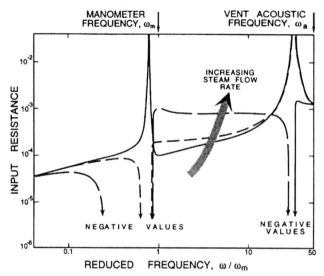

Figure 15.13. The real part of the input impedance (the input resistance) of the suppression pool as a function of the perturbation frequency for several steam flow rates. Adapted from Brennen (1979).

The frequency of the phenomenon tends to lock in on one of the natural modes of oscillation of the system in the absence of condensation. There are two obvious natural modes. The first, is the manometer mode of the liquid inside the end of the vent. In the absence of any steam flow, this manometer mode will have a typical small-amplitude frequency, $\omega_m = (g/h)^{\frac{1}{2}}$, where g is the acceleration due to gravity. This is usually a low frequency of the order of 1 Hz or less and, when the condensation instability locks into this low frequency, the phenomenon is known as chugging. The pressure oscillations resulting from chugging can be quite violent and can cause structural loads that are of concern to the safety engineer. Another natural mode is the first acoustic mode in the vent whose frequency, ω_a, is approximately given by $\pi c/\ell$, where c is the sound speed in the steam. There are also observations of lock-in to this higher frequency. The oscillations that result from this are known as condensation oscillations and tend to be of smaller amplitude than the chugging oscillations.

Figure 15.13 illustrates the results of a linear stability analysis of the suppression pool system (Brennen 1979) that was carried out using the transfer function methodology described in Section 15.7. Transfer functions were constructed for the vent or downcomer, for the phase change process, and for the manometer motions of the pool. Combining these, one can calculate the input impedance of the system viewed from the steam supply end of the vent. A positive input resistance implies that the system is absorbing fluctuation energy and is therefore stable; a negative input resistance implies an unstable system. In Figure 15.13, the input resistance is plotted against the perturbation frequency for several steam flow rates. Note that, at low steam flow rates, the system is stable for all frequencies. However, as the steam flow rate is increased, the system first becomes unstable over a narrow range of frequencies close to the manometer frequency, ω_m. Thus chugging is predicted to occur at some critical steam flow rate. At still higher flow rates, the system also becomes unstable over a narrow range of frequencies close to the first vent acoustic frequency, ω_a; thus the possibility of condensation oscillations is also predicted. Note that the quasistatic input resistance at small frequencies remains positive throughout and therefore the system is quasistatically stable for all steam flow rates. Thus, chugging and condensation oscillations are true, dynamic instabilities.

It is, however, important to observe that a linear stability analysis cannot model the highly nonlinear processes that occur during a chug and, therefore, cannot provide information on the subject of most concern to the practical engineer, namely the magnitudes of the pressure excursions and the structural loads that result from these condensation instabilities. Although models have been developed in an attempt to make these predictions (see, for example, Sargis *et al.* 1979) they are usually very specific to the particular problem under investigation. Often, they must also resort to empirical information on unknown factors such as the transient mixing and condensation rates.

Finally, we note that instabilities that are similar to chugging have been observed in other contexts. For example, when steam was injected into the wake of a streamlined underwater body to explore underwater jet propulsion, the flow became very unstable (Kiceniuk 1952).

15.7 Transfer Functions

15.7.1 Unsteady Internal Flow Methods

Although the details are beyond the scope of this book, it is nevertheless of value to conclude the present chapter with a brief survey of the transfer function methods referred to in Section 15.6. There are two basic approaches to unsteady internal flows, namely solution in the time domain or in the frequency domain. The traditional time domain or *water-hammer* methods for hydraulic systems can and should be used in many circumstances; these are treated in depth elsewhere (for example, Streeter and Wylie 1967, 1974, Amies *et al.* 1977). They have the great advantage that they can incorporate the nonlinear convective inertial terms in the equations of fluid flow. They are best suited to evaluating the transient response of flows in long pipes in which the equations of the flow and the structure are well established. However, they encounter great difficulties when either the geometry is complex (for example inside a pump) or the fluid is complex (for example in a multiphase flow). Under these circumstances, frequency domain methods have distinct advantages, both analytically and experimentally. Specifically, unsteady flow experiments are most readily conducted by subjecting the component or device to fluctuations in the flow over a range of frequencies and measuring the fluctuating quantities at inlet and discharge. The main disadvantage of the frequency domain methods is that the nonlinear convective inertial terms cannot readily be included and, consequently, these methods are accurate only for small perturbations from the mean flow. Although this permits evaluation of stability limits, it does not readily allow the evaluation of the amplitude of large unstable motions. However, there does exist a core of fundamental knowledge pertaining to frequency domain methods (see, for example, Pipes 1940, Paynter 1961, Brown 1967) that is summarized in Brennen (1994). A good example of the application of these methods is contained in Amies and Greene (1977).

15.7.2 Transfer Functions

As in the quasistatic analyses described at the beginning of this chapter, the first step in the frequency domain approach is to identify all the flow variables that are needed to completely define the state of the flow at each of the nodes of the system. Typical flow variables are the pressure, p (or total pressure, p^T), the velocities of the phases or components, the volume fractions, and so on. To simplify matters we count only those variables that are not related by simple algebraic expressions. Thus we do not count both the pressure and the density of a phase that behaves barotropically, nor do we count the mixture density, ρ, and the void fraction, α, in a mixture of two incompressible fluids. The minimum number of variables needed to completely define the flow at all of the nodes is called *the order of the system* and is denoted by N. Then the state of the flow at any node, i, is denoted by the vector of state variables, $\{q_i^n\}$, $n = 1, 2 \to N$. For example, in a homogeneous flow we could choose $q_i^1 = p$, $q_i^2 = u$, $q_i^3 = \alpha$, to be the pressure, velocity, and void fraction at the node i.

The next step in a frequency domain analysis is to express all the flow variables, $\{q_i^n\}$, $n = 1, 2 \to N$, as the sum of a mean component (denoted by an overbar) and a fluctuating component (denoted by a tilde) at a frequency, ω. The complex fluctuating component incorporates both the amplitude and phase of the fluctuation as follows:

$$\{q^n(s,t)\} = \{\bar{q}^n(s)\} + \text{Re}\{\{\tilde{q}^n(s,\omega)\}e^{i\omega t}\} \tag{15.5}$$

for $n = 1 \to N$ where i is $(-1)^{\frac{1}{2}}$ and Re denotes the real part. For example

$$p(s,t) = \bar{p}(s) + \text{Re}\{\tilde{p}(s,\omega)e^{i\omega t}\} \tag{15.6}$$

$$\dot{m}(s,t) = \bar{\dot{m}}(s) + \text{Re}\{\tilde{\dot{m}}(s,\omega)e^{i\omega t}\} \tag{15.7}$$

$$\alpha(s,t) = \bar{\alpha}(s) + \text{Re}\{\tilde{\alpha}(s,\omega)e^{i\omega t}\}. \tag{15.8}$$

Because the perturbations are assumed linear ($|\tilde{u}| \ll \bar{u}$, $|\tilde{\dot{m}}| \ll \bar{\dot{m}}$, $|\tilde{q}^n| \ll \bar{q}^n$) they can be readily superimposed, so a summation over many frequencies is implied in the above expressions. In general, the perturbation quantities, $\{\tilde{q}^n\}$, will be functions of the mean flow characteristics as well as position, s, and frequency, ω.

The utilization of transfer functions in the context of fluid systems owes much to the pioneering work of Pipes (1940). The concept is the following. If the quantities at

15.7 Transfer Functions

inlet and discharge are denoted by subscripts $m = 1$ and $m = 2$, respectively, then the transfer matrix, $[T]$, is defined as follows:

$$\{\tilde{q}_2^n\} = [T]\{\tilde{q}_1^n\}. \tag{15.9}$$

It is a square matrix of the order of N. For example, for an order $N = 2$ system in which the independent fluctuating variables are chosen to be the total pressure, $\tilde{p}^T$, and the mass flow rate, $\tilde{m}$, then a convenient transfer matrix is

$$\begin{Bmatrix} \tilde{p}_2^T \\ \tilde{m}_2 \end{Bmatrix} = \begin{bmatrix} T_{11} & T_{12} \\ T_{21} & T_{22} \end{bmatrix} \begin{Bmatrix} \tilde{p}_1^T \\ \tilde{m}_1 \end{Bmatrix}. \tag{15.10}$$

In general, the transfer matrix will be a function of the frequency, ω, of the perturbations and the mean flow conditions in the device. Given the transfer functions for each component one can then synthesize transfer functions for the entire system using a set of simple procedures described in detail in Brennen (1994). This allows one to proceed to a determination of whether a system is stable or unstable given the boundary conditions acting on it.

The transfer functions for many simple components are readily identified (see Brennen 1994) and are frequently composed of impedances due to fluid friction and inertia (that primarily contribute to the real and imaginary parts of T_{12} respectively) and compliances due to fluid and structural compressibility (that primarily contribute to the imaginary part of T_{21}). More complex components or flows have more complex transfer functions that can often be determined only by experimental measurement. For example, the dynamic response of pumps can be critical to the stability of many internal flow systems (Ohashi 1968, Greitzer 1981) and consequently the transfer functions for pumps have been extensively explored (Fanelli 1972, Anderson et al. 1971, Brennen and Acosta 1976). Under stable operating conditions (see Sections 15.3 and 16.4.2) and in the absence of phase change, most pumps can be modeled with resistance, compliance, and inertance elements and they are therefore dynamically passive. However, the situation can be quite different when phase change occurs. For example, cavitating pumps are now known to have transfer functions that can cause instabilities in the hydraulic system of which they are a part. Note that under cavitating conditions, the instantaneous flow rates at inlet and discharge will be different because of the rate of change of the total volume, V, of cavitation within the pump and this leads to complex transfer functions that are described in more detail in Section 16.4.2. These characteristics of cavitating pumps give rise to a variety of important instabilities such as cavitation surge (see Section 15.6.2) or the Pogo instabilities of liquid-propelled rockets (Brennen 1994).

Much less is known about the transfer functions of other devices involving phase change, for example boiler tubes or vertical evaporators. As an example of the transfer function method, in the next section we consider a simple homogeneous multiphase flow.

15.7.3 Uniform Homogeneous Flow

As an example of a multiphase flow that exhibits the solution structure described in Section 15.7.2, we explore the form of the solution for the inviscid, frictionless flow of a two-component gas and liquid mixture in a straight, uniform pipe. The relative motion between the two components is neglected so there is only one velocity, $u(s, t)$. Surface tension is also neglected so there is only one pressure, $p(s, t)$. Moreover, the liquid is assumed incompressible (ρ_L constant) and the gas is assumed to behave barotropically with $p \propto \rho_G^k$. Then the three equations governing the flow are the continuity equations for the liquid and for the gas and the momentum equation for the mixture which are, respectively:

$$\frac{\partial}{\partial t}(1 - \alpha) + \frac{\partial}{\partial s}[(1 - \alpha)u] = 0 \tag{15.11}$$

$$\frac{\partial}{\partial t}(\rho_G \alpha) + \frac{\partial}{\partial s}(\rho_G \alpha u) = 0 \tag{15.12}$$

$$\rho\left(\frac{\partial u}{\partial t} + u\frac{\partial u}{\partial s}\right) = -\frac{\partial p}{\partial s}, \tag{15.13}$$

where ρ is the usual mixture density. Note that this is a system of the order of $N = 3$ and the most convenient flow variables are p, u, and α. These relations yield the following equations for the perturbations:

$$-i\omega\tilde{\alpha} + \frac{\partial}{\partial s}[(1 - \bar{\alpha})\tilde{u} - \bar{u}\tilde{\alpha}] = 0 \tag{15.14}$$

$$i\omega\bar{\rho}_G\tilde{\alpha} + i\omega\bar{\alpha}\tilde{\rho}_G + \bar{\rho}_G\bar{\alpha}\frac{\partial \tilde{u}}{\partial s} + \bar{\rho}_G\bar{u}\frac{\partial \tilde{\alpha}}{\partial s} + \bar{\alpha}\bar{u}\frac{\partial \tilde{\rho}_G}{\partial s} = 0 \tag{15.15}$$

$$-\frac{\partial \tilde{p}}{\partial s} = \bar{\rho}\left[i\omega\tilde{u} + \bar{u}\frac{\partial \tilde{u}}{\partial s}\right], \tag{15.16}$$

where $\tilde{\rho}_G = \tilde{p}\bar{\rho}_G/k\bar{p}$. Assuming the solution has the simple form

$$\begin{Bmatrix} \tilde{p} \\ \tilde{u} \\ \tilde{\alpha} \end{Bmatrix} = \begin{Bmatrix} P_1 e^{i\kappa_1 s} + P_2 e^{i\kappa_2 s} + P_3 e^{i\kappa_3 s} \\ U_1 e^{i\kappa_1 s} + U_2 e^{i\kappa_2 s} + U_3 e^{i\kappa_3 s} \\ A_1 e^{i\kappa_1 s} + A_2 e^{i\kappa_2 s} + A_3 e^{i\kappa_3 s} \end{Bmatrix} \tag{15.17}$$

it follows from Eqs. (15.14), (15.15), and (15.16) that

$$\kappa_n(1 - \bar{\alpha})U_n = (\omega + \kappa_n \bar{u})A_n \tag{15.18}$$

$$(\omega + \kappa_n \bar{u})A_n + \frac{\bar{\alpha}}{k\bar{p}}(\omega + \kappa_n \bar{u})P_n + \bar{\alpha}\kappa_n U_n = 0 \tag{15.19}$$

$$\bar{\rho}(\omega + \kappa_n \bar{u})U_n + \kappa_n P_n = 0 \tag{15.20}$$

15.7 Transfer Functions

Eliminating A_n, U_n, and P_n leads to the following dispersion relation

$$(\omega + \kappa_n \bar{u})\left[1 - \frac{\bar{\alpha}\bar{\rho}}{k\bar{p}}\frac{(\omega + \kappa_n \bar{u})^2}{\kappa_n^2}\right] = 0. \tag{15.21}$$

The solutions to this dispersion relation yield the following wavenumbers and velocities, $c_n = -\omega/\kappa_n$, for the perturbations:

- $\kappa_1 = -\omega/\bar{u}$, which has a wave velocity, $c_0 = \bar{u}$. This is a purely kinematic wave, a concentration wave that from Eqs. (15.18) and (15.20) has $U_1 = 0$ and $P_1 = 0$ so that there are no pressure or velocity fluctuations associated with this type of wave. In other, more complex flows, kinematic waves may have some small pressure and velocity perturbations associated with them and their velocity may not exactly correspond with the mixture velocity but they are still called kinematic waves if the major feature is the concentration perturbation.

- $\kappa_2, \kappa_3 = -\omega/(\bar{u} \pm c)$, where c is the sonic speed in the mixture, namely $c = (k\bar{p}/\bar{\alpha}\bar{\rho})^{\frac{1}{2}}$. Consequently, these two modes have wave speeds $c_2, c_3 = \bar{u} \pm c$ and are the two acoustic waves traveling downstream and upstream respectively.

Finally, we list the solution in terms of three unknown, complex constants, P_2, P_3, and A_1:

$$\begin{Bmatrix} \tilde{p} \\ \tilde{u} \\ \tilde{\alpha} \end{Bmatrix} = \begin{bmatrix} 0 & e^{i\kappa_2 s} & e^{i\kappa_3 s} \\ 0 & -e^{i\kappa_2 s}/\bar{\rho}c & e^{i\kappa_3 s}/\bar{\rho}c \\ e^{i\kappa_1 s} & -(1-\bar{\alpha})e^{i\kappa_2 s}/\bar{\rho}c^2 & -(1-\bar{\alpha})e^{i\kappa_3 s}/\bar{\rho}c^2 \end{bmatrix} \begin{Bmatrix} A_1 \\ P_2 \\ P_3 \end{Bmatrix} \tag{15.22}$$

and the transfer function between two locations $s = s_1$ and $s = s_2$ follows by eliminating the vector $\{A_1, P_2, P_3\}$ from Eq. (15.22) for the state vectors at those two locations.

Transfer function methods for multiphase flow are nowhere near as well developed as they are for single-phase flows but, given the number and ubiquity of instability problems in multiphase flows (Ishii 1982), it is inevitable that these methods will gradually develop into a tool that is useful in a wide spectrum of applications.

16

Kinematic Waves

16.1 Introduction

The one-dimensional theory of sedimentation was introduced in a classic article by Kynch (1952), and the methods he used have since been expanded to cover a wide range of other multiphase flows. In Chapter 14 we introduced the concept of drift flux models and showed how these can be used to analyze and understand a class of steady flows in which the relative motion between the phases is determined by external forces and the component properties. The present chapter introduces the use of the drift flux method to analyze the formation, propagation, and stability of concentration (or kinematic) waves. For a survey of this material, the reader may wish to consult Wallis (1969).

The general concept of a kinematic wave was first introduced by Lighthill and Whitham (1955) and the reader is referred to Whitham (1974) for a rigorous treatment of the subject. Generically, kinematic waves occur when a functional relation connects the fluid density with the flux of some physically conserved quantity such as mass. In the present context a kinematic (or concentration) wave is a gradient or discontinuity in the volume fraction, α. We refer to such gradients or discontinuities as local *structure* in the flow; only multiphase flows with a constant and uniform volume fraction are devoid of such structure. Of course, in the absence of any relative motion between the phases or components, the structure is simply convected at the common velocity in the mixture. Such flows may still be nontrivial if the changing density at some Eulerian location causes deformation of the flow boundaries and thereby creates a dynamic problem. But we do not follow that path here. Rather, this chapter examines the velocity of propagation of the structure when there is relative motion between the phases. Then, inevitably, the structure will propagate at a velocity that does not necessarily correspond to the velocity of either of the phases or components. Thus it is a genuinely propagating wave. When the pressure gradients associated with the wave are negligible and its velocity of propagation is governed by mass conservation alone, we call the waves *kinematic* to help distinguish them from the *dynamic* waves in which the primary gradient or discontinuity is in the pressure rather than the volume fraction.

16.2 Two-Component Kinematic Waves

16.2.1 Basic Analysis

Consider the most basic model of two-component pipe flow (components A and B) in which the relative motion is nonnegligible. We assume a pipe of uniform cross section. In the absence of phase change the continuity equations become the following:

$$\frac{\partial \alpha_A}{\partial t} + \frac{\partial j_A}{\partial s} = 0; \quad \frac{\partial \alpha_B}{\partial t} + \frac{\partial j_B}{\partial s} = 0. \tag{16.1}$$

For convenience we set $\alpha = \alpha_A = 1 - \alpha_B$. Then, using the standard notation of Eqs. (15.5) to (15.8), we expand α, j_A, and j_B in terms of their mean values (denoted by an overbar) and harmonic perturbations (denoted by the tilde) at a frequency ω in the form used in Eq. (15.5). The solution for the mean flow is simply as follows:

$$\frac{d(\bar{j}_A + \bar{j}_B)}{ds} = \frac{d\bar{j}}{ds} = 0 \tag{16.2}$$

and therefore $\bar{j}$ is a constant. Moreover, the following equations for the perturbations emerge:

$$\frac{\partial \tilde{j}_A}{\partial s} + i\omega\tilde{\alpha} = 0; \quad \frac{\partial \tilde{j}_B}{\partial s} - i\omega\tilde{\alpha} = 0. \tag{16.3}$$

Now consider the additional information that is necessary to determine the dispersion equation and therefore the different modes of wave propagation that can occur in this flow. First, we note that

$$j_A = \alpha j + j_{AB}; \quad j_B = (1 - \alpha)j - j_{AB} \tag{16.4}$$

and it is convenient to replace the variables, j_A and j_B, with j, the total volumetric flux, and j_{AB}, the drift flux. Substituting these expressions into Eq. (16.3), we obtain the following:

$$\frac{\partial \tilde{j}}{\partial s} = 0; \quad \frac{\partial(\bar{j}\tilde{\alpha} + \tilde{j}_{AB})}{\partial s} + i\omega\tilde{\alpha} = 0 \tag{16.5}$$

The first of these yields a uniform and constant value of $\tilde{j}$ that corresponds to a synchronous motion in which the entire length of the multiphase flow in the pipe is oscillating back and forth in unison. Such motion is not of interest here and we assume for the purposes of the present analysis that $\tilde{j} = 0$.

The second equation [Eq. (16.5)] has more interesting implications. It represents the connection between the two remaining fluctuating quantities, $\tilde{j}_{AB}$ and $\tilde{\alpha}$. To proceed further it is therefore necessary to find a second relation connecting these same quantities. It now becomes clear that, from a mathematical point of view, there is considerable simplicity in the Drift Flux Model (Chapter 14), in which it is assumed that the relative motion is governed by a simple algebraic relation connecting j_{AB} and α,

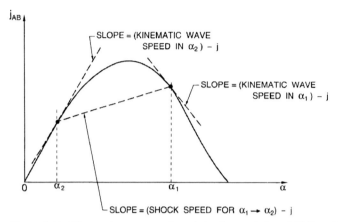

Figure 16.1. Kinematic wave speeds and shock speeds in a drift flux chart.

We utilize that model here and assume the existence of a known, functional relation, $j_{AB}(\alpha)$. Then the second Eq. (16.5) can be written as follows:

$$\left(\bar{j} + \frac{dj_{AB}}{d\alpha}\bigg|_{\bar{\alpha}}\right)\frac{\partial \tilde{\alpha}}{\partial s} + i\omega\tilde{\alpha} = 0, \tag{16.6}$$

where $dj_{AB}/d\alpha$ is evaluated at $\alpha = \bar{\alpha}$ and is therefore a known function of $\bar{\alpha}$. It follows that the dispersion relation yields a single wave type given by the wavenumber, κ, and wave velocity, c, where

$$\kappa = -\frac{\omega}{\bar{j} + \frac{dj_{AB}}{d\alpha}|_{\bar{\alpha}}} \quad \text{and} \quad c = \bar{j} + \frac{dj_{AB}}{d\alpha}\bigg|_{\bar{\alpha}}. \tag{16.7}$$

This is called a kinematic wave because its primary characteristic is the perturbation in the volume fraction and it travels at a velocity close to the velocity of the components. Indeed, in the absence of relative motion $c \to \bar{j} = u_A = u_B$.

Equation (16.7) [and, for the kinematic shock speed, Eq. (16.19)] reveal that the propagation speed of kinematic waves (and shocks) relative to the total volumetric flux, j, can be conveniently displayed in a drift flux chart as illustrated in Figure 16.1. The kinematic wave speed at a given volume fraction is the slope of the tangent to the drift flux curve at that point (plus j). This allows a graphical and comparative display of wave speeds that, as we demonstrate, is very convenient in flows that can be modeled using the drift flux methodology.

16.2.2 Kinematic Wave Speed at Flooding

In Section 14.3.1 (and Figure 14.2) we identified the phenomenon of flooding and drew the analogies to choking in gas dynamics and open-channel flow. Note that in these analogies, the choked flow is independent of conditions downstream because signals (small-amplitude waves) cannot travel upstream through the choked flow because the

16.2 Two-Component Kinematic Waves

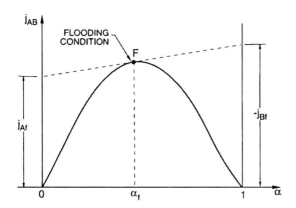

Figure 16.2. Conditions of flooding at a volume fraction of α_f and volume fluxes j_{Af} and j_{Bf}.

fluid velocity there is equal to the small-amplitude wave propagation speed relative to the fluid. Hence in the laboratory frame, the upstream traveling wave speed is zero. The same holds true in a flooded flow as illustrated in Figure 16.2, which depicts flooding at a volume fraction of α_f and volume fluxes, j_{Af} and j_{Bf}. From the geometry of this figure it follows that

$$j_{Af} + j_{Bf} = j_f = -\left.\frac{dj_{AB}}{d\alpha}\right|_{\alpha_f} \quad (16.8)$$

and therefore the kinematic wave speed at the flooding condition, c_f is as follows:

$$c_f = j_f + \left.\frac{dj_{AB}}{d\alpha}\right|_{\alpha_f} = 0. \quad (16.9)$$

Thus the kinematic wave speed in the laboratory frame is zero and small disturbances cannot propagate through flooded flow. Consequently, the flow is choked just as it is in the gas dynamic or open-channel flow analogies.

One way to visualize this limit in a practical flow is to consider countercurrent flow in a vertical pipe whose cross-sectional area decreases as a function of axial position until it reaches a throat. Neglecting the volume fraction changes that could result from the changes in velocity and therefore pressure, the volume flux intercepts in Figure 16.2, j_A and j_B, therefore increase with decreasing area. Flooding or choking will occur at a throat when the fluxes reach the flooding values, j_{Af} and j_{Bf}. The kinematic wave speed at the throat is then zero.

16.2.3 Kinematic Waves in Steady Flows

In many, nominally steady two-phase flows there is sufficient ambient *noise* or irregularity in the structure that the inhomogeneity instability analyzed in Section 7.4.1 leads to small-amplitude kinematic waves that propagate that structure (see, for example, El-Kaissy and Homsy, 1976). Although those structures may be quite irregular and sometimes short-lived, it is often possible to detect their presence by cross-correlating

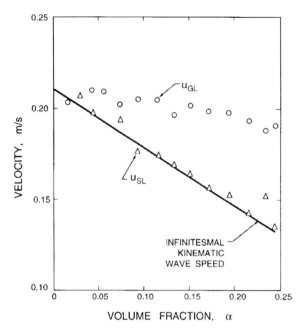

Figure 16.3. Kinematic wave speeds, $u_{SL}(\triangle)$, in nominally steady bubbly flows of an air/water mixture with $j_L = 0.169$ m/s in a vertical, 0.102-m-diameter pipe as obtained from cross-correlograms. Also shown is the speed of infinitesimal kinematic waves (solid line, calculated from the measured drift flux) and the measured bubble velocities relative to the liquid (u_{GL}, ⊙). Adapted from Bernier (1982).

volume fraction measurements at two streamwise locations a short distance apart. For example, Bernier (1982) cross-correlated the outputs from two volume fraction meters 0.108 m apart in a nominally steady vertical bubbly flow in a 0.102-m diameter pipe. The cross-correlograms displayed strong peaks that corresponded to velocities, u_{SL}, relative to the liquid that are shown in Figure 16.3. From that figure it is clear that u_{SL} corresponds to the infinitesimal kinematic wave speed calculated from the measured drift flux. This confirms that the structure consists of small-amplitude kinematic waves. Similar results were later obtained for solid/liquid mixtures by Kytomaa and Brennen (1990) and others.

It is important to note that, in these experiments, the cross-correlation yields the speed of the propagating structure and not the speed of individual bubbles (shown for contrast as u_{GL} in Figure 16.3) because the volume fraction measurement performed was an average over the cross section and therefore an average over a volume much larger than the individual bubbles. If the probe measuring volume were small relative to the bubble (or disperse phase) size and if the distance between the probes was also small, then the cross-correlation would yield the dispersed phase velocity.

16.3 Two-Component Kinematic Shocks

16.3.1 Kinematic Shock Relations

The results of Section 16.2.1 can be extended by considering the relations for a finite kinematic wave or shock. As sketched in Figure 16.4 the conditions ahead of

16.3 Two-Component Kinematic Shocks

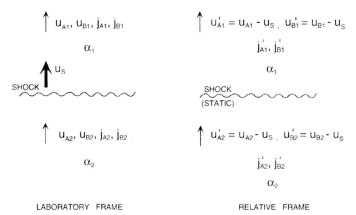

Figure 16.4. Velocities and volume fluxes associated with a kinematic shock in the laboratory frame (left) and in a frame relative to the shock (right).

the shock will be denoted by the subscript 1 and the conditions behind the shock by the subscript 2. Two questions must be asked. First, does such a structure exist and, if so, what is its propagation velocity, u_S? Second, is the structure stable in the sense that it will persist unchanged for a significant time? The first question is addressed in this section, the second question in the section that follows. For the sake of simplicity, any differences in the component densities across the shock are neglected; it is also assumed that no exchange of mass between the phases or components occurs within the shock. In Section 16.3.3, the role that might be played by each of these effects is considered.

To determine the speed of the shock, u_S, it is convenient to first apply a Galilean transformation to the situation on the left in Figure 16.4 so that the shock position is fixed (the diagram on the right in Figure 16.4). In this relative frame we denote the velocities and fluxes by the prime. By definition it follows that the fluxes relative to the shock are related to the fluxes in the original frame by the following:

$$j'_{A1} = j_{A1} - \alpha_1 u_s; \quad j'_{B1} = j_{B1} - (1 - \alpha_1)u_s \qquad (16.10)$$

$$j'_{A2} = j_{A2} - \alpha_2 u_s; \quad j'_{B2} = j_{B2} - (1 - \alpha_2)u_s. \qquad (16.11)$$

Then, because the densities are assumed to be the same across the shock and no exchange of mass occurs, conservation of mass requires that

$$j'_{A1} = j'_{A2}; \quad j'_{B1} = j'_{B2}. \qquad (16.12)$$

Substituting Eqs. (16.10) and (16.11) into Eq. (16.12) and replacing the fluxes, j_{A1}, j_{A2}, j_{B1}, and j_{B2}, using Eq. (16.4) involving the total flux, j, and the drift fluxes, j_{AB1} and j_{AB2}, we obtain the following expression for the shock propagation velocity, u_s:

$$u_s = j + \frac{j_{AB2} - j_{AB1}}{\alpha_2 - \alpha_1}, \qquad (16.13)$$

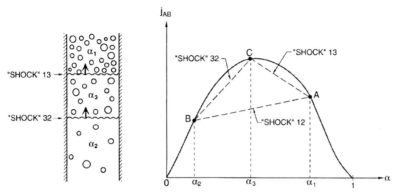

Figure 16.5. Shock stability for $\alpha_1 > \alpha_2$.

where the total flux, j, is necessarily the same on both sides of the shock. Now, if the drift flux is a function only of α it follows that this expression can be written as

$$u_s = j + \frac{j_{AB}(\alpha_2) - j_{AB}(\alpha_1)}{\alpha_2 - \alpha_1}. \tag{16.14}$$

Note that, in the limit of a small-amplitude wave ($\alpha_2 \to \alpha_1$), this reduces, as it must, to Eq. (16.7) for the speed of an infinitesimal wave.

So now we add another aspect to Figure 16.1 and indicate that, as a consequence of Eq. (16.14), the speed of a shock between volume fractions α_2 and α_1 is given by the slope of the line connecting those two points on the drift flux curve (plus j).

16.3.2 Kinematic Shock Stability

The stability of the kinematic shock waves analyzed in the last section is most simply determined by considering the consequences of the shock splitting into several fragments. Without any loss of generality we assume that component A is less dense than component B so that the drift flux, j_{AB}, is positive when the upward direction is defined as positive (as in Figures 16.4 and 16.1).

Consider first the case in which $\alpha_1 > \alpha_2$ as shown in Figure 16.5 and suppose that the shock begins to split such that a region of intermediate volume fraction, α_3, develops. Then the velocity of the shock fragment labeled Shock 13 is given by the slope of the line CA in the drift flux chart, whereas the velocity of the shock fragment labeled Shock 32 is given by the slope of the line BC. The former is smaller than the speed of the original Shock 12, whereas the latter fragment has a higher velocity. Consequently, even if such fragmentation were to occur, the shock fragments would converge and rejoin. Another version of the same argument is to examine the velocity of small perturbations that might move ahead of or be left behind the main Shock 12. A small perturbation that might move ahead would travel at a velocity given by the slope of the tangent to the drift flux curve at the point A. Because this velocity is much smaller than the velocity of the main shock such dispersion of the shock is

16.3 Two-Component Kinematic Shocks

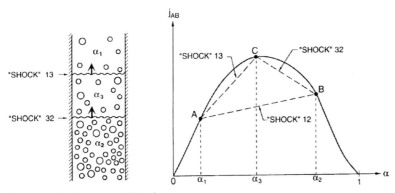

Figure 16.6. Shock instability for $\alpha_1 < \alpha_2$.

not possible. Similarly, a perturbation that might be left behind would travel with a velocity given by the slope of the tangent at the point B and because this is larger than the shock speed the perturbation would catch up with the shock and be reabsorbed. Therefore, the shock configuration depicted in Figure 16.5 is stable and the shock will develop a permanent form.

Conversely, a parallel analysis of the case in which $\alpha_1 < \alpha_2$ (Figure 16.6), clearly leads to the conclusion that, once initiated, fragmentation will continue because the velocity of the shock fragment Shock 13 will be greater than the velocity of the shock fragment Shock 32. Also the kinematic wave speed of small perturbations in α_1 will be greater than the velocity of the main shock and the kinematic wave speed of small perturbations in α_2 will be smaller than the velocity of the main shock. Therefore, the shock configuration depicted in Figure 16.6 is unstable. No such shock will develop and any imposed transient of this kind will disperse if $\alpha_1 < \alpha_2$.

Using the analogy with gas dynamic shocks, the case of $\alpha_1 > \alpha_2$ is a compression wave and develops into a shock, whereas the case of $\alpha_1 < \alpha_2$ is an expansion wave that becomes increasingly dispersed. All of this is not surprising because we defined A to be the less dense component and therefore the mixture density decreases with increasing α. Therefore, in the case of $\alpha_1 > \alpha_2$, the lighter fluid is on top of the heavier fluid and this configuration is stable, whereas, in the case of $\alpha_1 < \alpha_2$, the heavier fluid is on top and this configuration is unstable according to the Kelvin–Helmholtz analysis (see Section 7.5.1).

16.3.3 Compressibility and Phase-Change Effects

In this section the effects of the small pressure difference that must exist across a kinematic shock and the consequent effects of the corresponding density differences are explored. The effects of phase change are also explored.

By applying the momentum theorem to a control volume enclosing a portion of a kinematic shock in a frame of reference fixed in the shock, the following expression for the difference in the pressure across the shock is readily obtained:

$$p_2 - p_1 = \rho_A \left\{ \frac{(j'_{A1})^2}{\alpha_1} - \frac{(j'_{A2})^2}{\alpha_2} \right\} + \rho_B \left\{ \frac{(j'_{B1})^2}{(1-\alpha_1)} - \frac{(j'_{B2})^2}{(1-\alpha_2)} \right\}. \quad (16.15)$$

Here we have assumed that any density differences that might occur will be second-order effects. Because $j'_{A1} = j'_{A2}$ and $j'_{B1} = j'_{B2}$, it follows that

$$p_2 - p_1 = \frac{\rho_A(1-\alpha_1)(1-\alpha_2)}{(\alpha_1 - \alpha_2)} \left\{ \frac{j_{AB1}}{(1-\alpha_1)} - \frac{j_{AB2}}{(1-\alpha_2)} \right\}^2$$
$$- \frac{\rho_B \alpha_1 \alpha_2}{(\alpha_1 - \alpha_2)} \left\{ \frac{j_{AB1}}{\alpha_1} - \frac{j_{AB2}}{\alpha_2} \right\}^2. \quad (16.16)$$

Because the expressions inside the curly brackets are of the order of $(\alpha_1 - \alpha_2)$, the order of magnitude of $p_2 - p_1$ is given by the following:

$$p_2 - p_1 = O\left((\alpha_1 - \alpha_2)\rho u_{AB}^2\right), \quad (16.17)$$

provided neither α_1 nor α_2 are close to zero or unity. Here ρ is some representative density, for example, the mixture density or the density of the heavier component. Therefore, provided the relative velocity, u_{AB}, is modest, the pressure difference across the kinematic shock is small. Consequently, the dynamic effects on the shock are small. If, under unusual circumstances, $(\alpha_1 - \alpha_2)\rho u_{AB}^2$ were to become significant compared with p_1 or p_2, the character of the shock would begin to change substantially.

Consider, now, the effects of the differences in density that the pressure difference given by Eq. (16.17) imply. Suppose that the component B is incompressible but that the component A is a compressible gas that behaves isothermally so that

$$\frac{\rho_{A1}}{\rho_{A2}} - 1 = \delta = \frac{p_1}{p_2} - 1. \quad (16.18)$$

If the kinematic shock analysis of Section 16.3.1 is revised to incorporate a small density change ($\delta \ll 1$) in component A, the result is the following modification to Eq. (16.13) for the shock speed:

$$u_s = j_1 + \frac{j_{AB2} - j_{AB1}}{\alpha_2 - \alpha_1} - \frac{\delta(1-\alpha_2)(j_{AB1}\alpha_2 - j_{AB2}\alpha_1)}{(\alpha_2 - \alpha_1)(\alpha_2 - \alpha_1 - \alpha_2(1-\alpha_1)\delta)}, \quad (16.19)$$

where terms of the order of δ^2 have been neglected. The last term in Eq. (16.19) represents the first-order modification to the propagation speed caused by the compressibility of component A. From Eqs. (16.17) and (16.18), it follows that the order of the magnitude of δ is $(\alpha_1 - \alpha_2)\rho u_{AB}^2/p$, where p is a representative pressure. Therefore, from Eq. (16.19), the order of magnitude of the correction to u_s is $\rho u_{AB}^3/p$, which is the typical velocity, u_{AB}, multiplied by a Mach number. Clearly, this is usually a negligible correction.

Another issue that may arise concerns the effect of phase change in the shock. A different modification to the kinematic shock analysis allows some evaluation of this effect. Assume that, within the shock, mass is transfered from the more dense

16.4 Examples of Kinematic Wave Analyses

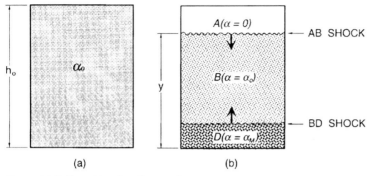

Figure 16.7. Type I batch sedimentation.

component B (the liquid phase) to the component A (or vapor phase) at a condensation rate equal to I per unit area of the shock. Then, neglecting density differences, the kinematic shock analysis leads to the following modified form of Eq. (16.13):

$$u_s = j_1 + \frac{j_{AB2} - j_{AB1}}{\alpha_2 - \alpha_1} + \frac{I}{(\alpha_1 - \alpha_2)} \left\{ \frac{(1-\alpha_2)}{\rho_A} + \frac{\alpha_2}{\rho_B} \right\}. \quad (16.20)$$

Because $\alpha_1 > \alpha_2$ it follows that the propagation speed increases as the condensation rate increases. Under these circumstances, it is clear that the propagation speed will become greater than j_{A1} or j_1 and that the flux of vapor (component A) will be down through the shock. Thus, we can visualize that the shock will evolve from a primarily kinematic shock to a much more rapidly propagating condensation shock (see Section 9.5.3).

16.4 Examples of Kinematic Wave Analyses

16.4.1 Batch Sedimentation

Because it presents a useful example of kinematic shock propagation, we consider the various phenomena that occur in batch sedimentation. For simplicity, it is assumed that this process begins with a uniform suspension of solid particles of volume fraction, α_0, in a closed vessel (Figure 16.7(a)). Conceptually, it is convenient to visualize gravity being switched on at time $t = 0$. Then the sedimentation of the particles leaves an expanding clear layer of fluid at the top of the vessel as indicated in Figure 16.7(b). This implies that at time $t = 0$ a kinematic shock is formed at the top of the vessel. This shock is the moving boundary between the region A of Figure 16.7(b) in which $\alpha = 0$ and the region B in which $\alpha = \alpha_0$. It travels downward at the shock propagation speed given by the slope of the line AB in Figure 16.8(left) (note that in this example $j = 0$).

Now consider the corresponding events that occur at the bottom of the vessel. Beginning at time $t = 0$, particles will start to come to rest on the bottom and a

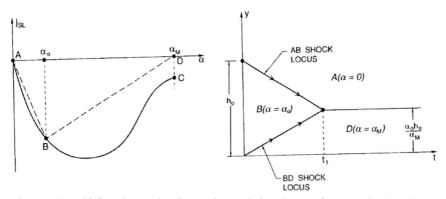

Figure 16.8. Drift flux chart and sedimentation evolution diagram for Type I batch sedimentation.

layer comprising particles in a packed state at $\alpha = \alpha_m$ will systematically grow in height (we neglect any subsequent adjustments to the packing that might occur as a result of the increasing overburden). A kinematic shock is therefore present at the interface between the packed region D (Figure 16.7(b)) and the region B; clearly this shock is also formed at the bottom at time $t = 0$ and propagates upward. Because the conditions in the packed bed are such that both the particle and liquid flux are zero and, therefore, the drift flux is zero, this state is represented by the point D in the drift flux chart, Figure 16.8(left) (rather than the point C). It follows that, provided that none of the complications discussed later occur, the propagation speed of the upward moving shock is given by the slope of the line BD in Figure 16.8(left). Note that both the downward moving AB shock and the upward moving BD shock are stable.

The progress of the batch sedimentation process can be summarized in a time evolution diagram such as Figure 16.8(right) in which the elevations of the shocks are plotted as a function of time. When the AB and BD shocks meet at time $t = t_1$, the final packed bed depth equal to $\alpha_0 h_0 / \alpha_m$ is achieved and the sedimentation process is complete. Note that

$$t_1 = \frac{h_0 \alpha_0 (\alpha_m - \alpha_0)}{\alpha_m j_{SL}(\alpha_0)} = \frac{h_0 (\alpha_m - \alpha_0)}{\alpha_m (1 - \alpha_0) u_{SL}(\alpha_0)}. \quad (16.21)$$

The simple batch evolution described above is known as Type I sedimentation. There are, however, other complications that can arise if the shape of the drift flux curve and the value of α_0 are such that the line connecting B and D in Figure 16.8(left) intersects the drift flux curve. Two additional types of sedimentation may occur under those circumstances and one of these, Type III, is depicted in Figures 16.9 and 16.10. In Figure 16.9, the line STD is tangent to the drift flux curve at the point T and the point P is the point of inflection in the drift flux curve. Thus are the volume fractions, α_S, α_P, and α_T defined. If α_0 lies between α_S and α_P the process is known as Type III sedimentation and this proceeds as follows (the line BQ is a tangent to the drift flux curve at the point Q and defines the value of α_Q). The first shock to form at the bottom is one in which the volume fraction is increased from α_0 to α_Q. As depicted in

16.4 Examples of Kinematic Wave Analyses

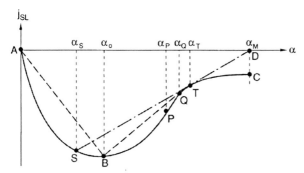

Figure 16.9. Drift flux chart for Type III sedimentation.

Figure 16.10 this is followed by a continuous array of small kinematic waves through which the volume fraction is increased from α_Q to α_T. Because the speeds of these waves are given by the slopes of the drift flux curve at the appropriate volume fractions, they travel progressively more slowly than the initial BQ shock. Finally this kinematic wave array is followed by a second, upward moving shock, the TD shock across which the volume fraction increases from α_Q to α_m. While this package of waves rises from the bottom, the usual AB shock moves down from the top. Thus, as depicted in Figure 16.10, the sedimentation process is more complex but, of course, arrives at the same final state as in Type I.

A third type, Type II, occurs when the initial volume fraction, α_0, is between α_P and α_T. This evolves in a manner similar to Type III except that the kinematic wave array is not preceded by a shock like the BQ shock in Type III.

16.4.2 Dynamics of Cavitating Pumps

Another very different example of the importance of kinematic waves and their interaction with dynamic waves occurs in the context of cavitating pumps. The dynamics of cavitating pumps are particularly important because of the dangers associated with the instabilities such as cavitation surge (see Section 15.6.2) that can result in very

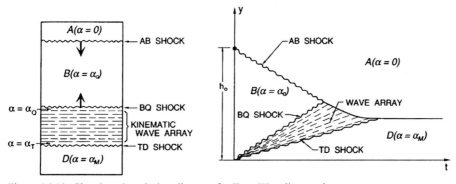

Figure 16.10. Sketch and evolution diagram for Type III sedimentation.

large pressure and flow rate oscillations in the entire system of which the pumps are a part (Brennen 1994). Therefore, in many pumping systems (for example, the fuel and oxidizer systems of a liquid-propelled rocket engine), it is very important to be able to evaluate the stability of that system and knowledge of the transfer function for the cavitating pumps is critical to that analysis (Rubin 1966).

For simplicity in this analytical model, the pump inlet and discharge flows are assumed to be purely incompressible liquid (density ρ_L). Then the inlet and discharge flows can be characterized by two flow variables; convenient choices are the total pressure, p_i^T, and the mass flow rate, $\dot{m}_i$, where the inlet and discharge quantities are given by $i = 1$ and $i = 2$ respectively as described in Section 15.7.2. Consider now the form of the transfer function [Eq. (15.10)] connecting these fluctuating quantities. As described in Section 15.7.2 the transfer function will be a function not only of frequency but also of the pump geometry and the parameters defining the mean flow (see Section 8.4.1). The instantaneous flow rates at inlet and discharge will be different because of the rate of change of the total volume, V, of cavitation within the pump. In the absence of cavitation, the pump transfer function is greatly simplified because (a) if the liquid and structural compressibilities are neglected then $\dot{m}_1 = \dot{m}_2$ and it follows that $T_{21} = 0$, $T_{22} = 1$ and (b) because the total pressure difference across the pump must be independent of the pressure level it follows that $T_{11} = 1$. Thus the noncavitating transfer function has only one nontrivial component, namely T_{12}, where $-T_{12}$ is known as the pump impedance. As long as the real part of $-T_{12}$ (the pump resistance) is positive, the pump is stable at all frequencies. Instabilities only occur at off-design operating points where the resistance becomes negative (when the slope of the total pressure rise against flow rate characteristic becomes positive). Measurements of T_{12} (which is a function of frequency) can be found in Anderson *et al.* (1971) and Ng and Brennen (1976).

A cavitating pump is much more complex because all four elements of $[T]$ are then nontrivial. The first complete measurements of $[T]$ were obtained by Ng and Brennen (1976) (see also Brennen *et al.* 1982). These revealed that cavitation could cause the pump dynamic characteristics to become capable of initiating instability in the system in which it operates. This helped explain the cavitation surge instability described in Section 15.6.2. Recall that cavitation surge occurs when the pump resistance (the real part of $-T_{12}$) is positive; thus it results from changes in the other elements of $[T]$ that come about as a result of cavitation.

A quasistatic approach to the construction of the transfer function of a cavitating pump was first laid out by Brennen and Acosta (1973, 1976) and proceeds as follows. The steady-state total pressure rise across the pump, $\Delta p^T(p_1^T, \dot{m})$, and the steady-state volume of cavitation in the pump, $V(p_1^T, \dot{m})$, will both be functions of the mean mass flow rate $\dot{m}$. They will also be functions of the inlet pressure (or, more accurately, the inlet pressure minus the vapor pressure) because this will change the cavitation number and the total pressure rise may depend on the cavitation number as discussed

16.4 Examples of Kinematic Wave Analyses

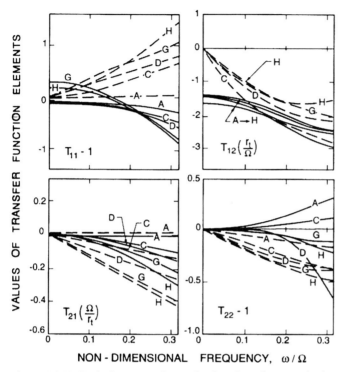

Figure 16.11. Typical measured transfer functions for a cavitating pump operating at five different cavitation numbers, σ = (A) 0.37, (C) 0.10, (D) 0.069, (G) 0.052, and (H) 0.044. Real and imaginary parts which are denoted by the solid and dashed lines respectively, are plotted against the nondimensional frequency, ω/Ω; r_t is the impeller tip radius. Adapted from Brennen et al. (1982).

in Section 8.4.1. Note that V is not just a function of cavitation number but also depends on $\dot{m}$ because changing $\dot{m}$ changes the angle of incidence on the blades and therefore the volume of cavitation bubbles produced. Given these two functions we could then construct the quasistatic or low-frequency form of the transfer function as follows:

$$[T] = \begin{bmatrix} 1 + \frac{d(\Delta p^T)}{dp_1^T}\big|_{\dot{m}} & \frac{d\Delta p^T}{d\dot{m}}\big|_{p_1^T} \\ i\omega\rho_L \frac{dV}{dp_1^T}\big|_{\dot{m}} & 1 + i\omega\rho_L \frac{dV}{d\dot{m}}\big|_{p_1^T} \end{bmatrix} \quad (16.22)$$

The constant $K^* = -\rho_L(dV/dp_1^T)_{\dot{m}}$ is known as the *cavitation compliance*, while the constant $M^* = -\rho_L(dV/d\dot{m})_{p_1^T}$ is called the *cavitation mass flow gain factor*. Later, we comment further on these important elements of the transfer function.

Typical measured transfer functions (in nondimensional form) for a cavitating pump are shown in Figure 16.11 for operation at four different cavitation numbers. Note that case (A) involved virtually no cavitation and that the volume of cavitation increases as σ decreases. In the figure, the real and imaginary parts of each of the

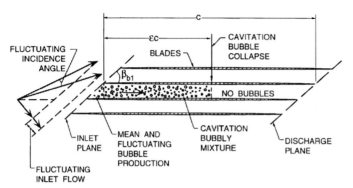

Figure 16.12. Schematic of the bubbly flow model for the dynamics of cavitating pumps (adapted from Brennen 1978).

elements are shown by the solid and dashed lines respectively and are plotted against a nondimensional frequency. Note that both the compliance, K^*, and the mass flow gain factor, M^*, increase monotonically as the cavitation number decreases.

To model the dynamics of the cavitation and generate some understanding of data such as that of Figure 16.11, we have generated a simple bubbly flow model (Brennen 1978) of the cavitating flow in the blade passages of the pump. The essence of this model is depicted schematically in Figure 16.12, which shows the blade passages as they appear in a developed, cylindrical surface within an axial-flow impeller. The cavitation is modeled as a bubbly mixture that extends over a fraction, ϵ, of the length of each blade passage before collapsing at a point where the pressure has risen to a value that causes collapse. This quantity, ϵ, will in practice vary inversely with the cavitation number, σ (experimental observations of the pump of Figure 16.11 indicate $\epsilon \approx 0.02/\sigma$), and therefore ϵ is used in the model as a surrogate for σ. The bubbly flow model then seeks to understand how this flow will respond to small, linear fluctuations in the pressures and mass flow rates at the pump inlet and discharge. Pressure perturbations at inlet will cause pressure waves to travel through the bubbly mixture and this part of the process is modeled using a mixture compressibility parameter, K^{**}, which essentially fixes the wave speed. In addition, fluctuations in the inlet flow rate produce fluctuations in the angle of incidence that cause fluctuations in the rate of production of cavitation at inlet. These disturbances would then propagate down the blade passage as kinematic or concentration waves that travel at the mean mixture velocity. This process is modeled by a factor of proportionality, M^{**}, which relates the fluctuation in the angle of incidence to the fluctuations in the void fraction. Neither of the parameters, K^{**} or M^{**}, can be readily estimated analytically; they are, however, the two key features in the bubbly flow model. Moreover, they respectively determine the cavitation compliance and the mass flow gain factor; see Brennen (1994) for the specific relationships between K^{**} and K^* and between M^{**} and M^*. Comparison of the model predictions with the experimental measurements indicate that $K^{**} = 1.3$ and

16.4 Examples of Kinematic Wave Analyses

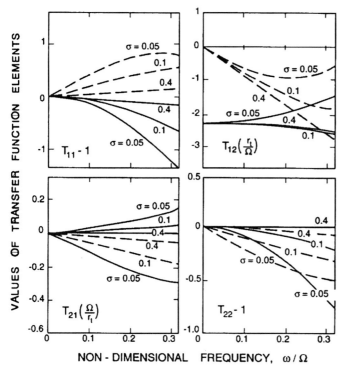

Figure 16.13. Theoretical transfer functions calculated from the bubbly flow model for comparison with the experimental results of Figure 16.11. The calculations use $K^{**} = 1.3$ and $M^{**} = 0.8$ (adapted from Brennen et al. 1982).

$M^{**} = 0.8$ are appropriate values and, with these, the complete theoretical transfer functions for various cavitation numbers are as depicted in Figure 16.13. This should be compared with the experimentally obtained transfer functions of Figure 16.11. Note that, with only a small number of discrepancies, the general features of the experimental transfer functions, and their variation with cavitation number, are reproduced by the model.

Following its verification, we must then ask how this knowledge of the pump transfer function might be used to understand cavitation-induced instabilities. In a given system, a stability analysis requires a complete model (transfer functions) of all the system elements; then a dynamic model must be constructed for the entire system. By interrogating the model, it is then possible to identify the key physical processes that promote instability. In the present case, such an interrogation leads to the conclusion that it is the formation and propagation of the kinematic waves that are responsible for those features of the transfer function (in particular the mass flow gain factor) that lead to cavitation-induced instability. In comparison, the acoustic waves and the cavitation compliance have relatively benign consequences. Hence a more complete understanding of the mass flow gain factor and the kinematic wave

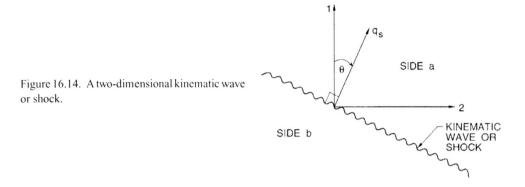

Figure 16.14. A two-dimensional kinematic wave or shock.

production processes that contribute to it will be needed to enhance our ability to predict these instabilities.

16.5 Two-Dimensional Kinematic Waves

Noting that all of the above analyses are for simple one-dimensional flow, we should add a footnote on the nature of kinematic waves and shocks in a more general three-dimensional flow. Although the most general analysis is quite complex, a relative simple extension of the results of the preceding sections is obtained when attention is restricted to those flows in which the direction of the relative velocity vector, u_{ABi}, is everywhere the same. Such would be the case, for example, for a relative velocity caused by buoyancy alone. Let that be the 1 direction (so that $u_{AB2} = 0$) and consider, therefore, a planar flow in the 12 plane in which, as depicted in Figure 16.14, the kinematic wave or shock is inclined at an angle, θ, to the 2 direction and is moving at a velocity, q_s, normal to itself. It is readily shown that the volume flux of any component, N, normal to and relative to the shock is as follows

$$\alpha_N(u_{N2} \sin\theta + u_{N1} \cos\theta - q_s) \qquad (16.23)$$

and the total volume flux, j_θ, relative to the shock is as follows:

$$j_\theta = j_2 \sin\theta + j_1 \cos\theta - q_s. \qquad (16.24)$$

Now consider a multiphase flow consisting of two components, A and B, with velocity vectors, u_{Ai}, u_{Bi}, with volume flux vectors, j_{Ai}, j_{Bi}, with volume fractions, $\alpha_A = \alpha$, $\alpha_B = 1 - \alpha$, and with a drift flux vector, u_{ABi}, where, in the present case, $u_{AB2} = 0$. If the indices a and b denote conditions on the two sides of the shock, and if the individual volume fluxes into and out of the shock are equated, we obtain, two relations. The first relation simply states that the total volume flux, j_θ, must be the

16.5 Two-Dimensional Kinematic Waves

same on the two sides of the shock. The second relation yields the following:

$$q_s = j_\theta + \cos\theta \left\{ \frac{j_{AB1a} - j_{AB1b}}{\alpha_a - \alpha_b} \right\}, \tag{16.25}$$

which, in the case of a infinitesimal wave, becomes the following:

$$q_s = j_\theta + \cos\theta \left. \frac{dj_{AB1}}{d\alpha} \right|_\alpha. \tag{16.26}$$

These are essentially the same as the one-dimensional results, Eqs. (16.13) and (16.7), except for the $\cos\theta$. Consequently, within the restricted class of flows considered here, the propagation and evolution of a kinematic wave or shock in two and three dimensions can be predicted if the drift flux function, $j_{AB1}(\alpha)$, is known and its direction is uniform.

Bibliography

Acosta, A.J. and Parkin, B.R. (1975). Cavitation inception – a selective review. *J. Ship Res.*, **19**, 193–205.

Adam, S. and Schnerr, G.H. (1997). Instabilities and bifurcation of non-equilibrium two-phase flows. *J. Fluid Mech.*, **348**, 1–28.

Amies, G., Levek, R. and Struesseld, D. (1977). Aircraft hydraulic systems dynamic analysis. Volume II. Transient analysis (HYTRAN). *Wright-Patterson Air Force Base Technical Report AFAPL-TR-76-43*, **II**.

Amies, G. and Greene, B. (1977). Aircraft hydraulic systems dynamic analysis. Volume IV. Frequency response (HSFR). *Wright-Patterson Air Force Base Technical Report AFAPL-TR-76-43*, **IV**.

Andeen, G.B. and Marks, J.S. (1978). Analysis and testing of steam chugging in pressure systems. *Electric Power Res. Inst. Report NP-908*.

Anderson, D.A., Blade, R.J. and Stevens, W. (1971). Response of a radial-bladed centrifugal pump to sinusoidal disturbances for non-cavitating flow. *NASA TN D-6556*.

Anderson, T.B. and Jackson, R. (1968). Fluid mechanical description of fluidized beds. *Ind. Eng. Chem. Fund.*, **7**, No. 1, 12–21.

Apfel, R.E. (1981). Acoustic cavitation prediction. *J. Acoust. Soc. Am.*, **69**, 1624–1633.

Arakeri, V.H. (1979). Cavitation inception. *Proc. Indian Acad. Sci.*, **C2**, Part 2, 149–177.

Arakeri, V.H. and Shangumanathan, V. (1985). On the evidence for the effect of bubble interference on cavitation noise. *J. Fluid Mech.*, **159**, 131–150.

Arnold, G.S., Drew, D.A. and Lahey, R.T. (1989). Derivation of constitutive equations for interfacial force and Reynolds stress for a suspension of spheres using ensemble cell averaging. *Chem. Eng. Comm.*, **86**, 43–54.

ASTM. (1967). *Erosion by cavitation or impingement*. American Society for Testing and Materials, ASTM STP408.

Atkinson, C.M. and Kytömaa, H.K. (1992). Acoustic wave speed and attenuation in suspensions. *Int. J. Multiphase Flow*, **18**, No. 4, 577–592.

Azbel, D. and Liapis, A.I. (1983). Mechanisms of liquid entrainment. In *Handbook of Fluids in Motion* (eds: N.P. Cheremisinoff and R. Gupta), 453–482.

Babic, M. and Shen, H.H. (1989). A simple mean free path theory for the stresses in a rapidly deforming granular material. *ASCE J. Eng. Mech.*, **115**, 1262–1282.

Bagnold, R.A. (1954). Experiments on a gravity-free dispersion of large solid particles in a Newtonian fluid under shear. *Proc. R. Soc. Lond.*, **A225**, 49–63.

Baker, O. (1954). Simultaneous flow of oil and gas. *Oil Gas J.*, **53**, 185.

Barber, B.P. and Putterman, S.J. (1991). Observations of synchronous picosecond sonoluminescence. *Nature*, **352**, 318.

Barnea, E. and Mizrahi, J. (1973). A generalized approach to the fluid dynamics of particulate systems. Part 1. General correlation for fluidization and sedimentation in solid multiparticle systems. *The Chem. Eng. J.*, **5**, 171–189.

Barnocky, G. and Davis, R.H. (1988). Elastohydrodynamic collision and rebound of spheres: experimental verification. *Phys. Fluids*, **31**, 1324–1329.

Baroczy, C.J. (1966). A systematic correlation for two-phase pressure drop. *Chem. Eng. Prog. Symp. Series*, **62**, No. 64, 232–249.

Basset, A.B. (1888). *A Treatise on Hydrodynamics*, volume II. Reprinted by Dover, NY, 1961.

Batchelor, G.K. (1967). *An Introduction to Fluid Dynamics*. Cambridge Univ. Press.

Batchelor, G.K. (1969). In *Fluid Dynamics Transactions*, 4 (eds: W. Fizdon, P. Kucharczyk, and W.J. Prosnak). Polish Sci. Publ., Warsaw.

Batcho, P.F., Moller, J.C., Padova, C. and Dunn, M.G. (1987). Interpretation of gas turbine response to dust ingestion. *ASME J. Eng. for Gas Turbines and Power*, **109**, 344–352.

Bathurst, R.J. and Rothenburg, L. (1988). Micromechanical aspects of isotropic granular assemblies with linear contact interactions. *J. Appl. Mech.*, **55**, 17–23.

Benjamin, T.B. and Ellis, A.T. (1966). The collapse of cavitation bubbles and the pressures thereby produced against solid boundaries. *Phil. Trans. Roy. Soc., London, Ser. A*, **260**, 221–240.

Bernier, R.J.N. (1982). *Unsteady two-phase flow instrumentation and measurement*. Ph.D. Thesis, Calif. Inst. of Tech., Pasadena, Cal.

Biesheuvel, A. and Gorissen, W.C.M. (1990). Void fraction disturbances in a uniform bubbly fluid. *Int. J. Multiphase Flow*, **16**, 211–231.

Billet, M.L. (1985). Cavitation nuclei measurement – a review. *Proc. 1985 ASME Cavitation and Multiphase Flow Forum*, 31–38.

Binnie, A.M. and Green, J.R. (1943). An electrical detector of condensation in high velocity steam. *Proc. Roy. Soc. A*, **181**, 134.

Birkhoff, G. (1954). Note on Taylor instability. *Quart. Appl. Math.*, **12**, 306–309.

Birkhoff, G. and Zarontonello, E.H. (1957). *Jets, Wakes and Cavities*. Academic Press, New York.

Bjerknes, V. (1909). *Die Kraftfelder*. Friedrich Vieweg and Sohn, Braunsweig.

Blake, F.G. (1949). The onset of cavitation in liquids. Acoustics *Res. Lab., Harvard Univ., Tech. Memo. No. 12*.

Blake, F.G. (1949). Bjerknes forces in stationary sound fields. *J. Acoust. Soc. Am.*, **21**, 551.

Blake, J.R. and Gibson, D.C. (1987). Cavitation bubbles near boundaries. *Ann. Rev. Fluid Mech.*, **19**, 99–124.

Blake, W.K., Wolpert, M.J. and Geib, F.E. (1977). Cavitation noise and inception as influenced by boundary-layer development on a hydrofoil. *J. Fluid Mech.*, **80**, 617–640.

Blake, W.K. and Sevik, M.M. (1982). Recent developments in cavitation noise research. *Proc. ASME Int. Symp. on Cavitation Noise*, 1–10.

Blake, W.K. (1986b). *Mechanics of Flow-Induced Sound and Vibration*. Academic Press, New York.

Blanchard, D.C. (1963). The electrification of the atmosphere by particles from bubbles in the sea. *Progr. Oceanogr.*, **1**, 72.

Blanchard, D.C. (1983). The production, distribution and bacterial enrichment of the sea-salt aerosol. In *Climate and Health Implications of Bubble-Mediated Sea-Air Exchange* (eds: E.C. Monahan and M.A. VanPatten). Connecticut Sea Grant College Program CT-SG-89-06, Groton, CT.

Boothroyd, R.G. (1971). *Flowing Gas–Solid Suspensions*. Chapman and Hall Ltd, New York.

Borishanski, V.M. (1956). An equation generalizing experimental data on the cessation of bubble boiling in a large volume of liquid. *Zh. Tekh. Fiz.*, **26**, No. 7, 452–456.

Borotnikova, M.I. and Soloukin, R.I. (1964). A calculation of the pulsations of gas bubbles in an incompressible liquid subject to a periodically varying pressure. *Sov. Phys. Acoust.*, **10**, 28–32.

Boure, J.A., Bergles, A.E. and Tong, L.S. (1973). Review of two-phase flow instability. *Nucl. Eng. Des.*, **25**, 165–192.

Boure, J.A. and Mercadier, Y. (1982). Existence and properties of structure waves in bubbly two-phase flows. *Appl. Sci. Res.*, **38**, 297–303.

Brady, J.F. and Bossis, G. (1988). Stokesian dynamics. *Ann. Rev. Fluid Mech.*, **20**, 111–157.

Brady, J.F. (2001). Computer simulations of viscous suspensions. *Chem. Eng. Sci.*, **56**, 2921–2926.

Braisted, D.M. and Brennen, C.E. (1980). Auto-oscillation of cavitating inducers. In *Polyphase Flow and Transport Technology* (ed: R.A. Bajura). ASME, New York, 157–166.

Brennen, C.E. (1970). Cavity surface wave patterns and general appearance. *J. Fluid Mech.*, **44**, 33–49.

Brennen, C.E. (1978). Bubbly flow model for the dynamic characteristics of cavitating pumps. *J. Fluid Mech.*, **89**, Part 2, 223–240.

Brennen, C.E. (1979). A linear, dynamic analysis of vent condensation stability. In *Basic Mechanisms in Two-Phase Flow and Heat Transfer* (eds. P.H. Rothe and R.T. Lahey, Jr.). ASME (No. G00179), New York.

Brennen, C.E. (1982). A review of added mass and fluid inertial forces. Naval Civil Eng. Lab., Port Hueneme, Calif., Report CR82.010.

Brennen, C.E. (1994). *Hydrodynamics of Pumps*. Concepts ETI and Oxford Univ. Press, Oxford, UK.

Brennen, C.E. (1995). *Cavitation and Bubble Dynamics*. Oxford Univ. Press, Oxford, UK.

Brennen, C.E. (2002). Fission of collapsing cavitation bubbles. *J. Fluid Mech.*, **472**, 153–166.

Brennen, C.E. and Acosta, A.J. (1973). Theoretical, quasistatic analyses of cavitation compliance in turbopumps. *J. Spacecraft Rockets*, **10**, No. 3, 175–180.

Brennen, C.E. and Acosta, A.J. (1976). The dynamic transfer function for a cavitating inducer. *ASME J. Fluids Eng.*, **98**, 182–191.

Brennen, C.E. and Braisted, D.M. (1980). Stability of hydraulic systems with focus on cavitating pumps. *Proc. 10th Symp. of IAHR, Tokyo*, 255–268.

Brennen, C.E., Meissner, C., Lo, E.Y. and Hoffman, G.S. (1982). Scale effects in the dynamic transfer functions for cavitating inducers. *ASME J. Fluids Eng.*, **104**, 428–433.

Brennen, C. and Pearce, J.C. (1978). Granular media flow in two-dimensional hoppers. *ASME J. Appl. Mech.*, **45**, No. 1, 43–50.

Brenner, H. (1961). The slow motion of a sphere through a viscous fluid towards a plane surface. *Chem. Eng. Sci.*, **16**, 242–251.

Brinkman, H.C. (1947). A calculation of the viscous force exerted by a flowing fluid on a dense swarm of particles. *Appl. Sci. Res.*, **A1**, 27–34.

Bromley, L.A. (1950). Heat transfer in stable film boiling. *Chem. Eng. Prog. Ser.*, **46**, 221–227.

Brown, F.T. (1967). A unified approach to the analysis of uniform one-dimensional distributed systems. *ASME J. Basic Eng.*, **89**, No. 6, 423–432.

Butterworth, D. (1977). Introduction to condensation. Filmwise condensation. In *Two-Phase Flow and Heat Transfer* (eds: D. Butterworth and G.F. Hewitt). Oxford Univ. Press, Oxford, UK.

Butterworth, D. and Hewitt, G.F. (1977). *Two-Phase Flow and Heat Transfer*. Oxford Univ. Press, Oxford, UK.

Campbell, C.S. (1993). Boundary interactions for two-dimensional granular flow. Parts 1 and 2. *J. Fluid Mech.*, **247**, 111–136 and **247**, 137–156.

Campbell, C.S. (1990). Rapid granular flows. *Ann. Rev. of Fluid Mech.*, **22**, 57–92.

Campbell, C.S. (2002). Granular shear flows at the elastic limit. *J. Fluid Mech.*, **465**, 261–291.

Campbell, C.S. (2003). Stress controlled elastic granular shear flows. Submitted for publication.

Campbell, C.S. and Brennen, C.E. (1985a). Computer simulation of granular shear flows. *J. Fluid Mech.*, **151**, 167–188.

Campbell, C.S. and Brennen, C.E. (1985b). Chute flows of granular material: some computer simulation. *ASME J. Appl. Mech.*, **52**, No. 1, 172–178.

Campbell, I.J. and Pitcher, A.S. (1958). Shock waves in a liquid containing gas bubbles. *Proc. Roy. Soc. London, A*, **243**, 534–545.

Carman, P.C. (1937). Fluid flow through a granular bed. *Trans. Inst. Chem. Engrs.*, **15**, 150.

Carrier, G.F. (1958). Shock waves in a dusty gas. *J. Fluid Mech.*, **4**, 376–382.

Carstensen, E.L. and Foldy, L.L. (1947). Propagation of sound through a liquid containing bubbles. *J. Acoust. Soc. Amer.*, **19**, 481–501.

Catipovic, N.M., Govanovic, G.N. and Fitzgerald, T.J. (1978). Regimes of fluidization for large particles. *AIChE J.*, **24**, 543–547.

Ceccio, S.L. and Brennen, C.E. (1991). Observations of the dynamics and acoustics of travelling bubble cavitation. *J. Fluid Mech.*, **233**, 633–660.

Ceccio, S.L., Gowing, S. and Shen, Y. (1997). The effects of salt water on bubble cavitation. *ASME J. Fluids Eng.*, **119**, 155–163.

Cha, Y.S. and Henry, R.E. (1981). Bubble growth during decompression of a liquid. *ASME J. Heat Transfer*, **103**, 56–60.

Chahine, G.L. (1977). Interaction between an oscillating bubble and a free surface. *ASME J. Fluids Eng.*, **99**, 709–716.

Chahine, G.L. (1982). Cloud cavitation: theory. In *Proc. 14th ONR Symp. on Naval Hydrodynamics*. 165–194. Natl. Academies Press, Wash. D.C.

Chahine, G.L. and Duraiswami, R. (1992). Dynamical interactions in a multibubble cloud. *ASME J. Fluids Eng.*, **114**, 680–686.

Chapman, R.B. and Plesset, M.S. (1971). Thermal effects in the free oscillation of gas bubbles. *ASME J. Basic Eng.*, **93**, 373–376.

Cheremisinoff, N.P. and Cheremisinoff, P.N. (1984). *Hydrodynamics of Gas–Solids Fluidization*. Gulf, Houston, TX.

Chiu, H.H. and Croke, E.J. (1981). *Group combustion of liquid fuel sprays*. Univ. of Illinois at Chicago, Energy Technology Lab. Report 81-2.

Class, G. and Kadlec, J. (1976). Survey of the behavior of BWR pressure suppression systems during the condensation phase of LOCA. Paper presented at Amer. Nucl. Soc. Int. Conf., Washington, D.C., Nov. 1976.

Cole, R.H. (1948). *Underwater Explosions*. Princeton Univ. Press (reprinted by Dover, 1965), Princeton, NJ.

Collier, J.G. and Thome, J.R. (1994). *Convective Boiling and Condensation*. Clarendon, Oxford.

Crespo, A. (1969). Sound and shock waves in liquids containing bubbles. *Phys. Fluids*, **12**, 2274–2282.

Crowe, C.T. (1982). Review – numerical models for dilute gas–particle flow. *ASME J. Fluids Eng.*, **104**, 297–303.

Crowe, C.T., Sommerfeld, M. and Tsuji, Y. (1998). *Multiphase Flows with Droplets and Particles*. CRC Press, Boca Raton, FL.

Crum, L.A. (1979). Tensile strength of water. *Nature*, **278**, 148–149.

Crum, L.A. (1980). Measurements of the growth of air bubbles by rectified diffusion. *J. Acoust. Soc. Am.*, **68**, 203–211.

Crum, L.A. (1983). The polytropic exponent of gas contained within air bubbles pulsating in a liquid. *J. Acoust. Soc. Am.*, **73**, 116–120.

Crum, L.A. (1984). Rectified diffusion. *Ultrasonics*, **22**, 215–223.

Crum, L.A. and Eller, A.I. (1970). Motion of bubbles in a stationary sound field. *J. Acoust. Soc. Am.*, **48**, 181–189.

Cundall, P.A. and Strack, O.D.L. (1979). A discrete numerical model for granular assemblies. *Geotechnique*, **29**, 47–65.

Cunningham, E. (1910). On the velocity of steady fall of spherical particles through fluid medium. *Proc. Roy. Soc. A*, **83**, 357–365.

d'Agostino, L. and Brennen, C.E. (1983). On the acoustical dynamics of bubble clouds. *ASME Cavitation and Multiphase Flow Forum*, 72–75.

d'Agostino, L. and Brennen, C.E. (1988). Acoustical absorption and scattering cross-sections of spherical bubble clouds. *J. Acoust. Soc. Am.*, **84**, 2126–2134.

d'Agostino, L., Brennen, C.E. and Acosta, A.J. (1988). Linearized dynamics of two-dimensional bubbly and cavitating flows over slender surfaces. *J. Fluid Mech.*, **192**, 485–509.

d'Agostino, L. and Brennen, C.E. (1989). Linearized dynamics of spherical bubble clouds. *J. Fluid Mech.*, **199**, 155–176.

d'Agostino, L., d'Auria, F. and Brennen, C.E. (1997). On the inviscid stability of parallel bubbly flows. *J. Fluid Mech.*, **339**, 261–274.

Dai, Z., Hsiang, L.-P. and Faeth, G. (1997). Spray formation at the free surface of turbulent bow sheets. In *Proc. 21st ONR Symp. on Naval Hydrodynamics*. 490–505. Natl. Academies Press, Wash. D.C.

Dai, Z., Sallam, K. and Faeth, G. (1998). Turbulent primary breakup of plane-free bow sheets. In *Proc. 22nd ONR Symp. on Naval Hydrodynamics*. Natl. Academies Press, Wash. D.C.

Daily, J.W. and Johnson, V.E., Jr. (1956). Turbulence and boundary layer effects on cavitation inception from gas nuclei. *Trans. ASME*, **78**, 1695–1706.

Davenport, W.G., Bradshaw, A.V. and Richardson, F.D. (1967). Behavior of spherical-cap bubbles in liquid metals. *J. Iron Steel Inst.*, **205**, 1034–1042.

Davidson, J.F. and Harrison, D. (1963). *Fluidized Particles*. Cambridge Univ. Press, New York.

Davidson, J.F. and Harrison, D. (1971). *Fluidization*. Academic Press, New York.

Davidson, J.F., Clift, R. and Harrison, D. (1985). Fluidization, second edition. Academic Press, New York.

Davies, C.N. (1966). *Aerosol Science*. Academic Press, New York.

Davies, R.M. and Taylor, G.I. (1942). The vertical motion of a spherical bubble and the pressure surrounding it. In *The Scientific Papers of G.I. Taylor* (ed: G.K. Batchelor), **III**, 320–336, Cambridge Univ. Press, New York.

Davies, R.M. and Taylor, G.I. (1943). The motion and shape of the hollow produced by an explosion in a liquid. In *The Scientific Papers of G.I. Taylor* (ed: G.K. Batchelor), **III**, 337–353. Cambridge Univ. Press, New York.

Davies, R.M. and Taylor, G.I. (1950). The mechanics of large bubbles rising through extended liquids and through liquids in tubes. *Proc. Roy. Soc. A*, **200**, 375–390.

Davis, R.H., Serayssol, J.M. and Hinch, E.J. (1986). The elastohydrodynamic collison of two spheres. *J. Fluid Mech.*, **163**, 479–497.

Delale, C.F., Schnerr, G.H. and Zierep, J. (1995). Asymptotic solution of shock-tube flows with homogeneous condensation. *J. Fluid Mech.*, **287**, 93–118.

de Leeuw, G. (1987). Near-surface particle size distribution profiles over the North Sea. *J. Geophys. Res.*, **92**, 14, 631.

Dergarabedian, P. (1953). The rate of growth of vapor bubbles in superheated water. *ASME J. Appl. Mech.*, **20**, 537–545.

Didwania, A.K. and Homsy, G.M. (1981). Flow regimes and flow transitions in liquid fluidized beds. *Int. J. Multiphase Flow*, **7**, 563–580.

Dowling, A.P. and Ffowcs Williams, J.E. (1983). *Sound and Sources of Sound*. Ellis Horwood Ltd. and John Wiley and Sons, New York.

Drescher, A. and De Josselin de Jong, G. (1972). Photoelastic verification of a mechanical model for the flow of a granular material. *J. Mech. Phys. Solids*, **20**, 337–351.

Drew, D.A. (1983). Mathematical modelling of two-phase flow. *Ann. Rev. Fluid Mech.*, **15**, 261–291.

Drew, D.A. (1991). Effect of particle velocity fluctuations in particle fluid flows. *Physica*, **179A**, 69–80.

Dunn, M.G., Baran, A.J. and Miatech, J. (1996). Operation of gas turbine engines in volcanic ash clouds. *ASME J. Eng. for Gas Turbines and Power*, **118**, 724–731.

Durand, R. and Condolios, E. (1952). Experimental study of the hydraulic transport of coal and solid material in pipes. *Proc. Colloq. on the Hydraulic Transport of Coal, Natl. Coal Board, U.K.*, Paper IV, 39–55.

Eaton, J.K. (1994). Experiments and simulations on turbulence modification by dispersed particles. *Appl. Mech. Rev.*, **47**, No. 6, Part 2, S44–S48.

Einstein, A. (1906). *Annalen der Physik (4)*, **19**, 289–306. [For translation see Einstein, A. (1956), chapter III, 36–62.]

Einstein, A. (1956). *Investigations on the Theory of Brownian Movement*. Dover, New York.

Elghobashi, S.E. and Truesdell, G.C. (1993). On the two-way interaction between homogeneous turbulence and dispersed solid particles. I: Turbulence modification. *Phys. Fluids*, **A5**, 1790–1801.

El-Kaissy, M.M. and Homsy, G.M. (1976). Instability waves and the origin of bubbles in fluidized beds. *Int. J. Multiphase Flow*, **2**, 379–395.

Eller, A.I. and Flynn, H.G. (1965). Rectified diffusion during non-linear pulsation of cavitation bubbles. *J. Acoust. Soc. Am.*, **37**, 493–503.

Emmons, H.W., Pearson, C.E. and Grant, H.P. (1955). Compressor surge and stall propagation. *Trans. ASME*, **79**, 455–469.

Epstein, P.S. and Plesset, M.S. (1950). On the stability of gas bubbles in liquid–gas solutions. *J. Chem. Phys.*, **18**, 1505–1509.

Ervine, D.A. and Falvey, H.T. (1987). Behavior of turbulent water jets in the atmosphere and in plunge pools. *Proc. Inst. Civil Eng.*, **83**, Part 2, 295–314.

Esche, R. (1952). Untersuchung der Schwingungskavitation in Flüssigkeiten. *Acustica*, **2**, AB208–AB218.

Esipov, I.B. and Naugol'nykh, K.A. (1973). Collapse of a bubble in a compressible liquid. *Akust. Zh.*, **19**, 285–288.

Faeth, G.M. and Lazar, R.S. (1971). Fuel droplet burning rates in a combustion gas environment. *AIAA J.*, **9**, 2165–2171.

Faeth, G.M. (1983). Evaporation and combustion of sprays. *Prog. Energy Combust. Sci.*, **9**, 1–76.

Fanelli, M. (1972). Further considerations on the dynamic behaviour of hydraulic machinery. *Water Power*, June 1972, 208–222.

Fessler, J.R., Kulick, J.D. and Eaton, J.K. (1994). Preferential concentration of heavy particles in a turbulent channel flow. *Phys. Fluids*, **6**, 3742–3749.

Fitzpatrick, H.M. and Strasberg, M. (1956). Hydrodynamic sources of sound. *Proc. First ONR Symp. on Naval Hydrodynamics*, 241–280.

Flagan, R.C. and Seinfeld, J.H. (1988). *Fundamentals of Air Pollution Engineering*. Prentice Hall, New York.

Flint, E.B. and Suslick, K.S. (1991). The temperature of cavitation. *Science*, **253**, 1397–1399.

Flynn, H.G. (1964). Physics of acoustic cavitation in liquids. *Physical Acoustics*, **1B**. Academic Press, New York.

Foerster, S.F., Louge, M.Y., Chang, A.H. and Allia, K. (1994). Measurements of the collisional properties of small spheres. *Phys. Fluids*, **6**, 1108–1115.

Foldy, L.L. (1945). The multiple scattering of waves. *Phys. Rev.*, **67**, 107–119.

Forster, H.K. and Zuber, N. (1954). Growth of a vapor bubble in a superheated liquid. *J. Appl. Phys.*, **25**, No. 4, 474–478.

Fortes, A.F., Joseph, D.D. and Lundgren, T.S. (1987). Nonlinear mechanics of fluidization of beds of spherical particles. *J. Fluid Mech.*, **177**, 469–483.

Fox, F.E., Curley, S.R. and Larson, G.S. (1955). Phase velocity and absorption measurements in water containing air bubbles. *J. Acoust. Soc. Am.*, **27**, 534–539.

Franklin, R.E. and McMillan, J. (1984). Noise generation in cavitating flows, the submerged jet. *ASME J. Fluids Eng.*, **106**, 336–341.

Franz, R., Acosta, A.J., Brennen, C.E. and Caughey, T.C. (1990). The rotordynamic forces on a centrifugal pump impeller in the presence of cavitation. *ASME J. Fluids Eng.*, **112**, 264–271.

Friedlander, S.K. (1977). *Smoke, Dust amd Haze: Fundamentals of Aerosol Behavior*. John Wiley and Sons, New York.

Fritz, W. (1935). Berechnung des Maximal Volume von Dampfblasen. *Phys. Z.*, **36**, 379.

Frost, D. and Sturtevant, B. (1986). Effects of ambient pressure on the instability of a liquid boiling explosively at the superheat limit. *ASME J. Heat Transfer*, **108**, 418–424.

Fujikawa, S. and Akamatsu, T. (1980). Effects of the non-equilibrium condensation of vapour on the pressure wave produced by the collapse of a bubble in a liquid. *J. Fluid Mech.*, **97**, 481–512.

Furuya, O. (1985). An analytical model for prediction of two-phase (non-condensable) flow pump performance. *ASME J. Fluids Eng.*, **107**, 139–147.

Furuya, O. and Maekawa, S. (1985). An analytical model for prediction of two-phase flow pump performance – condensable flow case. *ASME Cavitation and Multiphase Flow Forum*, **FED-23**, 74–77.

Garabedian, P.R. (1964). *Partial Differential Equations*. John Wiley and Sons, New York.

Gardner, G.C. (1963) Events leading to erosion in the steam turbine. *Proc. Inst. Mech. Eng.*, **178**, Part 1, No. 23, 593–623.

Gates, E.M. and Bacon, J. (1978). Determination of cavitation nuclei distribution by holography. *J. Ship Res.*, **22**, No. 1, 29–31.

Gavalas, G.R. (1982). *Coal Pyrolysis*. Elsevier, New York.

Geldart, D. (1973). *Types of gas fluidization*. *Powder Tech.*, **7**, 285–292.

Gibilaro, L.G. (2001). *Fluidization – dynamics*. Butterworth-Heinemann, New York.

Gibson, D.C. and Blake, J.R. (1982). The growth and collapse of bubbles near deformable surfaces. *Appl. Sci. Res.*, **38**, 215–224.

Gidaspow, D. (1994). *Multiphase flow and fluidization*. Academic Press.

Gill, L.E., Hewitt, G.F., Hitchon, J.W. and Lacey, P.M.C. (1963). Sampling probe studies of the gas core distribution in annular two-phase flow. I. The effect of length on phase and velocity distribution. *Chem. Eng. Sci.*, 18, 525–535.

Gilmore, F.R. (1952). The collapse and growth of a spherical bubble in a viscous compressible liquid. *Calif. Inst. of Tech. Hydrodynamics Lab. Rep. No. 26–4*.

Ginestra, J.C., Rangachari, S. and Jackson, R. (1980). A one-dimensional theory of flow in a vertical standpipe. *Powder Tech.*, **27**, 69–84.

Glassman, I. (1977). *Combustion*. Academic Press, New York.

Goldsmith, W. (1960). *Impact*. E. Arnold, New York.

Gore, R.A. and Crowe, C.T. (1989). The effect of particle size on modulating turbulent intensity. *Int. J. Multiphase Flow*, **15**, 279–285.

Gouesbet, G. and Berlemont, A. (1993). *Instabilities in Multiphase Flows*. Plenum, New York/London.

Gouse, S.W. and Brown, G.A. (1964). A survey of the velocity of sound in two-phase mixtures. *ASME Paper 64-WA/FE-35*.

Green, H.L. and Lane, W.R. (1964). *Particulate Clouds: Dusts, Smokes and Mists*. E. and F.N. Spon Ltd., London.

Gregor, W. and Rumpf, H. (1975). Velocity of sound in two-phase media. *Int. J. Multiphase Flow*, **1**, 753–769.

Greitzer, E.M. (1976). Surge and rotating stall in axial flow compressors. Part I: Theoretical compression system model. Part II: Experimental results and comparison with theory. *ASME J. Eng. for Power*, **98**, 190–211.

Greitzer, E.M. (1981). The stability of pumping systems – the 1980 Freeman Scholar Lecture. *ASME J. Fluids Eng.*, **103**, 193–242.

Griffith, P. and Wallis, J.D. (1960). The role of surface conditions in nucleate boiling. *Chem. Eng. Prog. Symp.*, Ser. 56, **30**, 49–63.

Haberman, W.L. and Morton, R.K. (1953). An experimental investigation of the drag and shape of air bubbles rising in various liquids. *David Taylor Model Basin, Washington, Report No. 802*.

Hadamard, J. (1911). Movement permanent lent d'une sphere liquide et visqueuse dans un liquide visqueux. *Comptes Rendus*, **152**, 1735.

Haff, P.K. (1983). Grain flow as a fluid-mechanical phenomenon. *J. Fluid Mech.*, **134**, 401–430.

Hampton, L. (1967). Acoustic properties of sediments. *J. Acoust. Soc. Am.*, **42**, 882–890.

Hanes, D.M. and Inman, D.L. (1985). Observations of rapidly flowing granular-fluid flow. *J. Fluid Mech.*, **150**, 357–380.

Hanson, I., Kedrinskii, V.K. and Mørch, K.A. (1981). On the dynamics of cavity clusters. *J. Appl. Phys.*, **15**, 1725–1734.

Happel, J. and Brenner, H. (1965). *Low Reynolds Number Hydrodynamics*. Prentice Hall, New York.

Harper, J.F., Moore, D.W. and Pearson, J.R.A. (1967). The effect of the variation of surface tension with temperature on the motion of bubbles and drops. *J. Fluid Mech.*, **27**, 361–366.

Hartunian, R.A. and Sears, W.R. (1957). On the instability of small gas bubbles moving uniformly in various liquids. *J. Fluid Mech.*, **3**, 27–47.

Hayden, J.W. and Stelson, T.E. (1971). Hydraulic conveyance of solids in pipes. In *Advances in Solid-Liquid Flow in Pipes and Its Applications* (ed: I. Zandi), 149–163. Pergamon Press, New York.

Henry, R.E. and Fauske, H.K. (1971). The two-phase critical flow of one-component mixtures in nozzles, orifices, and short tubes. *ASME J. Heat Transfer*, **93**, 179–187.

Herring, C. (1941). The theory of the pulsations of the gas bubbles produced by an underwater explosion. *US Nat. Defence Res. Comm. Report*.

Hetsroni, G. (1989). Particles–turbulence interaction. *Int. J. Multiphase Flow*, **15**, 735.

Hewitt, G.F., and Roberts, D.N. (1969). Studies of two-phase flow patterns by simultaneous X-ray and flash photography. *U.K.A.E.A. Rep. No. AERE-M2159*.

Hewitt, G.F. and Hall-Taylor, N.S. (1970). *Annular Two-Phase Flow*. Pergamon, New York.

Hewitt, G.F. (1982). Flow regimes. In *Handbook of Multiphase Systems* (ed: G. Hetsroni). McGraw-Hill, New York.

Hewitt, H.C. and Parker, J.D. (1968). Bubble growth and collapse in liquid nitrogen. *ASME J. Heat Transfer*, **90**, 22–26.

Hickling, R. (1963). The effects of thermal conduction in sonoluminescence. *J. Acoust. Soc. Am.*, **35**, 967–974.

Hickling, R. and Plesset, M.S. (1964). Collapse and rebound of a spherical bubble in water. *Phys. Fluids*, **7**, 7–14.

Hill, M.J.M. (1894). On a spherical vortex. *Phil. Trans. Roy. Soc., London, Ser. A.*, **185**, 213–245.

Hill, P.G. (1966). Condensation of water vapour during supersonic expansion in nozzles. *J. Fluid Mech.*, **25**, 593–620.

Hinze, J.O. (1959). *Turbulence*. McGraw-Hill, New York.

Hinze, J.O. (1961). Momentum and mechanical energy balance equations for a flowing homogeneous suspension with slip between the phases. *Appl. Sci. Res.*, **11**, 33.

Ho, B.P. and Leal, L.G. (1974). Inertial migration of rigid spheres in two-dimensional unidirectional flows. *J. Fluid Mech.*, **65**, 365.

Homsy, G.M., El-Kaissy, M.M. and Didwania, A.K. (1980). Instability waves and the origin of bubbles in fluidized beds. II. Comparison of theory and experiment. *Int. J. Multiphase Flow*, **6**, 305–318.

Homsy, G.M. (1983). A survey of some results in the mathematical theory of fluidization. In *Theory of Disperse Multiphase Flow* (ed: R.E. Meyer). Academic Press, New York.

Hoyt, J.W. and Taylor, J.J. (1977a). Waves on water jets. *J. Fluid Mech.*, **83**, 119–123.

Hoyt, J.W. and Taylor, J.J. (1977b). Turbulence structure in a water jet discharging in air. *Phys. Fluids*, **20**, Part 2, S253–S257.

Hsieh, D.-Y. and Plesset, M.S. (1961). Theory of rectified diffusion of mass into gas bubbles. *J. Acoust. Soc. Am.*, **33**, 206–215.

Hsu, Y.-Y. and Graham, R.W. (1976). *Transport Processes in Boiling and Two-Phase Systems*. Hemisphere and McGraw-Hill, New York.

Hubbard, N.G. and Dukler, A.E. (1966). *The Characterization of Flow Regimes in Horizontal Two-Phase Flow.* Heat Transfer and Fluid Mechanics Inst., Stanford Univ., CA.

Hui, K., Haff, P.K., Ungar, J.E. and Jackson, R. (1984). Boundary conditions for high-shear grain flows. *J. Fluid Mech.*, **145**, 223–233.

Hunt, M.L., Zenit, R., Campbell, C.S. and Brennen, C.E. (2002). Revisiting the 1954 suspension experiments of R.A. Bagnold. *J. Fluid Mech.*, **452**, 1–24.

Hutchinson, P. and Whalley, P.B. (1973). A possible characterization of entrainment in annular flow. *Chem. Eng. Sci.*, **28**, 974–975.

Hutter, K. (1993). Avalanche dynamics. In *Hydrology of Disasters* (ed: V.P. Singh). Kluwer, Amsterdam.

Ishii, M. (1982). Two-phase flow instabilities. In *Handbook of Multiphase Systems* (ed. G. Hetsroni). McGraw-Hill, New York.

Ivany, R.D. and Hammitt, F.G. (1965). Cavitation bubble collapse in viscous, compressible liquids – numerical analysis. *ASME J. Basic Eng.*, **87**, 977–985.

Iverson, R.M. (1997). The physics of debris flows. *Rev. Geophys.*, **35**, 245–296.

Jackson, R. (1963). The mechanics of fluidized beds. Part I: The stability of the state of uniform fluidization. Part II: The motion of fully developed bubbles. *Trans. Inst. Chem. Eng.*, **41**, 13–28.

Jackson, R. (1985). Hydrodynamic instability of fluid-particle systems. In *Fluidization*, second edition (eds: J.F. Davidson, R. Clift, and D. Harrison). Academic Press, New York.

Jaeger, H.M., Nagel, S.R. and Behringer, R.P. (1996). The physics of granular materials. *Phys. Today*, April 1996, 32–38.

Janssen, H.A. (1895). Versuche über Getreidedruck in Silozellen. *Zeit. Ver. Deutsch. Ing.*, **39**, 1045–1049.

Jarman, P. (1960). Sonoluminescence: a discussion. *J. Acoust. Soc. Am.*, **32**, 1459–1462.

Jenike, A.W. and Shield, R.T. (1959). On the plastic flow of Coulomb solids beyond original failure. *ASME J. Appl. Mech.*, **26**, 599–602.

Jenike, A.W. (1964). Storage and flow of solids. *Bull. No. 123, Utah Eng. Expt. Stat., U. of Utah.*

Jenkins, J.T. and Richman, M.W. (1985). Kinetic theory for plane flows of a dense gas of identical, rough, inelastic, circular disks. *Phys. Fluids*, **28**, 3485–3494.

Jenkins, J.T. and Richman, M.W. (1986). Boundary conditions for plane flows of smooth, nearly elastic, circular disks. *J. Fluid Mech.*, **171**, 53–69.

Jenkins, J.T. and Savage, S.B. (1983). A theory for the rapid flow of identical, smooth, nearly elastic particles. *J. Fluid Mech.*, **130**, 187–202.

Johanson, J.R. and Colijin, H. (1964). New design criteria for hoppers and bins. *Iron and Steel Eng.*, Oct. 1964, 85.

Johnson, P.C. and Jackson, R. (1987). Frictional-collisional constitutive relations for granular materials, with application to plane shearing. *J. Fluid Mech.*, **176**, 67–93.

Johnson, P.C., Nott, P. and Jackson, R. (1990). Frictional-collisional equations of motion for particulate flows and their application to chutes. *J. Fluid Mech.*, **210**, 501–535.

Jones, O.C. and Zuber, N. (1974). Statistical methods for measurement and analysis of two-phase flow. *Proc. Int. Heat Transfer Conf., Tokyo.*

Jones, O.C. and Zuber, N. (1978). Bubble growth in variable pressure fields. *ASME J. Heat Transfer*, **100**, 453–459.

Jorgensen, D.W. (1961). Noise from cavitating submerged jets. *J. Acoust. Soc. Am.*, **33**, 1334–1338.

Joseph, D. (1993). Finite size effects in fluidized suspension experiments. In *Particulate Two-Phase Flow* (ed: M.C. Roco), Butterworth-Heinemann, New York.

Joseph, G.G., Zenit, R., Hunt, M.L. and Rosenwinkel, A.M. (2001). Particle–wall collisions in a viscous fluid. *J. Fluid Mech.*, **433**, 329–346.

Joseph, G.G. (2003). Collisional dynamics of macroscopic particles in a viscous fluid. Ph.D. Thesis, California Institute of Technology.

Kamijo, K., Shimura, T. and Watanabe, M. (1977). An experimental investigation of cavitating inducer instability. *ASME Paper 77-WA/FW-14.*

Kaplun, S. and Lagerstrom, P.A. (1957). Asymptotic expansions of Navier–Stokes solutions for small Reynolds numbers. *J. Math. Mech.*, **6**, 585–593.

Karplus, H.B. (1958). The velocity of sound in a liquid containing gas bubbles. *Illinois Inst. Tech. Rep. COO-248.*

Katz, J. (1978). Determination of solid nuclei and bubble distributions in water by holography. *Calif. Inst. of Tech., Eng. and Appl. Sci. Div. Rep. No. 183-3.*

Kawaguchi, T. and Maeda, M. (2003). Measurement technique for analysis in two-phase flows involving distributed size of droplets and bubbles using interferometric method. In preparation.

Keller, J.B. and Kolodner, I.I. (1956). Damping of underwater explosion bubble oscillations. *J. Appl. Phys.*, **27**, 1152–1161.

Kennard, E.M. (1967). Irrotational flow of frictionless fluid, mostly of invariable density. *David Taylor Model Basin, Washington, Report No. 2299.*

Kenning, V.M. and Crowe, C.T. (1997). On the effect of particles on carrier phase turbulence in gas-particle flows. *Int. J. Multiphase Flow*, **23**, 403–408.

Kermeen, R.W. (1956). Water tunnel tests of NACA 4412 and Walchner Profile 7 hydrofoils in non-cavitating and cavitating flows. *Calif. Inst. of Tech. Hydro. Lab. Rep. 47–5.*

Keulegan, G.H. and Carpenter, L.H. (1958). Forces on cylinders and plates in an oscillating fluid. *U.S. Nat. Bur. Standards J. Res.*, **60**, No. 5, 423–440.

Kiceniuk, T. (1952). A preliminary investigation of the behavior of condensible jets discharging in water. *Calif. Inst. of Tech. Hydro. Lab. Rept.*, E-24.6.

Kimoto, H. (1987). An experimental evaluation of the effects of a water microjet and a shock wave by a local pressure sensor. *Int. ASME Symp. on Cavitation Res. Facilities and Techniques*, **FED 57**, 217–224.

Klyachko, L.S. (1934). Heating and ventilation. *USSR Journal Otopl. i Ventil.*, No. 4.

Knapp, R.T. and Hollander, A. (1948). Laboratory investigations of the mechanism of cavitation. *Trans. ASME*, **70**, 419–435.

Knapp, R.T., Daily, J.W. and Hammitt, F.G. (1970). *Cavitation.* McGraw-Hill, New York.

Koch, E. and Karwat, H. (1976). Research efforts in the area of BWR pressure suppression containment systems. *Proc. 4th Water Reactor Safety Research Meeting, Gaithersburg, MD, Sept. 1976.*

Kozeny, J. (1927). *Sitzber. Akad. Wiss. Wien, Math-naturnw. kl (Abt. Ha)*, **136**, 271.

Kraus, E.B. and Businger, J.A. (1994). *Atmosphere–Ocean Interaction.* Oxford Univ. Press, Oxford, UK.

Kraynik, A.M. (1988). Foam flows. *Ann. Rev. Fluid Mech.*, **20**, 325–357.

Kuhn de Chizelle, Y., Ceccio, S.L., Brennen, C.E., and Gowing, S. (1992a). Scaling experiments on the dynamics and acoustics of travelling bubble cavitation. In *Proc. 3rd I. Mech. E. Int. Conf. on Cavitation, Cambridge, England*, 165–170.

Kuhn de Chizelle, Y., Ceccio, S.L., Brennen, C.E., and Shen, Y. (1992b). Cavitation scaling experiments with headforms: bubble acoustics. *Proc. 19th ONR Symp. on Naval Hydrodynamics*, 72–84.

Kulick, J.D., Fessler, J.R. and Eaton, J.K. (1994). Particle response and turbulence modification in fully developed channel flow. *J. Fluid Mech.*, **277**, 109–134.

Kumar, S. and Brennen, C.E. (1991). Non-linear effects in the dynamics of clouds of bubbles. *J. Acoust. Soc. Am.*, **89**, 707–714.

Kumar, S. and Brennen, C.E. (1992). Harmonic cascading in bubble clouds. *Proc. Int. Symp. on Propulsors and Cavitation, Hamburg*, 171–179.

Kumar, S. and Brennen, C.E. (1993). Some nonlinear interactive effects in bubbly cavitation clouds. *J. Fluid Mech.*, **253**, 565–591.

Kumar, S. and Brennen, C.E. (1993). A study of pressure pulses generated by travelling bubble cavitation. *J. Fluid Mech.*, **255**, 541–564.

Kuo, K.K. (1986). *Principles of Combustion*. Wiley Interscience, New York.

Kutateladze, S.S. (1948). On the transition to film boiling under natural convection. *Kotloturbostroenie*, **3**, 10.

Kutateladze, S.S. (1952). Heat transfer in condensation and boiling. *U.S. AEC Rep. AEC-tr-3770*.

Kynch, G.J. (1952). A theory of sedimentation. *Trans. Faraday Soc.*, **48**, 166–176.

Kytömaa, H.K. (1974). Stability of the Structure in Multicomponent Flows. Ph.D. Thesis, Calif. Inst. of Tech.

Kytömaa, H.K. and Brennen, C.E. (1990). Small amplitude kinematic wave propagation in two-component media. *Int. J. Multiphase Flow*, **17**, No. 1, 13–26.

Lamb, H. (1932). *Hydrodynamics*. Cambridge Univ. Press, New York.

Landau, L.E. and Lifshitz, E.M. (1959). *Fluid Mechanics*. Pergamon, New York.

Lauterborn, W. and Bolle, H. (1975). Experimental investigations of cavitation bubble collapse in the neighborhood of a solid boundary. *J. Fluid Mech.*, **72**, 391–399.

Lauterborn, W. (1976). Numerical investigation of nonlinear oscillations of gas bubbles in liquids. *J. Acoust. Soc. Am.*, **59**, 283–293.

Lauterborn, W. and Suchla, E. (1984). Bifurcation superstructure in a model of acoustic turbulence. *Phys. Rev. Lett.*, **53**, 2304–2307.

Law, C.K. (1982). Recent advances in droplet vaporization and combustion. *Progr. Energy Combustion Sci.*, **8**, No. 3, 169–199.

Lazarus, J.H. and Neilson, I.D. (1978). A generalized correlation for friction head losses of settling mixtures in horizontal smooth pipelines. *Hydrotransport 5. Proc. 5th Int. Conf. on Hydraulic Transport of Solids in Pipes.*, **1**, B1-1-B1-32.

Ledinegg, M. (1983). Instabilität der Strömung bei Natürlichen und Zwangumlaut. *Warme*, **61**, No. 8, 891–898.

Lee, D.W. and Spencer, R.C. (1933). Photomicrographic studies of fuel sprays. *NACA Tech. Note*, 454.

Lee, J., Cowin, S.C. and Templeton, J.S. (1974). An experimental study of the kinematics of flow through hoppers. *Trans. Soc. Rheol.*, **18**, No. 2, 247.

Lienhard, J.H. and Sun, K. (1970). Peak boiling heat flux on horizontal cylinders. *Int. J. Heat Mass Trans.*, **13**, 1425–1440.

Lighthill, M.J. and Whitham, G.B. (1955). On kinematic waves. I. Flood movement in long waves. II. Theory of traffic flow along crowded roads. *Proc. Roy. Soc. A*, **229**, 281–345.

Liss, P.S. and Slinn, W.G.N. (eds.) (1983). *Air–Sea Exchange of Gases and Particles*. D. Reidel, New York.

Lockhart, R.W. and Martinelli, R.C. (1949). Proposed correlation of data for isothermal two-phase two-component flow in pipes. *Chem. Eng. Progress*, **45**, 39–48.

Lun, C.K.K., Savage, S.B., Jeffrey, D.J. and Chepurniy, N. (1984). Kinetic theories for granular flow: inelastic particles in Couette flow and slightly inelastic particles in a general flow field. *J. Fluid Mech.*, **140**, 223–256.

Lun, C.K.K. and Savage, S.B. (1986). The effects of an impact velocity dependent coefficient of restitution on stresses developed by sheared granular materials. *Acta Mechanica*, **63**, 15–44.

Lush, P.A. and Angell, B. (1984). Correlation of cavitation erosion and sound pressure level. *ASME. J. Fluids Eng.*, **106**, 347–351.

Macpherson, J.D. (1957). The effect of gas bubbles on sound propagation in water. *Proc. Phys. Soc. London*, **70B**, 85–92.

Magnaudet, J. and Legendre, D. (1998). The viscous drag force on a spherical bubble with a time-dependent radius. *Phys. Fluids*, **10**, No. 3, 550–554.

Makay, E. and Szamody, O. (1978). Survey of feed pump outages. *Electric Power Res. Inst. Rep. FP-754*.

Mandhane, J.M., Gregory, G.A. and Aziz, K.A. (1974). A flow pattern map for gas-liquid flow in horizontal pipes. *Int. J. Multiphase Flow*, **1**, 537–553.

Maneely, D.J. (1962). A study of the expansion process of low quality steam through a de Laval nozzle. *Univ. of Calif. Radiation Lab. Rep. UCRL-6230.*

Marble, F.E. (1970). Dynamics of dusty gases. *Ann. Rev. Fluid Mech.*, **2**, 397–446.

Marble, F.E. and Wooten, D.C. (1970). Sound attenuation in a condensing vapor. *Phys. Fluids*, **13**, No. 11, 2657–2664.

Marinesco, M. and Trillat, J.J. (1933). Action des ultrasons sur les plaques photographiques. *Compt. Rend.*, **196**, 858–860.

Martin, C.S., Medlarz, H., Wiggert, D.C. and Brennen, C. (1981). Cavitation inception in spool valves. *ASME. J. Fluids Eng.*, **103**, 564–576.

Martinelli, R.C. and Nelson, D.B. (1948). Prediction of pressure drop during forced circulation boiling of water. *Trans. ASME*, **70**, 695–702.

Maw, N., Barber, J.R. and Fawcett, J.N. (1976). The oblique impact of elastic spheres. *Wear*, **38**, 101–114. See also (1981). The role of elastic tangential compliance in oblique impact. *ASME J. Lubr. Tech.*, **103**, 74–80.

McCoy, D.D. and Hanratty, T.J. (1977). Rate of deposition of droplets in annular two-phase flow. *Int. J. Multiphase Flow*, **3**, 319–331.

McTigue, D.F. (1978). A model for stresses in shear flows of granular materials. *Proc. US-Japan Sem. on Cont.-Mech. and Stat. Approaches to Mech. Granular Mater.*, 266–271.

Mellen, R.H. (1954). Ultrasonic spectrum of cavitation noise in water. *J. Acoust. Soc. Am.*, **26**, 356–360.

Mertes, T.S. and Rhodes, H.B. (1955). Liquid–particle behavior (Part 1). *Chem. Eng. Prog.*, **51**, 429–517.

Meyer, E. and Kuttruff, H. (1959). Zur Phasenbeziehung zwischen Sonolumineszenz und Kavitations-vorgang bei periodischer Anregung. *Zeit angew. Phys.*, **11**, 325–333.

Miller, C.D. and Gross, L.A. (1967). A performance investigation of an eight-inch hubless pump inducer in water and liquid nitrogen. *NASA TN D-3807*.

Minnaert, M. (1933). Musical air bubbles and the sound of running water. *Phil. Mag.*, **16**, 235–248.

Moller, W. (1938). Experimentelle Untersuchungen zur Hydrodynamik der Kugel. *Physik. Z.*, **39**, 57–80.

Monahan, E.C. and Zietlow, C.R. (1969). Laboratory comparisons of fresh-water and salt-water whitecaps. *J. Geophys. Res.*, **74**, 6961–6966.

Monahan, E.C. and Van Patten, M.A. (eds.) (1989). *Climate and Health Implications of Bubble-Mediated Sea–Air Exchange.* Connecticut Sea Grant College Program CT-SG-89-06, Groton, CT.

Monahan, E.C. (1989). From the laboratory tank to the global ocean. In *Climate and Health Implications of Bubble-Mediated Sea–Air Exchange* (editors, E.C. Monahan and M.A. VanPatten). Connecticut Sea Grant College Program CT-SG-89-06, Groton, CT.

Mørch, K.A. (1980). On the collapse of cavity cluster in flow cavitation. *Proc. First Int. Conf. on Cavitation and Inhomogeneities in Underwater Acoustics, Springer Series in Electrophysics*, **4**, 95–100.

Mørch, K.A. (1981). Cavity cluster dynamics and cavitation erosion. *Proc. ASME Cavitation and Polyphase Flow Forum*, 1–10.

Morison, J.R., O'Brien, M.P., Johnson, J.W. and Schaaf, S.A. (1950). The forces exerted by surface waves on piles. *AIME Trans., Petroleum Branch*, **189**, 149–154.

Morrison, F.A. and Stewart, M.B. (1976). Small bubble motion in an accelerating liquid. *ASME J. Appl. Mech.*, **43**, 399–403.

Morrison, H.L. and Richmond, O. (1976). Application of Spencer's ideal soil model to granular materials flow. *ASME J. Appl. Mech.*, **43**, 49–53.

Muir, T.F. and Eichhorn, R. (1963). Compressible flow of an air–water mixture through a vertical two-dimensional converging-diverging nozzle. Proc. 1963 Heat Transfer and Fluid Mechanics Institute, Stanford Univ. Press, 183–204.

Murakami, M. and Minemura, K. (1977). Flow of air bubbles in centrifugal impellers and its effect on the pump performance. *Proc. 6th Australasian Hydraulics and Fluid Mechanics Conf.*, **1**, 382–385.

Murakami, M. and Minemura, K. (1978). Effects of entrained air on the performance of a horizontal axial-flow pump. In *Polyphase Flow in Turbomachinery* (eds: C.E. Brennen, P. Cooper and P.W. Runstadler, Jr.), ASME, 171–184.

Musmarra, D., Poletta, M., Vaccaro, S. and Clift, R. (1995). Dynamic waves in fluidized beds. *Powder Tech.*, **82**, 255–268.

Nakagawa, M. (1988). Kinetic theory for plane flows of rough, inelastic circular discs. Ph.D. Dissertation, Cornell University, Ithaca, NY.

NASA. (1970). Prevention of coupled structure-propulsion instability. *NASA SP-8055*.

Naude, C.F. and Ellis, A.T. (1961). On the mechanism of cavitation damage by non-hemispherical cavities in contact with a solid boundary. *ASME. J. Basic Eng.*, **83**, 648–656.

Neppiras, E.A. and Noltingk, B.E. (1951). Cavitation produced by ultrasonics: theoretical conditions for the onset of cavitation. *Proc. Phys. Soc., London*, **64B**, 1032–1038.

Neppiras, E.A. (1969). Subharmonic and other low-frequency emission from bubbles in sound-irradiated liquids. *J. Acoust. Soc. Am.*, **46**, 587–601.

Neppiras, E.A. (1980). Acoustic cavitation. *Phys. Rep.*, **61**, 160–251.

Neusen, K.F. (1962). Optimizing of flow parameters for the expansion of very low quality steam. *Univ. of Calif. Radiation Lab. Rep. UCRL-6152*.

Newitt, D.M., Richardson, H.F., Abbot, M. and Turtle, R.B. (1955). Hydraulic conveying of solids in horizontal pipes. *Trans. Inst. Chem. Engrs.*, **33**, 93.

Ng, S.L. and Brennen, C. (1978). Experiments on the dynamic behavior of cavitating pumps. *ASME J. Fluids Eng.*, **100**, No. 2, 166–176.

Nguyen, T.V., Brennen, C. and Sabersky, R.H. (1979). Gravity flow of granular materials in conical hoppers. *ASME J. Appl. Mech.*, **46**, No. 3, 529–535.

Nguyen, T.V., Brennen, C. and Sabersky, R.H. (1980). Funnel flow in hoppers. *ASME J. Appl. Mech.*, **102**, No. 4, 729–735.

Nigmatulin, R.I. (1979). Spatial averaging in the mechanics of heterogeneous and dispersed systems. *Int. J. Multiphase Flow*, **5**, 353–385.

Noltingk, B.E. and Neppiras, E.A. (1950). Cavitation produced by ultrasonics. *Proc. Phys. Soc., London*, **63B**, 674–685.

Noordzij, L. (1973). Shock waves in mixtures of liquid and air bubbles. Ph.D. Thesis, Technische Hogeschool, Twente, Netherlands.

Noordzij, L. and van Wijngaarden, L. (1974). Relaxation effects, caused by relative motion, on short waves in gas-bubble/liquid mixtures. *J. Fluid Mech.*, **66**, 115–143.

Ogawa, S. (1978). Multitemperature theory of granular materials. In *Proc. US-Japan Sem. on Cont.-Mech. and Stat. Approaches to Mech. Granular Mater.*, 208–217. Gakujutsu Bunken Fukyakai, Tokyo.

Ogawa, S., Umemura, A. and Oshima, N. (1980). On the equations of fully fluidized granular materials. *J. Appl. Math. Phys. (ZAMP)*, **31**, 483–493.

Ohashi, H. (1968). Analytical and experimental study of dynamic characteristics of turbopumps. *NASA TN D-4298*.

Ohashi, H., Matsumoto, Y., Ichikawa, Y. and Tsukiyama, T. (1990). Air/water two-phase flow test tunnel for airfoil studies. *Expts. Fluids*, **8**, 249–256.

O'Hern, T.J., Katz, J. and Acosta, A.J. (1985). Holographic measurements of cavitation nuclei in the sea. *Proc. ASME Cavitation and Multiphase Flow Forum*, 39–42.

Omta, R. (1987). Oscillations of a cloud of bubbles of small and not so small amplitude. *J. Acoust. Soc. Am.*, **82**, 1018–1033.

Oseen, C.W. (1910). Über die Stokessche Formel und über die verwandte Aufgabe in der Hydrodynamik. *Arkiv Mat., Astron. och Fysik*, **6**, No. 29.

Owens, W.L. (1961). Two-phase pressure gradient. International developments in heat transfer. *ASME Paper no. 41*, **2**, 363–368.

Pan, Y. and Banerjee, S. (1997). Numerical investigation of the effects of large particles on wall-turbulence. *Phys. Fluids*, **9**, 3786–3807.

Paris, A.D. and Eaton, J.K. (2001). Turbulence attenuation in a particle laden channel flow. Mech. Eng. Dept., Stanford Univ. Report.

Parkin, B.R. (1952). Scale effects in cavitating flow. Ph.D. Thesis, Calif. Inst. of Tech.

Parlitz, U., Englisch, V., Scheffczyk, C., and Lauterborn, W. (1990). Bifurcation structure of bubble oscillators. *J. Acoust. Soc. Am.*, **88**, 1061–1077.

Parsons, C.A. (1906). The steam turbine on land and at sea. Lecture to the Royal Institution, London.

Parthasarathy, R.N. and Faeth, G.M. (1987). Structure of particle-laden turbulent water jets in still water. *Int. J. Multiphase Flow*, **13**, 699–716.

Patel, B.R. and Runstadler, P.W., Jr. (1978). Investigations into the two-phase flow behavior of centrifugal pumps. In *Polyphase Flow in Turbomachinery* (eds: C.E. Brennen, P. Cooper and P.W. Runstadler, Jr.), ASME, 79–100.

Patton, K.T. (1965). Tables of hydrodynamic mass factors for translational motion. *ASME Paper, 65-WA/UNT-2*.

Paynter, H.M. (1961). *Analysis and Design of Engineering Systems*. MIT Press, Cambridge, MA.

Pearcey, T. and Hill, G.W. (1956). The accelerated motion of droplets and bubbles. *Austr. J. Phys.*, **9**, 19–30.

Peterson, F.B., Danel, F., Keller, A.P., and Lecoffre, Y. (1975). Comparative measurements of bubble and particulate spectra by three optical methods. *Proc. 14th Int. Towing Tank Conf.*

Phinney, R.E. (1973). The breakup of a turbulent jet in a gaseous atmosphere. *J. Fluid Mech.*, **60**, 689–701.

Pipes, L.A. (1940). The matrix theory for four terminal networks. *Phil. Mag.*, **30**, 370.

Plesset, M.S. (1949). The dynamics of cavitation bubbles. *ASME J. Appl. Mech.*, **16**, 228–231.

Plesset, M.S. and Zwick, S.A. (1952). A nonsteady heat diffusion problem with spherical symmetry. *J. Appl. Phys.*, **23**, No. 1, 95–98.

Plesset, M.S. and Mitchell, T.P. (1956). On the stability of the spherical shape of a vapor cavity in a liquid. *Quart. Appl. Math.*, **13**, No. 4, 419–430.

Plesset, M.S. and Chapman, R.B. (1971). Collapse of an initially spherical vapor cavity in the neighborhood of a solid boundary. *J. Fluid Mech.*, **47**, 283–290.

Plesset, M.S. and Prosperetti, A. (1977). Bubble dynamics and cavitation. *Ann. Rev. Fluid Mech.*, **9**, 145–185.

Poritsky, H. (1952). The collapse or growth of a spherical bubble or cavity in a viscous fluid. *Proc. First Nat. Cong. in Appl. Math.*, 813–821.

Prosperetti, A. (1977). Thermal effects and damping mechanisms in the forced radial oscillations of gas bubbles in liquids. *J. Acoust. Soc. Am.*, **61**, 17–27.

Prosperetti, A. (1982). Bubble dynamics: a review and some recent results. *Appl. Sci. Res.*, **38**, 145–164.

Prosperetti, A. (1984). Bubble phenomena in sound fields: part one and part two. *Ultrasonics*, **22**, 69–77 and 115–124.

Prosperetti, A. and Jones, A.V. (1985,7). The linear stability of general two-phase flow models – I and II. *Int. J. Multiphase Flow*, **11**, 133–148 and **13**, 161–171.

Prosperetti, A. and Lezzi, A. (1986). Bubble dynamics in a compressible liquid. Part 1. First-order theory. *J. Fluid Mech.*, **168**, 457–478.

Proudman, I. and Pearson, J.R.A. (1957). Expansions at small Reynolds number for the flow past a sphere and a circular cylinder. *J. Fluid Mech.*, **2**, 237–262.

Ranz, W.E. and Marshall, W.R. (1952). Evaporation from drops, I and II. *Chem. Eng. Prog.*, **48**, 141 and 173.

Rayleigh, Lord (Strutt, J.W.). (1917). On the pressure developed in a liquid during the collapse of a spherical cavity. *Phil. Mag.*, **34**, 94–98.

Reeks, M.W. (1992). On the continuum equations for dispersed particles in nonuniform flows. *Phys. Fluids A*, **4**, 1290–1303.

Reynolds, A.B. and Berthoud, G. (1981). Analysis of EXCOBULLE two-phase expansion tests. *Nucl. Eng. and Design*, **67**, 83–100.

Reynolds, O. (1873). The causes of the racing of the engines of screw steamers investigated theoretically and by experiment. *Trans. Inst. Naval Arch.*, **14**, 56–67.

Richman, M.W. (1988). Boundary conditions based upon a modified Maxwellian velocity distribution for flows of identical, smooth, nearly elastic spheres. *Acta Mechanica*, **75**, 227–240.

Rohsenow, W.M. and Hartnett, J.P. (1973). *Handbook of Heat Transfer*, Section 13. McGraw-Hill, New York.

Rubin, S. (1966). Longitudinal instability of liquid rockets due to propulsion feedback (POGO). *J. Spacecraft and Rockets*, **3**, No. 8, 1188–1195.

Rudinger, G. (1969). Relaxation in gas–particle flow. In *Nonequilibrium Flows, Part 1* (ed: P.P. Wegener). Marcel Dekker, New York and London.

Rybzynski, W. (1911). Über die fortschreitende Bewegung einer flüssigen Kugel in einem zähen Medium. *Bull. Acad. Sci. Cracovie*, **A**, 40.

Ryskin, G. and Rallison, J.M. (1980). On the applicability of the approximation of material frame-indifference in suspension mechanics. *J. Fluid Mech.*, **99**, 525–529.

Sack, L.E. and Nottage, H.B. (1965). System oscillations associated with cavitating inducers. *ASME J. Basic Eng.*, **87**, 917–924.

Saffman, P.G. (1962). On the stability of laminar flow of a dusty gas. *J. Fluid Mech.*, **13**, 120–128.

Sangani, A.S., Zhang, D.Z. and Prosperetti, A. (1991). The added mass, Basset, and viscous drag coefficients in nondilute bubbly liquids undergoing small-amplitude oscillatory motion. *Phys. Fluids A*, **3**, 2955–2970.

Sangani, A.S. and Didwania, A.K. (1993a). Dispersed phase stress tensor in flows of bubbly liquids at large Reynolds number. *J. Fluid Mech.*, **248**, 27–54.

Sangani, A.S. and Didwania, A.K. (1993b). Dynamic simulations of flows of bubbly liquids at large Reynolds number. *J. Fluid Mech.*, **250**, 307–337.

Sangani, A.S., Mo, G., Tsao, H.-K., and Koch, D.L. (1996). Simple shear flows of dense gas-solid suspensions at finite Stokes numbers. *J. Fluid Mech.*, **313**, 309–341.

Sargis, D.A., Stuhmiller, J.H. and Wang, S.S. (1979). Analysis of steam chugging phenomena, Volumes 1, 2 and 3. *Electric Power Res. Inst. Report NP-962*.

Sarpkaya, T. and Isaacson, M. (1981). *Mechanics of Wave Forces on Offshore Structures*. Van Nostrand Reinhold Co., NY.

Sarpkaya, T. and Merrill, C. (1998). Spray formation at the free surface of liquid wall jets. In *Proc. 22nd ONR Symp. on Naval Hydrodynamics*, Natl. Academie Press, Wash. D.C.

Savage, S.B. (1965). The mass flow of granular materials derived from coupled velocity-stress fields. *Brit. J. Appl. Phys.*, **16**, 1885–1888.

Savage, S.B. (1967). Gravity flow of a cohesionless bulk solid in a converging conical channel. *Int. J. Mech. Sci.*, **19**, 651–659.

Savage, S.B. (1983). Granular flows down rough inclines – review and extension. In *Mechanics of Granular Materials: New Models and Constitutive Relations* (ed: J.T. Jenkins and M. Satake), 261–282. Elsevier, New York.

Savage, S.B. and Jeffrey, D.J. (1981). The stress tensor in a granular flow at high shear rates. *J. Fluid Mech.*, **110**, 255–272.

Savage, S.B. and McKeown, S. (1983). Shear stress developed during rapid shear of dense concentrations of large spherical particles between concentric cylinders. *J. Fluid Mech.*, **127**, 453–472.

Savage, S.B. and Sayed, M. (1984). Stresses developed by dry cohesionless granular materials in an annular shear cell.

Schicht, H.H. (1969). Flow patterns for an adiabatic two-phase flow of water and air within a horizontal tube. *Verfahrenstechnik*, **3**, No. 4, 153–161.

Schofield, A. and Wroth, P. (1968). *Critical State Soil Mechanics*. McGraw-Hill, New York.

Schrage, R.W. (1953). *A Theoretical Study of Interphase Mass Transfer*. Columbia Univ. Press, New York.

Scriven, L.E. (1959). On the dynamics of phase growth. *Chem. Eng. Sci.*, **10**, 1–13.

Shepherd, J.E. and Sturtevant, B. (1982). Rapid evaporation near the superheat limit. *J. Fluid Mech.*, **121**, 379–402.

Sherman, D.C. and Sabersky, R.H. (1981). Natural convection film boiling on a vertical surface. *Lett. Heat Mass Trans.*, **8**, 145–153.

Shima, A., Takayama, K., Tomita, Y. and Ohsawa, N. (1983). Mechanism of impact pressure generation from spark-generated bubble collapse near a wall. *AIAA J.*, **21**, 55–59.

Shook, C.A. and Roco, M.C. (1991). *Slurry Flow: Principles and Practice*. Butterworth-Heinemann, New York.

Shorr, M. and Zaehringer, A.J. (1967). *Solid Rocket Technology*. John Wiley and Sons, New York.

Silberman, E. (1957). Sound velocity and attenuation in bubbly mixtures measured in standing wave tubes. *J. Acoust. Soc. Am.*, **18**, 925–933.

Simmons, H.C. (1977). The correlation of drop-size distributions in fuel nozzle sprays. Parts I and II. *ASME J. Eng. for Power*, **99**, 309–319.

Sirignano, W.A. and Mehring, C. (2000). Review of theory of distortion and disintegration of liquid streams. *Prog. Energy Combust. Sci.*, **26**, 609–655.

Slattery, J.C. (1972). *Momentum, Energy, and Mass Transfer in Continua*. McGraw-Hill, New York.

Smereka, P., Birnir, B. and Banerjee, S. (1987). Regular and chaotic bubble oscillations in periodically driven pressure fields. *Phys. Fluids*, **30**, 3342–3350.

Smialek, J.L., Archer, F.A. and Garlick, R.G. (1994). Turbine airfoil degradation in the Persian Gulf war. *J. Minerals, Metals Materials Soc.*, **46**, 39–41.

Smith, A., Kent, R.P. and Armstrong, R.L. (1967). Erosion of steam turbine blade shield materials. In *Erosion by Cavitation and Impingement*, 125–151. ASTM STP 408.

Sondergaard, R., Chaney, K. and Brennen, C.E. (1990). Measurements of solid spheres bouncing off flat plates. *ASME J. Appl. Mech.*, **57**, 694–699.

Soo, S.L. (1983). Pneumatic transport. In *Handbook of Fluids in Motion* (eds: N.P. Cheremisinoff and R. Gupta), Ann Arbor Science, Ann Arbor, MI.

Squires, K.D. and Eaton, J.K. (1990). Particle response and turbulence modification in isotropic turbulence. *Phys. Fluids*, **A2**, 1191–1203.

Stokes, G.G. (1851). On the effect of the internal friction of fluids on the motion of pendulums. *Trans. Camb. Phil. Soc.*, **9**, Part II, 8–106.

Strasberg, M. (1961). Rectified diffusion: comments on a paper of Hsieh and Plesset. *J. Acoust. Soc. Am.*, **33**, 359.

Streeter, V.L. and Wylie, E.B. (1967). *Hydraulic Transients*. McGraw-Hill, New York.

Streeter, V.L. and Wylie, E.B. (1974). Waterhammer and surge control. *Ann. Rev. Fluid Mech.*, **6**, 57–73.

Stripling, L.B. and Acosta, A.J. (1962). Cavitation in turbopumps – Parts I and II. *ASME J. Basic Eng.*, **84**, 326–338 and 339–350.

Symington, W.A. (1978). Analytical studies of steady and non-steady motion of a bubbly liquid. Ph.D. Thesis, Calif. Inst. of Tech.

Tabakoff, W. and Hussein, M.F. (1971). Effects of suspended solid particles on the properties in cascade flow. *AIAA J.*, 1514–1519.

Tabakoff, W. and Hamed, A. (1986). The dynamics of suspended solid particles in a two-stage gas turbine. *ASME J. Turbomach.*, **108**, 298–302.

Tae-il, K., Okamoto, K. and Madarame, H. (1993). Study of the effective parameters for the geysering period. In *Instabilities in Multiphase Flows* (eds: G. Gouesbet and A. Berlemont), 125–136. Plenum, New York.

Taitel, Y. and Dukler, A.E. (1976). A model for predicting flow regime transitions in horizontal and near horizontal gas-liquid flow. *AIChE J.*, **22**, 47–55.

Tam, C.K.W. (1969). The drag on a cloud of spherical particles in low Reynolds number flow. *J. Fluid Mech.*, **38**, 537–546.

Taneda, S. (1956). Studies on wake vortices (III). Experimental investigation of the wake behind a sphere at low Reynolds number. *Rep. Res. Inst. Appl. Mech., Kyushu Univ.*, **4**, 99–105.

Tangren, R.F., Dodge, C.H. and Seifert, H.S. (1949). Compressibility effects in two-phase flow. *J. Appl. Phys.*, **20**, No. 7, 637–645.

Taylor, K.J. and Jarman, P.D. (1970). The spectra of sonoluminescence. *Aust. J. Phys.*, **23**, 319–334.

Theofanous, T., Biasi, L., Isbin, H.S. and Fauske, H. (1969). A theoretical study on bubble growth in constant and time-dependent pressure fields. *Chem. Eng. Sci.*, **24**, 885–897.

Thiruvengadam, A. (1967). *The concept of erosion strength*. In *Erosion by Cavitation or Impingement*, 22–35. Am. Soc. Testing Mats, STP 408.

Thiruvengadam, A. (1974). Handbook of cavitation erosion. Tech. Rep. 7301-1, Hydronautics, Inc., Laurel, MD.

Thomas, D.G. (1962). Transport characteristics of suspensions, part VI. Minimum transport velocity for large particle size suspensions in round horizontal pipes. *AIChE J.*, **8**, 373–378.

Tomita, Y. and Shima, A. (1977). On the behaviour of a spherical bubble and the impulse pressure in a viscous compressible liquid. *Bull. JSME*, **20**, 1453–1460.

Tomita, Y. and Shima, A. (1990). High-speed photographic observations of laser-induced cavitation bubbles in water. *Acustica*, **71**, No. 3, 161–171.

Torobin, L.B. and Gauvin, W.H. (1959). Fundamental aspects of solids–gas flow. Part II. The sphere wake in steady laminar fluids. *Can. J. Chem. Eng.*, **37**, 167–176.

Tsao, H.-K., and Koch, D.L. (1995). Rapidly sheared, dilute gas-solid suspensions. *J. Fluid Mech.*, **296**, 211–245.

Tulin, M.P. (1964). Supercavitating flows – small perturbation theory. *J. Ship Res.*, **7**, No. 3, 16–37.

Turner, J.M. and Wallis, G.B. (1965). The separate-cylinders model of two-phase flow. *AEC Report NYO-3114-6*.

Urick, R.J. (1948). The absorption of sound in suspensions of irregular particles. *J. Acoust. Soc. Am.*, **20**, 283–289.

van Wijngaarden, L. (1964). On the collective collapse of a large number of gas bubbles in water. In *Proc. 11th Int. Cong. Appl. Mech.*, 854–861. Springer-Verlag, Berlin.

van Wijngaarden, L. (1972). One-dimensional flow of liquids containing small gas bubbles. *Ann. Rev. Fluid Mech.*, **4**, 369–396.

van Wijngaarden, L. (1976). Hydrodynamic interaction between gas bubbles in liquid. *J. Fluid Mech.*, **77**, 27–44.

Vernier, P. and Delhaye, J.M. (1968). General two-phase flow equations applied to the thermodynamics of boiling water reactors. *Energ. Primaire*, **4**, 5–46.

Wade, G.E. (1974). Evolution and current status of the BWR containment system. *Nuclear Safety*, **15**, No. 2.

Wallis, G.B. (1969). *One-Dimensional Two-Phase Flow*. McGraw-Hill, New York.

Wallis, G.B. (1991). The averaged Bernoulli equation and macroscopic equations of motion for the potential flow of a two-phase dispersion. *Int. J. Multiphase Flow*, **17**, 683–695.

Walton, O.R. and Braun, R.L. (1986a). Viscosity, granular-temperature, and stress calculations for shearing assemblies of inelastic, frictional disks. *J. Rheology*, **30(5)**, 949–980.

Walton, O.R. and Braun, R.L. (1986b). Stress calculations for assemblies of inelastic spheres in uniform shear. *Acta Mechanica*, **63**, 73–86.

Wang, Q. and Monahan, E.C. (1995). The influence of salinity on the spectra of bubbles formed in breaking wave simulations. In *Sea Surface Sound '94* (eds: M.J. Buckingham and J.R. Potter), 312–319. World Scientific, Singapore.

Wassgren, C.R., Brennen, C.E. and Hunt, M.L. (1996). Vertical vibration of a deep bed of granular material in a container. *ASME J. Appl. Mech.*, **63**, 712–719.

Weaire, D. and Hutzler, S. (2001). *The physics of foams*. Oxford Univ. Press.

Wegener, P.P. and Mack, L.M. (1958). Condensation in supersonic and hypersonic wind tunnels. *Adv. Appl. Mech.*, **5**, 307–447.

Wegener, P.P., Sundell, R.E. and Parlange, J.-Y. (1971). Spherical-cap bubbles rising in liquids. *Z. Flugwissenschaften*, **19**, 347–352.

Wegener, P.P. and Parlange, J.-Y. (1973). Spherical-cap bubbles. *Ann. Rev. Fluid Mech.*, **5**, 79–100.

Weighardt, K. (1975). Experiments in granular flow. *Ann. Rev. Fluid Mech.*, **7**, 89–114.

Weir, G.J. (2001). Sound speed and attenuation in dense, non-cohesive air-granular systems. *Chem. Eng. Sci.*, **56**, 3699–3717.

Weisman, J. (1983). Two-phase flow patterns. In *Handbook of Fluids in Motion* (eds: N.P. Cheremisinoff and R. Gupta), 409–425. Ann Arbor Science, Ann Arbor, MI.

Weisman, J. and Kang, S.Y. (1981). Flow pattern transitions in vertical and upwardly inclined lines. *Int. J. Multiphase Flow*, **7**, 27.

Westwater, J.W. (1958). Boiling of liquids. *Adv. Chem. Eng.*, **2**, 1–56.

Whalley, P.B. (1987). *Boiling, Condensation and Gas–Liquid Flow*. Oxford Science, Oxford, UK.

Whitham, G.B. (1974). *Linear and Non-Linear Waves*. John Wiley and Sons, New York.

Williams, F.A. (1965). *Combustion Theory*. Addison-Wesley, New York.

Witte, J.H. (1969). Mixing shocks in two-phase flow. *J. Fluid Mech.*, **36**, 639–655.

Wood, I.R. (ed.) (1991). *Air Entrainment in Free-Surface Flows*. Balkema, Rotterdam.

Woods, L.C. (1961). *Theory of Subsonic Plane Flow*. Cambridge Univ. Press, New York.

Wu, P.-K., Miranda, R.F. and Faeth, G.M. (1995). Effects of initial flow conditions on primary breakup of nonturbulent and turbulent round jets. *Atomization and Sprays*, **5**, 175–196.

Wu, P.-K. and Faeth, G.M. (1993). Aerodynamic effects on primary breakup of turbulent liquids. *Atomization and Sprays*, **3**, 265–289.

Wu, P.-K. and Faeth, G.M. (1995). Onset and end of drop formation along the surface of turbulent liquid jets in still gases. *Phys. Fluids A*, **7**, 2915–2917.

Wu, T.Y. (1972). Cavity and wake flows. *Ann. Rev. Fluid Mech.*, **4**, 243–284.

Yamamoto, K. (1991). Instability in a cavitating centrifugal pump. *JSME Int. J., Ser. II*, **34**, 9–17.

Yih, C.-S. (1969). *Fluid Mechanics*. McGraw-Hill, New York.

Young, F.R. (1989). *Cavitation*. McGraw-Hill, New York.

Young, N.O., Goldstein, J.S. and Block, M.J. (1959). The motion of bubbles in a vertical temperature gradient. *J. Fluid Mech.*, **6**, 350–356.

Yuan, Z. and Michaelides, E.E. (1992). Turbulence modulation in particulate flows – a theoretical approach. *Int. J. Multiphase Flow*, **18**, 779–785.

Zandi, I. (1971). Hydraulic transport of bulky materials. In *Advances in Solid–Liquid Flow in Pipes and Its Applications* (ed: I. Zandi), 1–34. Pergamon, New York.

Zandi, I. and Govatos, G. (1967). Heterogeneous flow of solids in pipelines. *ASCE J. Hyd. Div.*, **93**, 145–159.

Zeininger, G. and Brennen, C.E. (1985). Interstitial fluid effects in hopper flows of granular materials. In *ASME Cavitation and Multiphase Flow Forum, 1985*, 132–136.

Zenit, R., Hunt, M.L. and Brennen, C.E. (1997). Collisional particle pressure measurements in solid–liquid flows. *J. Fluid Mech.*, **353**, 261–283.

Zenz, F.A. and Othmer, D.F. (1960). *Fluidization and Fluid-Particle Systems*. Reinhold, New York.

Zhang, D.Z. and Prosperetti, A. (1994). Averaged equations for inviscid disperse two-phase flow. *J. Fluid Mech.*, **267**, 185–219.

Zick, A. and Homsy, G.M. (1982). Stokes flow through periodic arrays of spheres. *J. Fluid Mech.*, **115**, 13–26.

Bibliography

Zuber, N. (1959). Hydrodynamic aspects of boiling heat transfer. Ph.D. Thesis, UCLA.

Zuber, N., Tribus, M. and Westwater, J.W. (1961). The hydrodynamic crisis in pool boiling of saturated and subcooled liquids. *Proc. 2nd Int. Heat Tranf. Conf.*, Section A, Part II, 230–237.

Zuber, N. (1964). On the dispersed two-phase flow in the laminar flow regime. *Chem. Eng. Sci.*, **19**, 897–917.

Zuber, N., and Findlay, J. (1965). Average volumetric concentration in 2-phase flow systems. *ASME J. Heat Trans.*, **87**, 453.

Zung, L.B. (1967). Particle–fluid mechanics in shear flows, acoustic waves and shock waves. Ph.D. thesis, California Institute of Technology.

Index

acoustic absorption, 215
acoustic attenuation, 181, 203, 224–227
acoustic damping, 224–227
acoustic impedance, 179
acoustic impulse, 113
acoustic pressure, 110
added mass matrix, 38
annular flow, 164
   instability, 154
atomizing nozzle, 237
attached cavitation, 108
   tails, 108
auto-oscillation, 292
avalanches, 252
averaging, 8–9, 25–26, 27, 282–283

Bagnold number, 268
barotropic relation, 176, 186–187
Basset term, 47, 50
binary collision time, 254
Bjerknes forces, 38, 68, 94
Blake critical pressure, 83
Blake critical radius, 83
boiling, 116–125
   vertical surfaces, 122–125
boiling crisis, 119, 278
Brownian motion, 38
bubble
   acceleration, 48–52
   cloud, 103, 210–216
   collapse, 99–115
   damping, 93, 203
   deformation, 60–65
   fission, 107
   migration, 69–72
   natural frequency, 91–93, 201, 209
   stability, 82
   translation, 30–59, 60–72

bubbly flow, 140, 148, 199–216
   limits, 142–144
   shock waves, 205–210, 216
bulk modulus, 179
burning rate, 248

cavitating pump dynamics, 313–318
cavitation, 97
   bubble cloud, 103
   bubble collapse, 99
   bubble shape, 101, 106–109
   damage, 102, 104, 110
   event rate, 113
   events, 106, 113
   inception, 97
   inception number, 97
   luminescence, 115
   noise, 109–115
   nuclei, 98
   number, 97
   patch, 108
   scaling, 114
   stable acoustic, 94
   transient acoustic, 94
cavitation surge, 292–293, 313, 314
   frequency, 293
   onset, 293
charge separation, 144
choked flow, 189, 231, 275, 304
chugging, 293–297
Clausius–Clapeyron equation, 76
cloud natural frequency, 212
coefficient of restitution, 254, 269
component
   characteristic, 285
   resistance, 285
compressibility, 309
compressor surge, 288, 293
concentration waves, 136, 290–292, 302

condensation, 125–126
condensation oscillations, 293–297
condensation shocks, 195–198
conjugate states, 274
conservation of mass, 9–10
continuous phase, 134
Coulomb friction, 253, 260
Coulomb yield criterion, 260
critical gas volume fraction, 189, 192
critical heat flux, 119, 281
critical mass flow rate, 190, 193
critical pressure ratio, 190, 193
critical radius, 83
critical solids fraction, 258, 261
critical vapor volume fraction, 193

D'Alembert's paradox, 32
debris flow, 252
density wave, 291
disperse flow, 2, 134
    friction, 155–163
    limits, 145–148
disperse phase, 134
    separation, 136–139
dispersion, 137
drag coefficient, 32
drift flux, 5
drift flux models, 272–283, 303
drift velocity, 5
droplet
    combustion, 245–249
    concentration, 235
    deposition, 235
    entrainment, 235
    evaporation, 243–245
    mechanics, 243–249
    size, 239
dusty gases, 217–231
dynamic instability, 292–297

ebullition cycle, 119
effective viscosity, 158
elastic-quasistatic regime, 256
electromagnetic forces, 144
energy equation, 16–19
energy interaction, 18, 28–29
enthalpy, 5
entropy, 5
equations of motion, 8–21

far-wake, 32, 37
Fick's law, 11
film boiling, 120, 122–125, 278
film condensation, 125–126

flame front, 246, 248
flexible coating, 103
flooding, 275, 278, 304
flow patterns, 127–154
flow regimes, 127–154
    annular flow, 129, 132
    bubble flow, 129
    churn flow, 132
    churn-turbulent flow, 133
    disperse flow, 129, 132
    fluidized bed, 149
    Geldart chart, 150
    granular flow, 255–259
    heterogeneous flow, 132
    homogeneous flow, 131
    map, 128
    saltation flow, 132
    slug flow, 129, 132
    stratified flow, 129
    wave flow, 129
fluidized bed, 145, 182, 252, 276–278
    bubble, 148
foam flow, 144
force chains, 252, 255, 256
force interaction, 13, 15, 28–29
free streamline theory, 3
frequency dispersion, 181
frequency domain methods, 297
friction coefficient, 156
Froude number, 52, 139
fully separated flow, 135

gas turbines, 217
Geldart classification, 150
geyser instability, 289–290
grain elevators, 144
granular energy, 264
granular flow, 252–271
    boundary conditions, 267
    computer simulations, 267–268
    kinetic theory, 264–268
granular heat flux, 264
granular temperature, 263, 264

Haberman–Morton number, 61, 281
Hadamard–Rybczynski flow, 34, 47, 51
hard particle model, 253, 255
harmonic cascading, 216
heterogeneous flow
    friction, 159–161
homogeneous flow, 134, 176–198, 199, 219–220, 300–301
    equilibrium model, 183, 194
    friction, 157–159

Index

frozen model, 183, 193
  in nozzles, 187–195
hopper flows, 261–262
  funnel flow, 261
  mass flow, 261
hydraulic diameter, 163
hydraulic gradient, 155

imposed vibration, 255
inertia tensor, 38
inertial regime, 256
inhomogeneity instability, 144–148, 305
intermittency, 136, 148
intermolecular forces, 144
internal friction angle, 260
interstitial fluid effects, 268–271
isotropy assumption, 260
ITTC headform, 112

Jakob number, 86
jet breakup, 237–243

Kelvin impulse, 71
Kelvin–Helmholtz instability, 151–153
Keulegan–Carpenter number, 40
kinematic shocks, 274, 306–311
  stability, 308–309
kinematic waves, 136, 148, 291, 302–319
  speed, 304
  two-dimensional, 318–319
Knudsen number, 37
Kolmogorov scales, 21

laminar boundary layer, 106
landslides, 252
Ledinegg instability, 288–289
Leidenfrost effect, 121–122
Lewis number, 246
limit of fluidization, 278
liquid compressibility, 100, 186
Lockhart–Martinelli correlation, 163–168

Marangoni effects, 66–68
Martinelli correlations, 163–172
Martinelli parameter, 165
Martinelli–Nelson correlation, 168–172
mass diffusion, 89–90
mass flux, 4
mass fraction, 4
mass interaction, 9, 28–29
mass mean diameter, 8, 239
mass quality, 4
microjet, 102–106

microlayer, 118
mist flow
  limits, 142–144
mixing shock, 144
mixture density, 5
Mohr–Coulomb models, 260–261
momentum equation, 12–15
Monte Carlo methods, 266
Morison's equation, 39
multiphase flow
  models, 2–3
  notation, 3–6

natural convection, 117, 122
near-wake, 32, 37
nomenclature, xv–xxi
nozzle flow, 187–198
nucleate boiling, 117, 119–120, 278
nucleation, 117, 196
  sites, 119

oblique collisions, 269
ocean spray, 233–234
one-way coupling, 53
operating point, 286
Oseen flow, 35, 48

particle
  acceleration, 48–52
  added mass, 38–41, 55
  charge, 144
  collisions, 268–270
  drag, 55–58
  fission, 140–142
  heat transfer, 19–21
  interactions, 252, 253–255, 270
  loading, 218
  size, 140–142, 175, 239
  slip, 220
  stiffness, 254
  turbulence interaction, 21–25
patch cavitation, 108
photocopiers, 144
photophoresis, 38
pipe friction, 139, 155–172
Plesset–Zwick equation, 77
plunge pools, 232
polytropic constant, 78
pool boiling, 117–119, 278–282
porous media flow, 270
pressure suppression systems, 294

pumps
- axial, 174, 292
- bubbly flow, 141–142
- cavitation, 174
- cavitation number, 174, 293
- centrifugal, 174, 292
- dredge pump, 172
- dynamics, 313–318
- energy conversion, 172–175
- flow coefficient, 172, 293
- head coefficient, 172
- head degradation, 174
- multiphase flow, 172–175

quality, 4

rapid granular flow, 256, 263–268
Rayleigh collapse time, 81
Rayleigh–Plesset equation, 73, 200
Rayleigh–Taylor instability, 120, 152, 153, 280
rectified diffusion, 69, 90, 94, 95–96
relative motion, 176
relative velocity, 4, 16
relaxation time, 16, 49
remnant cloud, 105, 108
reservoir conditions, 186
Reynolds number, 32, 156
Reynolds stresses, 26–27
ring frequency, 209
rocket engines, 217, 314

salt water, 234
sand storms, 217
Sauter mean diameter, 8, 240
scattering cross-section, 216
Schiebe headform, 112
sedimentation, 302, 311–313
segregation, 137, 159
separated flow, 2, 134
- friction, 163–172
- limits, 151–153

shape distortion, 101
shock wave, 100, 104, 115, 205–210, 216, 221–224
size distribution, 6–8
slow granular flow, 259–262
slurry flow, 131, 139, 156, 252
small slip perturbation, 229–231
soft particle model, 253, 255
soil liquefaction, 252
sonic speed, 177–186, 201
sonophoresis, 38
spherical-cap bubble, 62
spillways, 232

spray, 232–251
- combustion, 249–251
- formation, 232–243

stability
- of laminar flow, 227–228
- of multiphase flows, 284–301

steam turbines, 217
stoichiometry, 246
Stokes flow, 32–37, 44–48, 49
Stokes number, 269
Stokes streamfunction, 32
stratified flow, 153–154
Strouhal number, 37
subharmonics, 94
subscripts, xx
super-resonant flow, 216
superficial velocity, 4
superscripts, xx
surface contaminants, 68
surface roughening, 108
surface tension, 66
system characteristic, 286
system components, 284–286
system stability, 286–297

temperature relaxation, 220–221
terminal velocity, 49, 55
thermal conductivity, 5
thermal effects, 84
thermocapillary effects, 66
thermodynamic equilibrium, 177
thermophoresis, 38
throat conditions, 188
time domain methods, 297
Tollmein–Schlicting waves, 238
trajectory models, 2
transfer functions, 297–301
- cavitating pumps, 317

transfer matrices, 297–301
- homogeneous flow, 300–301
- pumps, 299

turbines
- energy conversion, 172–175

turbomachine surge, 288
turbulent jets, 239–243
two-fluid models, 2
two-way coupling, 53

units, xxi
unsteady internal flow, 297–298

velocity relaxation, 220–221
ventilation, 97
vertical flow
- friction, 161–163

Index

vertical pipe flow, 274–276
viscosity, 5
volcanic dust, 217
volume flux, 3
volume fraction, 4
volumetric flux, 4
volumetric quality, 4
vortex shedding, 32, 37, 41

water-hammer methods, 297
wavy wall flow, 228–229
Weber number, 142, 241
white caps, 234
Whitehead paradox, 34

yield criterion, 260
Young's modulus, 254